AF691485

The professions of architect and engineer, which had maintained very close links since the time of the Renaissance, became increasingly isolated from one another in France during the course of the eighteenth century, the Age of Enlightenment. This book analyses the meaning of this gradual mutual isolation, the consequences of which can still be felt today at a variety of different levels, and offers a unique insight in English into the teaching and practice of architects such as Jacques-François Blondel and Pierre Patte, and engineers such as Jean-Rodolphe Perronet and Gaspard Riche de Prony.

The architects and engineers who form the core of Picon's thesis are major figures in the history of French architecture, on whom very little can be found in English. The special interest of the battle between the architects and the engineers, particularly at the Ecole des Ponts et Chaussées, and the eventual victory of the engineers, can be seen in terms of the gradual replacement of the ideals of the *Ancien Régime* by those of nineteenth-century rationalism. The issues raised are central to eighteenth-century thought, including nature and reason, Newtonian science, the *Philosophes* and the *Encyclopédistes*. Picon also describes the centrality of the new interest in the town, and the debate over the construction of the church of Sainte-Geneviève (the Panthéon) which encapsulated many of the issues in the struggle between the architects and the engineers.

The text of the book is clear and easily comprehensible, and presents a fully accessible account of this key period in the development of architectural achievement and debate. It will also shed light on the current debate on the appropriate form for modern buildings, as well as offering a full survey of the social, technical and aesthetic representations of a profession that was undergoing rapid and permanent transformation.

CAMBRIDGE STUDIES IN THE HISTORY OF ARCHITECTURE

Edited by

ROBIN MIDDLETON
Professor of Art History, Columbia University

JOSEPH RYKWERT
Paul Philippe Cret Professor of Architecture, University of Philadelphia

and DAVID WATKIN
Lecturer in the History of Art, University of Cambridge and Fellow of Peterhouse

This is a new series of historical studies intended to embrace a wide chronological range, from antiquity to the twentieth century, and to become a natural counterpart to Cambridge Studies in the History of Art. Volumes in the series are intended primarily for professional historians of architecture and their students, but it is also intended to include a number of volumes for course work or of interest to the general reader.

CAMBRIDGE STUDIES IN THE HISTORY OF ARCHITECTURE

FRENCH ARCHITECTS AND ENGINEERS IN THE AGE OF ENLIGHTENMENT

FRENCH ARCHITECTS AND ENGINEERS

IN THE AGE OF ENLIGHTENMENT

ANTOINE PICON

TRANSLATED BY

MARTIN THOM

CAMBRIDGE
UNIVERSITY PRESS

CAMBRIDGE UNIVERSITY PRESS
Cambridge, New York, Melbourne, Madrid, Cape Town, Singapore,
São Paulo, Delhi, Dubai, Tokyo

Cambridge University Press
The Edinburgh Building, Cambridge CB2 8RU, UK

Published in the United States of America by Cambridge University Press, New York

www.cambridge.org
Information on this title: www.cambridge.org/9780521123693

French edition © Parenthèses, Marseilles, 1988
English translation © Cambridge University Press 1992

This publication is in copyright. Subject to statutory exception
and to the provisions of relevant collective licensing agreements,
no reproduction of any part may take place without the written
permission of Cambridge University Press.

Originally published in French as
Architectes et ingénieurs au siècle des lumières
by Parenthèses, Marseilles, 1988
First published in English by Cambridge University Press as
French architects and engineers in the Age of Enlightenment 1992
This digitally printed version 2009

A catalogue record for this publication is available from the British Library

Library of Congress Cataloguing in Publication data

Picon, Antoine.
[Architectes et ingénieurs au siècle des Lumières. English]
French architects and engineers in the Age of Enlightenment/
Antoine Picon: translated by Martin Thom.
p. cm. – (Cambridge studies in the history of architecture)
Translation of: Architectes et ingénieurs au siècle des Lumières.
Includes bibliographical references and index.
ISBN 0 521 38253 X
1. Neoclassicism (Architecture) – France. 2. Architecture–
Philosophy. 3. Engineering – Philosophy. France – Intellectual
life – 18th century. I. Title. II. Series.
NA1046.5.N4P513 1991
720.'944'09033–dc20 90–23827 CIP

ISBN 978-0-521-38253-3 Hardback
ISBN 978-0-521-12369-3 Paperback

Cambridge University Press has no responsibility for the persistence or
accuracy of URLs for external or third-party internet websites referred to in
this publication, and does not guarantee that any content on such websites is,
or will remain, accurate or appropriate.

CONTENTS

FIGURES

ACKNOWLEDGEMENTS

I owe much to the works of Bruno Fortier, Jacques Guillerme, Monique Mosser and Werner Szambien on classicism, the Enlightenment and neo-classicism. Bernard Haumont has helped me to clarify certain aspects of the professional debate between architects and engineers. Bénédicte Leclerc was kind enough to put the results of his study of Jacques-François Blondel's work at Strasburg at my disposal. Jean-Pierre Epron showed me his notes on the Academy of Architecture. Philippe Duboy, François Véry and Michel Vernes also supplied me with much information. Philippe Gresset and Pierre Saddy were kind enough to read through the manuscript, while Jean-Louis Cohen has proved an unfailing source of support, his advice proving of particular value on more than one occasion. Finally, I wish to thank the Centre Pédagogique de Documentation et de Communication de l'Ecole Nationale des Ponts et Chaussées, and Jean Michel and Michel Yvon in particular, for their help.

INTRODUCTION

In architecture, as in many other domains, the eighteenth century was a period of transition. The 'fine old taste' of Mansart, Lemercier and Le Vau was clearly no longer in favour, while research in progress was leading to the architecture of the revolutionary period, to neo-classicism and to eclecticism. Nevertheless, it is no easy matter to advance an interpretation of these developments, for the old and the new were inextricably interwoven. One can, for example, list the innovatory qualities of De Wailly, Boullée and Ledoux, while at the same time investigating what it was that bound them to the French tradition. Architectural forms develop in a manner that is resistant to any attempt at over-simplification.

The situation is perhaps a little clearer with regard to theory. The Enlightenment saw the demise of classical architectural theory, in spite of the attempts by Jacques-François Blondel in particular, from the 1750s to the 1770s, to revive it. The architects of the revolutionary period introduced a new concept, in which the rules presupposed by the ideal of *convenance* were rejected.

A transitional period may be characterised at least as much by what disappears, and by the reasons for that disappearance, as by the new configurations which emerge. The exhaustion of the architectural theory inherited from the *Grand Siècle* may thus serve as a guiding thread in our study of the changes that took place in the course of the eighteenth century.

One should not suppose, however, that the theoretical debate will by itself serve to explain very much, not least because it must reflect, albeit with distortions, the burning issues of the day. Among such issues, the new division of labour between architects and engineers may well have played an important role. Engineers were distinguished from architects in the eighteenth century, and began then to exert a considerable influence upon the design of space. By deploying a network of new institutions, by codifying their knowledge and practice, they developed a system which was to pose an increasingly serious threat to the architects and, indeed, to wrest some of their traditional prerogatives from them. The theoretical renewal of architecture in the revolutionary period may also be said to constitute an attempt to recover lost ground.

One of the main thrusts of the present study is therefore the link between the demise of classical theory and the relationship between architects and engineers. To be more precise, I have sought to compare the theoretical statements made by architects with writings of a more narrowly technical kind, so as to identify significant discrepancies.

My interpretation of these differences refers to changes on the social and economic level and to the effects of such changes upon the organisation of space. The division between architects and engineers increased as the state's control over planning was consolidated, as economic exchanges intensified and as the rate of technological change accelerated. The concepts of engineers bear the mark of these general tendencies, just as the doctrines of classicism reflected the desire for stability in the hierarchies of the society of the *Ancien Régime*.

The reader should be warned, however, against any too facile an opposition between architect and engineer, for the two professions were still complementary. Since there was no real breach until the very end of the century, the situation was characterised rather by the emergence of an independent but imaginary technology deployed by engineers, whose task it was to plan a territory. Their actual practice remained in fact close to that of the architects. In many respects, the latter also give the impression of being heirs to the architects of the *Grand Siècle* and to their efforts of rationalisation. In order to arrive at a more accurate assessment of the influence of this inheritance, one has to go back to the founding, in 1671, of the Royal Academy of Architecture. Colbert had designed the Academy as an instrument of technical control, which, given the influence of Perrault's doctrine, would seem to have blazed a trail for the engineers. Imbued from the start with contradictions, classical architectural theory presented a variety of approaches, its very diversity accounting for its longevity and its use, in part, by the technicians of the Enlightenment.

A retrospective survey is also required if one wishes to understand the attempt made by Jacques-François Blondel, undoubtedly the greatest teacher of architecture of his day, a convinced rationalist and a theorist of planning, to revive the classical ideal. His monumental *Cours d'Architecture*, which was published between 1771 and his death in 1774, combined original insights with the consolidation of retrogressive positions, and a highly refined aesthetic sensibility with the systematisations of stultifying academic rules. The failure of his endeavour was to take on a very particular meaning in later years. In his desire to re-establish links with tradition, he merely succeeded in highlighting its shortcomings, and thereby helped unwittingly to hasten its decline.

Blondel's ideal was a hierarchical architecture, reflecting the social conditions of the *Ancien Régime*, though rationalised as far as might be. In his mind, the notion of *convenance* was concerned mainly with the reflection of status made apparent through controlled architectural magnificence. But opposed to this was another rationality, one based on notions of utility rather than on any formal considerations. Founded in 1716, the corps of the Ponts et Chaussées was consolidated under Perronet. The engineers of the Ponts et Chaussées, who were entrusted with the building of roads and bridges, and also with architectural projects, developed an approach of undeniable efficacy, concerned as it was with issues of the moment. Utility, allied to the excellence and disinterestedness of the engineer, the general will and the policy of these norms and their relation to architectural theory will be the subject of the chapters that follow.

This comparison must be formulated in terms of the concrete issues of the day. It was because architectural theory had ceased to keep step with such issues that it finally fell apart. I have chosen to focus upon problems posed by the design and the construction of buildings and, also, those raised by the planning of town and country, subjects of irreconcilable oppositions.

Apart from Blondel, a number of other figures emerge at this period, for although he may have proved to be the most 'classical' theoretician of architecture he was by no means the only one. Where the decoration of buildings was concerned, Patte had much in common with Blondel, although he continued to be more concerned with problems of construction and advanced town planning. Perronet, who was chief engineer to the Ponts et Chaussées and director of the Ecole des Ponts et Chaussées from 1747 to 1794, was the embodiment of the engineer as artisan and administrator. Finally, Prony, who collaborated with Perronet on the building of the pont de Neuilly and on the pont Louis XVI, now the pont de la Concorde, sought to promote a standard based upon mathematics. In this respect, he seems to prefigure the approach of the nineteenth century. These figures will crop up again and again in the course of this study. They belonged to very much the same circles and, in the course of their development, they seem to have conducted a kind of conversation, through which the crucial questions of the day – architectural theory and practice, teaching and the diffusion of knowledge, planning and design – were debated.

In the end, this debate encountered some serious obstacles. Blondel's wish for spatial and functional coherence was at odds with a system involving the free disposition of fractures and reassemblages. One party favoured continuity pure and simple, whereas the other desired a combination of the continuous and the discontinuous. After a certain time, a confrontation became inevitable.

Such concepts were linked, or so it has been claimed, to differing perceptions of society and of the process of building. Whereas Blondel sought to rationalise what already existed, by harmonising architecture with the social structure, the thinking of the engineers evolved in an egalitarian context which may be described as pre-revolutionary. Their thinking was not therefore divorced from the paradoxes of state and citizen, of an intention to resort to planning and architectural design. In the area which concerns us here, the revolution begins with the publication in 1788 of Lagrange's *Mécanique analytique*, the closure of the old Academies, the foundation of the Polytechnique in 1794, the mathematisation of the science of engineering and the passage from the classical world to the universe of the machine and of precision. This passage would not have been a real possibility without the paradoxes upon which the engineers' system was based. The tensions within their discourse, and the often risky nature of their venture, seem to anticipate a universe of process, of flux and of movement, quantified by means of calculation.

Admittedly, the first intimations of this change only occurred around 1750, and it had not even assumed a very definite shape between 1770 and 1780. The French Revolution hastened the process to a quite spectacular degree. It was anyway obscured by contrary tendencies, involving the union of the arts and the sciences, the natural quality of technological intervention, and even the pronounced taste of engineers for aesthetics. However, the important thing to grasp is that the architecture of the closing years of the eighteenth century was already part of a new world. Through their rejection of classical formulae and their quest for a natural basis to architecture, Boullée and Ledoux belong to the same realm as the engineers. But did architecture yet represent a form of knowledge susceptible to being theorised? Did it yet have a place within the decision-making processes?

It was in order to answer such questions that I set about writing the present book. The study of the demise of classical theory and of the emergence of an independent technological discourse actually enables one to retrace the origins of the discord in the relationship between architecture and scientific and technological rationality which was often denounced subsequently. The crisis of modern architecture is linked in one respect to that affecting the relations between this rationality and the sphere of the visible, which no longer seem to be linked by any form of organic necessity. From this point of view, classicism offers the paradoxical example of a fertile relation between architectural debate and scientific investigation. If classical architecture in fact depended upon a network of knowledge and practice which was earlier than any scientific systematisation, the questions which it prompted were a constant source of stimulus for the scientists. The interest shown by a man like Claude Perrault, a doctor and a member of the Academy of Sciences, in the theory of Orders and its links with phenomena in the domain of perception was by no means unique. Architecture served as more than just a metaphor for man's domination of the world, for it constituted a field for experimentation, and a model for the comparison of already highly formalised theories with looser determinations, linked to aesthetics and to taste. This complicity dissolved in the course of the eighteenth century, giving rise to the emergence of opposed pairs, namely, sensation and calculation, solidity and construction, tradition and innovation. Theoreticians then saw it as their task to reconcile inexorably divergent terms. Blondel therefore perfected architectural criticism, Prony posited an intimate connection between poetic inspiration and the analytic spirit, while Patte tried to base construction on a number of propositions which were certain and self-evident to all. We know the subsequent fate of the union of the arts and the sciences, a theme which was nevertheless built into the programme of the Revolution. In this respect, the Enlightenment venture, and its failure, are still relevant to us today.

But it is also worth reflecting upon the links between theory, aesthetic sensibility and the pleasure of looking, a pleasure arising in large part out of habit but one which nevertheless does not necessarily rule out the invention of new combinations, nor even attempts at rationalisation. From this angle too, classical theory turns out to be full of interest. The details of the various orders, the cornices and the mouldings which feature on every page of the treatises, represent the basic elements of a genuine apprenticeship in perception. The architects were also quite willing to explore unexpected solutions, or at times to simplify their orders to make it easier for the workmen to execute them well. Classical theory offered a model open to change. This flexibility was gradually eroded in the course of the eighteenth century, a development which also corresponded to a far-reaching change in sensibility, making it possible for architectural detail to be reduced to ornament. I needed to understand this shift, which is as revealing as the general evolution of design. I therefore thought it appropriate to devote some space to the problem of detail in architecture, before studying Blondel's aesthetic more closely. If I were asked to define the qualities of classicism and of Enlightenment architectural theory which are of relevance to us today, I would suggest that the former's apprenticeship in the pleasure of looking, and the tension between invention and rigour, apparent in every treatise, are still worth retaining, and that the latter's ambiguous combination of loyalty to the past with an urgent sense of the need for renewal is a quality that we have not yet rediscovered.

1

GENERAL REMARKS

Every division into historical periods poses a problem, for one is involved in isolating coherent sequences within a confused mass of facts and hypotheses. The choice of such sequences generally reflects the author's basic preoccupations. I have noted my own preconceptions in the course of the present chapter, beginning with the notion of an eighteenth century in which the part played by tradition and by innovation is sometimes difficult to grasp. The collapse of classical architectural theory must be seen in the context of a general shift from an 'approximate world' to a 'universe of precision'.[1] I will then consider the connections between architectural theory and professional disputes, before, finally, discussing the key figures around whom I have chosen to organise this study.

THE TRADITION OF ENLIGHTENMENT

Since it is poised between the classical age and the industrial era, the eighteenth century represents an intermediate period, and one that is still difficult to interpret. Yet our view of Enlightenment architecture depends on the kind of interpretation that is advanced. If it is still legitimate, on the basis of the pioneering studies of Emil Kaufmann,[2] to emphasise the innovatory concepts of Ledoux, Boullée and Lequeu, one can by the same token draw inspiration from Allan Braham's analyses, and therefore reject the 'obsessive search for underlying principles' which, according to him, has long been in favour in art history.[3] Indeed, the problem revolves around the importance of the seventeenth and nineteenth centuries in the interpretation one advances of the Enlightenment.

I shall return to the question of classicism in chapter 2. Here I want simply to observe that, if its definition, architecturally speaking, dates back to the last century and is still hedged around with uncertainties, one can nevertheless point to a society and a mode of production that were classical – a fragmented universe, a society that was still traditional and in the grip of custom, and a technological context in which water and wooden machinery predominated.[4] In spite of scientific progress, the works of Newton, Cartesian philosophy or the bold speculations of Pierre Bayle, seventeenth-century life was still conducted at the leisurely pace dictated by the available means of transport, in a daily struggle against nature, liable at any moment to annihilate the boldest ventures. This latter aspect is typified by Holland's patient conquest of the sea. The development of its trade and the dynamism of its publishing houses also bear witness to innovatory currents which emerge in a framework which one might have supposed was immutable. The nineteenth century offers us another perspective from which to interpret the Enlightenment. Industrial expansion, a series of more or less violent spurts and periods of growth, progress, the mathematisation of knowledge, are phenomena too familiar to need describing. We still have to account, however, for the eighteenth century, and for the transition from the world of gesture and of approximation to that of the machine and of precision.

There is a great temptation to interpret this transition in terms of the succeeding period, and to place more emphasis upon innovatory concepts and 'pioneers' than upon tradition and its upholders. Through its confidence in the virtues of experimentation, through its advocacy of social reforms based upon reason, the philosophy of the Enlightenment portends a work of conscious elaboration.

However, the eighteenth-century mode of production resembles to a confusing degree that of the previous period. Although coal heating was then used in London, and Newcomen's machine had made its appearance in the mines,[5] hydraulic works were still well to the fore, especially in France.[6] Nevertheless, what the Enlightenment *Philosophes* understood by reason was not what was to be understood by the term around 1800, and they themselves were often disconcerted by a Revolution whose possibility and extent they had been far from suspecting. Although it sometimes seems to anticipate the nineteenth century, the eighteenth century remained above all else the heir of the classical age, and its main problems derived from those of the previous century. The thinkers of the *Grand Siècle*[7] were already haunted by questions concerning nature and reason, the individual and society, experiment and calculation, just as men such as D'Alembert and Lavoisier were to follow the trail blazed by Descartes, Mariotte or Huygens.

We therefore need to consider the range of possible interpretations of the Age of Enlightenment. One can of course choose to emphasise what was inherited from the past or what seemed to anticipate the future, but in either case one would undoubtedly be guilty of over-simplification. What in fact seems to have occurred is a kind of critical mass phenomenon, a slow rise culminating in a qualitative leap around the year 1800.[8] In the eighteenth century, the classical edifice finally cracked; in a hesitant and contradictory fashion, the possibility of a new equilibrium was gradually discernible.

In architecture, the new developments are often difficult to assess. The main theoretical concepts, for example, date from the seventeenth century: *bienséance*, *convenance* and even taste itself had been the subject of debate within the Academy since the time of Colbert.[9] There was relatively little development, however, in what was actually built, and one wonders whether the forms of someone like Ledoux are genuinely 'revolutionary'. From the rope manufactory of Rochefort to the observatory of Paris, proceeding by way of the study of the art of fortifications, the bareness of walls, the contrast in volume and emphasis upon geometry are not alien to classical order. What occurs in Ledoux is perhaps a simple displacement of the ideal of *convenance*, which leads him to generalise solutions which had formerly been restricted to utilitarian projects. But this displacement refers to the problems posed by architectural theory, and by its institutions, around the year 1800. Since grand treatises, such as those by Durand and Rondelet, are to be found after this date, what exactly does one mean by the demise of classical theory?

What undoubtedly ceased was a certain kind of dialogue between the various treatises. As Werner Szambien emphasises, 'Classical theory only had possible lines of development as a collective venture for as long as there was still scope for conflict.'[10] The unity of classical theory therefore consisted in such conflicts, which were always liable to erupt, and which in reality reflected the existence of a broad consensus. In challenging each other over the question of the Orders, were architects not implicitly recognising the validity of research undertaken in their subject since the Renaissance? Conversely, the demise of classical theory was characterised by the increasingly tame nature of a number of debates. Thus, the proportions of the Orders or the rules of *convenance* became less and less important, and controversies on such topics abated. It remains for us to identify the transformations of which this disaffection was the symptom.

These transformations quite obviously reflect those occurring in society. The last years of the eighteenth century saw the collapse of the *Ancien Régime*, a hierarchised 'society of orders'[11] for which the court at Versailles constituted to some extent the archetype.[12] In place of the regulated play of oppositions and of parties, the Revolution established a new type of relation between state and citizen, and a new, institutionalised violence. At the same time, a technological rationality based upon physico-mathematical calculation had just been consolidated. In this context, the bonds between architecture and scientific and technological rationality were gradually dissolved. Soon architecture would constitute nothing more than an ineluctably subordinate domain for the application of technology. In spite of the fact that Louis Hautecœur continued his *Histoire de l'architecture classique en France* up to 1900,[13] there was nevertheless a distinct break at the beginning of the nineteenth century. As a new understanding of the nature of society arose, the order of knowledge ceased to reflect the need, which had still inspired the *Encyclopédistes*, for a visual and spatial equivalent. While 'court society' was able to recognise itself in the gradations of classical architecture, the rising bourgeoisie would increasingly identify itself with engineers whose task it was to organise the territory and to adapt it to the needs of industry and of profit. The accuracy of their calculations would become synonymous with progress. From the world of gesture and approximation to the universe of precision, the evolution of technology came up against that of society.

By relating architecture and technology to each other, I hope in this study to isolate some of the factors which made this evolution possible. The collapse of classical theory may also be interpreted in terms of the growing technological domination of the engineers, a domination whose structure and functioning I will explore below.

THEORY, PRINCIPLES AND APPLICATION

I propose to begin this study of theory by considering a number of questions arising out of the link between architectural theory, the execution of a particular project and professional debate. In *L'architecture et la règle*, Jean-Pierre Epron defines theory as an edifice with several floors, comprising a corpus of reference, the set of architectural principles which the corpus serves to illustrate and, finally, rules governing the execution of a particular project.[14] I shall generally take theory to consist of the exposition which the treatises give of it. The transformation of written knowledge into an actual design very probably involved an elaborate discourse, but it is extremely difficult for us to reconstruct it. Likewise, we can generally do little more than hazard a guess as to the complex relations between theory and the building-site. This is one of the basic shortcomings of the present study.

In architectural treatises, however, we encounter a corpus of reference, 'the most beautiful buildings in France', and this serves to define the ontological or formal principles behind the rules governing a particular project. But these elements must be distinguished from the norms implicit in professional practice. Engraving, in particular, made it possible for alterations to be made to designs which had in fact already been realised. Such 'corrections' were commonplace, the most striking example being Ledoux's thoroughgoing rearrangement of his château at Benouville.[15] Of his precursors, François Blondel might also be mentioned in this context. One would never suspect him of such irregularities, and yet he clearly did 'correct' the Porte Saint-Denis when he published his *Cours d'architecture*.[16] Indeed, treatises offer ideal interpretations of projects whereas these were executed as if at one remove from grand principles, with a considerable leeway being tolerated. A similar toleration was granted the practice of speculation among architects, although it was roundly condemned in their writings. In spite of peremptory declarations, the ideal of *convenance* was above all a matter for negotiation; supposedly inflexible principles yielded to more accommodating approaches.

Even though the recommendations contained in the treatises cannot be taken literally, they do nevertheless provide us with valuable evidence. They serve to delimit the domain of architecture and of its practice. This delimitation, however, is not unproblematic, insofar as authors often contradicted themselves regarding these boundaries. In the course of the eighteenth century, the emergence of independent architectural criticism tended, however, to stabilise the field covered by the treatises. Decoration and distribution were its backbone. Construction, still claimed as the preserve of architects, was subject to different approaches, as we shall see in the case of Jacques-François Blondel and Pierre Patte. Be this as it may, architectural treatises help to define a discipline, together with the profession which aspired to practise it.

This latter aspect is never clearer than in the context of teaching, which represents one of the main advances of the Enlightenment. Countless institutions were founded, both in Paris and in the provinces, from the Ecole des Arts of Jacques-François Blondel to the Ecole des Ponts et Chaussées run by Perronet, from the free Schools of Drawing to the institutions sponsored by the Academies of Dijon, Toulouse or Rennes.[17] In this context, the treatises were seen as pedagogic instruments, and Jacques-François Blondel would justify the publication of his *Cours* on the grounds that he wished to provide students with a useful tool for their work.[18] In the eighteenth century, teaching was closely tied to professional strategies, as is shown by the mode of recruitment of the Academy of Architecture[19] or by the *esprit de corps* of the Ecole des Ponts et Chaussées.[20] It was through teaching that theory confronted the concrete problems of the period, which were thereby more precisely delineated. Thus *convenance*, in Blondel's interpretation, did not merely represent an aesthetic aim but involved the gamut of relations between architecture and society. Theory and doctrine were therefore inextricably intertwined.

Information gleaned from treatises needs nevertheless to be supplemented, especially with respect to the engineers, whose works lay claim, significantly enough, to technical objectivity. I shall seek below to go beyond this apparent objectivity, by having recourse to other documents belonging in the main to the Ecole des Ponts et Chaussées.

NOTABLE INDIVIDUALS

I want finally to say a word or two about the main figures encountered in the course of this study, namely, Jacques-François Blondel, Pierre Patte, Jean-Rodolphe Perronet, Gaspard Riche de Prony, Etienne-Louis Boullée and Claude-Nicolas Ledoux, architects and engineers who have been judged to be sufficiently representative of the theoretical debate.

Jacques-François Blondel (1705–74) springs to mind first, since he defined himself as above all else a teacher and theoretician of architecture.[21] After a fairly long period of training, he founded the Ecole des Arts around 1740, which was intended to provide young 'artists' with a more exhaustive instruction than was available at the Academy of Architecture.[22] The Ecole des Arts was to be one of the most famous institutions of the eighteenth century, and its founder was to be given the task of writing most of the articles on architecture in the *Encyclopédie*.[23]

Appointed architect to the King in 1755, and Royal Professor in 1762, Blondel was the quintessential defender of tradition. Yet, in spite of his attacks upon the bolder flights of young architects, Ledoux in particular,[24] he tried to adapt this tradition to the changes that were under way. Through his influence upon characters as diverse as Leroy, De Wailly and Boullée, Blondel stands as a central figure, whose shortcomings reflect to a certain extent those of classical theory itself.

Virtually nothing is known of the life of Pierre Patte.[25] He was born in Paris in 1723 and died in 1814, but his main claim to fame was his polemic with the authors of the *Encyclopédie*, whom he accused of having plagiarised Réaumur,[26] and his part in the dispute over the Panthéon.[27] His name also survives because of the innovatory character of his concepts regarding town planning, mainly expounded in his *Monuments érigés à la gloire de Louis XV* (1765), and in his *Mémoires sur les objets les plus importants de l'architecture* (1769).[28] In contrast to Blondel, Patte was never a member of the Academy, although he applied upon several occasions, in 1769, in 1774 and again in 1803.[29]

Although an architect by training, Jean-Rodolphe Perronet epitomises the figure of the engineer.[30] Born in 1708, his humble origins prevented him from entering the corps of the Génie. He worked for an architect in Paris from 1725 to 1745, before entering the corps of the Ponts et Chaussées and being nominated for a post in Alençon.[31] Singled out by Trudaine for his talents as an engineer, he became director of the Ecole des Ponts et Chaussées, a position he held until his death in 1794.[32] Appointed chief engineer of the corps of the Ponts et Chaussées in 1747 also, he helped to formulate a training specific to engineers. His achievements as a structural engineer made his administration famous.[33] Indeed, Perronet's contemporaries were struck by the boldness of the bridges which he had built. Neuilly, Sainte-Maxence and the pont Louis XVI all reveal, as monuments, the dual allegiance of their creator, for in many respects Perronet was still an architect, whereas his work on structures transcends that field. We will return to this aspect of Perronet's career below, especially with respect to the system of 'stretched arches' to which he gave his name.[34]

Gaspard Riche de Prony was the youngest of the quartet, since he was born in 1758.[35] His death in 1839 makes him every bit as much a man of the nineteenth century as an engineer of the Enlightenment.[36] If, unlike Perronet, he never became an architect, his scientific capacities earned him the patronage of the older man and the post of director of the Ecole des Ponts et Chaussées in 1798, which he held until his death. He also taught mechanics at the Ecole Polytechnique from 1795 to 1815.[37] From the 1780s, Prony epitomised the disputes over the mathematisation of knowledge and the rationalisation of conceptions. This is why he features in the present study.

In spite of their diverse careers, Blondel, Patte, Perronet and Prony moved in the same circles, which is why their paths crossed so often. Like Blondel, Perronet was a member of the Academy of Architecture, whose lectures Patte had attended around 1745 and which he had unsuccessfully tried to join. Prony was Perronet's assistant when he was building the pont Louis XVI,[38] but he was also interested in architecture and the fine arts early on in his career.[39] Blondel, like Patte, contributed to the *Encyclopédie*; Patte also completed the older man's *Cours d'architecture*.[40] Any list of institutions around which these four figures gravitated would be virtually endless, as would a rehearsal of the occasions upon which their paths had crossed. Fréron was a friend of Blondel, and Patte was a contributor to the *Année littéraire*. Perronet drew inspiration from the Ecole des Arts when he was defining the pedagogical approach of the Ecole des Ponts et Chaussées. Blondel's Ecole would also serve subsequently as a preparatory school for the Ponts et Chaussées, with Perronet's pupils also coming to study architecture there.[41]

In the wake of Emil Kaufmann's pioneering works,[42] numerous studies have been devoted to Etienne-Louis Boullée (1728–99) and to Claude-Nicolas Ledoux (1736–1806),[43] so there is no need to devote too much space to them here. If it is a question of studying the demise of classical theory, in relation to the institutional and technological foundations of the engineers, Boullée and Ledoux allow a better understanding of the extent of the change. They were members of the Academy of Architecture, but their links with the world of technology were likewise close. Thus, Ledoux began his career by practising as an engineer in the employ of the administration of the Eaux et Forêts.[44] He was then architect for the Ferme Générale and Perronet's assistant as inspector of the saltworks in the eastern region of France.[45] Under the *Ancien Régime*, Ledoux's influence is discernible in the architectural projects of the pupils of the Ecole des Ponts et Chaussées.[46] During the Revolution, he was a member of the Institute, putting his signature, alongside those of Prony, Coulomb and Périer, to a *Rapport sur les moyens de sauver les personnes renfermées dans les maisons incendiées*.[47] Subsequently he became professor at the Ecole Centrale du Panthéon alongside Leplace and Monge.[48]

Blondel, Prony, Boullée and Ledoux thus constitute a homogeneous milieu of technical men, theoreticians and practitioners aligned with the dominant institutions. It is worth emphasising in this respect the restricted nature of this milieu, and the consequences of this for the scope of the present study. These architects and engineers, working for the most part in Paris, cannot in fact be held to reflect the situation in the country as a whole.

Conversely, it must be admitted that architecture was predominantly Parisian on the eve of the Revolution, at any rate as far as its theoretical definition and its claims to be exemplary were concerned. If construction in the provinces still represented, according to Pierre Chaunu, from 80% to 90% of what was built at the end of the eighteenth century,[49] then the finest buildings of the Ile-de-France nevertheless embody the national genius. The Ecole des Beaux-Arts and the Conseil des Bâtiments Civils also revived some of the traditions of the Academy of Architecture. The problems which architecture was to face, and the difficulties it experienced in adapting to the conditions created by the Revolution, were in part associated with its limited geographic extension, in contrast to the local involvement of engineers.

This objective difference between architects and engineers was to have repercussions, as we shall see below, at the theoretical level, even though the two professions were still closely linked. The architects' tackling of their projects in isolation was thus in contrast to the engineers' aim to organise things overall. More generally, the latter tended more and more to inscribe what was built within a dynamic system of tensions and flows, a circulation of men and commodities, for example, of which the design was supposed to be the expression. Where Blondel thought still in terms of spatial coherence and a direct relation between social demand and architectural response, one may discern the gradual introduction of new forces, and an interplay of disjunctions and connections, serving as a prelude to a much more general reorganisation of knowledge and practice. The crisis of classical architectural theory may therefore be regarded as having exemplary value.

2

THE CLASSICAL TRADITION AND ITS AMBIGUITIES

THE CLASSICAL 'MANNER'

Because Jacques-François Blondel would seem to have aligned himself with the inheritance of the *Grand Siècle* and with François Mansart,[1] at a time when new solutions were being advanced, we need to consider the problem of tradition and its contemporary relevance. What was the nature of the tradition? And what determined the attitude of eighteenth-century architects towards that tradition?

In his *Cours d'architecture*, Blondel repeatedly refers to buildings such as the Château of Maisons, the Val-de-Grâce, or the chapel of the Sorbonne. Tradition thus seems here to be identified with a classical architecture which might provide a point of departure for a study of the Enlightenment. Yet this point of departure seems itself to resist all attempts at analysis. A few years ago, Jacques Guillerme devoted an article to the semantic ambiguities surrounding the term 'classical', together with the frequent confusions between the art of Antiquity, the idea of an Apollonian perfection or, again, the exemplary nature of a particular work.[2] French classicism reflects this arbitrariness for, whereas its apogee occurred during the reign of Louis XIV, its boundaries remain ill-defined, since an authority such as Hautecœur, as we have seen, understands it to have extended from the Renaissance up until the nineteenth century, whereas other historians stop short at the eighteenth century. In opting for the term 'manner', which is the preferred usage of the architectural treatises, it may be that Jean-Marie Pérouse de Montclos has gone some way towards resolving this problem.[3]

Yet eighteenth-century architects were still not sure what the 'manner' of the *Grand Siècle* was. To be more precise, they felt that it was possible to identify tensions which might be turned to their own advantage. These tensions were predominantly concerned with architectural production, and Jacques-François Blondel thus contrasted the sobriety of François Mansart with the liberties taken by Jules Hardouin, as if to symbolise an ever-present alternative in his own practice as an architect.[4] Classical theory, from which Blondel borrowed a certain number of principles, represented yet another tradition. This too should be treated as a sign of ambiguities that were still operative, in spite of the inevitable displacement of the architectural debate. It is therefore feasible to try and trace a number of these ambiguities back to their source, regarding them as so many ruptures by means of which the classical edifice contrives some possible lines of development. This is effectively what we have attempted to do in the present chapter, which does not contain an exhaustive description of classical doctrine, but simply describes a number of the tensions which may be said to constitute it.

FROM THE TOWN TO THE DETAIL: THE LIMITS OF THE CARTESIAN MODEL

The paradigmatic character of Cartesian thought in the seventeenth century has often been emphasised. By establishing order and measure in the very heart of the world, this thought specified discontinuities and relations which offered the possibility of attaining a mastery over the real. The intelligible, as in mathematics, might prevail over the sensible, with the ultimate ambition being the founding of a general science of order,[5] the 'universal characteristic' of which Leibniz was later to dream.[6]

Intellectual pride of this sort had impinged upon architecture also, and the attempt was made to define a series of rules governing its production. The majority of authors of the architectural treatises of the period attempted to establish a form of architectural knowledge based upon the immutable figures of geometry, upon a combinatory which might regulate the interplay of the parts and subordinate them to the whole. Rational judgement thus supplanted the unsystematic invention of the Renaissance; the doctrine of proportions was the commonest expression, in the domain of architecture, of this rationalism.[7] Roland Fréart de Chambray, in his *Parallèle de l'architecture antique avec la moderne*, which was published in 1659, attempted to harmonise the measurements given by the classical authors, so as to systematise the use of proportions; François Blondel, in his *Cours d'architecture*, went on to compile an inventory of the solutions advanced to the problem since Vitruvius. In the seventeenth century, an *esprit de système* seemed often to prevail over the evidence of one's senses, as much in architecture as in any other domain.

This interpretation of classical architecture, developed by François Fichet in particular,[8] requires some qualification. Although the imagination of the *Grand Siècle* was very probably to some extent Cartesian, the towns of the period were still largely medieval, and the imprecision of machines constructed out of wood seems to belie the rigour of Descartes' mechanistic conceptions. In a domain like architecture, where the sheer weight of material factors is so strongly felt, one is bound to wonder what classical theory omitted to mention, touched upon in passing or was obliged to tolerate. A study of such limitations allows one a better understanding of the aspirations that found expression in the ordering of architectural knowledge, and of the purchase they had upon the world. The architects cannot in fact have been solely concerned with *mathesis universalis* or taxinomia.

THE TOWN

The town, which rarely appears as an entity, was very probably the most glaring of the omissions evident in the classical treatises, for it seemed to disappear behind the rhetoric of the Orders and the rules of *convenance*. Yet the town was invariably implicit in such discussions. The architect's knowledge of geometry was meant to allow him to come to terms with the impacted tissue of the town, in which he might have, for example, to recreate a symmetry in circumstances which suggested that no such thing was possible[9] – as was the case with Le Pautre's *tour de force* at the hôtel de Beauvais. Being both absent and present at the same time, the town was located at the very edge of architectural theory, while the operations to which it was subject were hard to articulate. It is in this context that one can best understand Le Muet's pragmatism, his 'realistic utopia'.[10] The town, which had not yet been fully tamed politically, was subjected to a typology whose rules were almost always obscured in the treatises.[11] Indeed, the town had later to be reintroduced into architectural discourse, even if at the risk of highlighting its shortcomings.[12]

ARCHITECTURAL DETAIL

The question of detail was absolutely crucial to the architecture of the period.[13] Treatises devoted lengthy expositions to this problem, for the coherence of architectural theory depended upon it. From the volute of the Ionic capital to a simple Astragal, arcs of a circle, conic sections or clusters of tangents were supposed to allow one to determine quite precisely the profile of each architectural member, by linking it to the general system of the Orders (Fig. 1). The works of Palladio, the Scamozzi and Vignola had presented methods of lay-out of varying degrees of complexity. Drawing their inspiration from Italian sources, French authors also formulated rules based upon the same principles.

Nevertheless, architects would seem to have been free to improvise to a degree, if one can judge their work fairly from the sections used in their buildings. In his *Cours d'architecture*, Jacques-François Blondel summarised some of the liberties that might in fact be taken, from the design of the casement windows in the Louvre to the mouldings on the collège des Quatres-Nations. As in the case of the town, so too here, the rules seemed not to take account of actual architectural detail.

Every author, however, testified to the importance of detail, starting with the mouldings. A good architect might be recognised by the elegance of his mouldings; irrespective of any large-scale regularisation, the quality of a building depended upon the design of its tori and of its cymas, upon the delicacy of its ornamentation. No doubt the subordination of the parts to the whole determined the general ordinance of the building, the relation of the frontispieces to the rest of the façade, the respective heights of the base, the main storey and the attic storey, yet one had still to decide upon the outline and dimensions of the various details, the section of the mouldings, together with that of the balustrades. One would then be in a position to identify the Order and the proportions of the column, and of its more discrete 'expression',[14] which made its effects more specific.

The attention paid to mouldings was especially noteworthy. As d'Aviler observed, 'mouldings are to architecture what letters are to writing'.[15] The 'expression' of the Order was organised by means of a kind of alphabet which the architect was absolutely required to master. This alphabet offered an almost infinite number of possible combinations: 'Now, just as through the combination of characters, an almost infinite quantity of words in various languages may be created, so too by the mixture of mouldings, one can invent a quantity of different sections for every kind of Order and for regular and irregular compositions.'[16]

Alongside that theory which prescribed imitation of the ancients, the art of sectioning established a complex kind of freedom. Admittedly, this freedom, like the invention of the architects, had its limits. But liberties were in the end justified in terms of the pleasure they afforded the eye; classicism was viewed in the light of a sensuality that was at once veiled and aroused by the exposition of its leading principles. Beneath the structure of pure reason, the spectator could enjoy, through the observation of subtle nuances, and of the learned play of elisions and transitions, an endlessly renewed source of pleasure.

It was very probably this same play of appearances which determined the nature of classical space, namely, the internal volumes whose dimensions nevertheless depended upon number or upon simple arithmetical relations.[17] One may in fact discern a series of modulations achieved through, for example, the positioning of mouldings where floors emerge, a device which served to modify the apparent height of the rooms.[18] Whether one was concerned with the interior of buildings or with their general ordinance, detail served in either case to render the overall effect more supple or more fluid. Kaufmann was probably not incorrect in contrasting the sensuality of the classical constructional material with neo-classical expressivity.[19] When placed before a building, the spectator would observe in turn the major balances, details of ornamentation, the rhythm of the frontispieces, then the refinement of design in the capitals. This passage of the eye was irreducible to a mechanistic approach; similarly, the mouldings added further nuances of light and shadow, mixing and dissolving them in ways that defy discursive analysis. Beneath the ideal of a scientific mastery of proportions, the possibility of taste, to which the eighteenth century would refer so much in the future, was thereby introduced.

But the art of detailing lends itself to other sorts of analysis. To begin with, one should stress its links with current thinking on geometrical optics and its limitations. In the treatises, the representation of details, the outline of the mouldings in particular, was not subject to the rules of elevational representation, and therefore seemed not to refer to the viewpoint of an unrealistically motionless spectator, enjoying an ideal view of the building. The delicacy of the architectural members was appreciated by changing one's position. With the profiles, there therefore appeared a kind of figuration which stood midway between realism and abstraction.

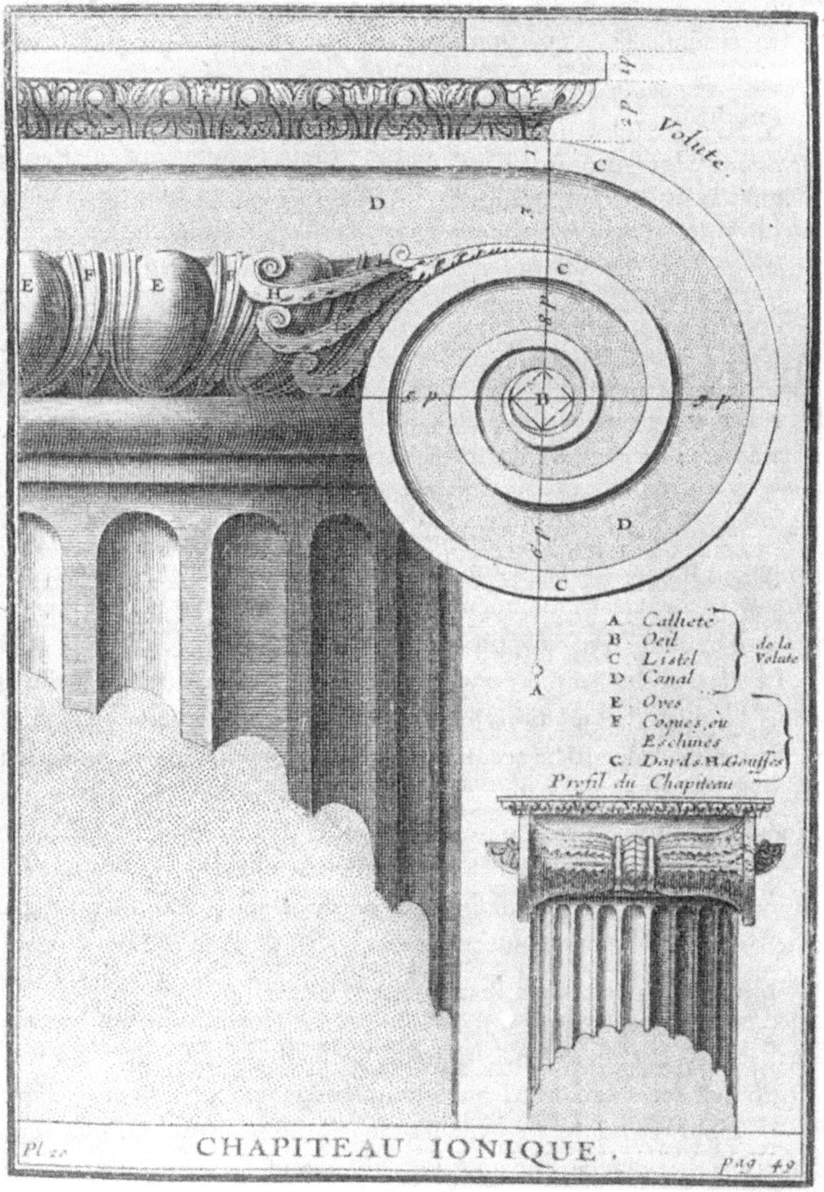

FIG. 1 C.-A. d'Aviler, *Cours d'architecture* (1691, second edition 1720), Ionic capital.

The problem of detail was also linked to that of optical correction in relation to height. This question, which had been a preoccupation of Vitruvius himself, and which Serlio had resolved by introducing a requisite distortion in the upper parts of the building, was raised once again at the end of the seventeenth century, at the time of the famous dispute between Perrault and Blondel in particular.[20] Optical adjustment concerns more particularly those architectural details which, at too great a height, might seem stunted. Whichever solution the authors of the various treatises adopted, it is worth noting the connection between this debate and the anamorphic experiments conducted by friars such as Father Du Breuil or Father Niceron.[21] It is but a short step to the development of projective geometry and the art of projection, to which Desargues was to devote his *Brouillon projet d'une atteinte aux événements des rencontres du cone avec un plan* (1639) and his *Exemple d'une manière universelle . . . touchant la pratique du trait à preuves pour la coupe des pierres en l'architecture* (1640), works which anticipate those of Monge and Poncelet.[22] Falling short of the bold formulations of Desargues, treatment of these questions usually involved abandonment of the rules of traditional perspective, which were, by and large, inappropriate. Through the investigation of the fringes of vision, of the frontier at which the privileged link between the eye and reason dissolves, the question that was actually being addressed was that of experience and the value to be accorded to it. Being irreducible to rules, detail was often of its nature experimental; similarly, anamorphoses or the art of projection seemed to undermine theory, whether architectural or optical.

One then has to investigate what it is, in theory, which gives rise to detail: the division into parts of the building, a kind of deconstruction whose main principles ought perhaps to be specified here. This division did not of course correspond to the real method of building, to the sequences of tasks to be performed on site. Nevertheless, if we look more closely into this question, we find that links between conception and realisation were still operative in the background. Although the architect might in general be at liberty to make his own decisions regarding the overall scheme, the realisation of his original intentions was heavily dependent upon the good will of the workmen, especially as regards the details of the design. The successful execution of the sections depended upon the workman's actions, which the imagination of the architect had to take into account. It was therefore crucial to arrive at some sort of understanding with the workforce, and theory was supposed to specify the modalities of such an understanding. Detail was a particularly sensitive issue in this respect, and it is therefore not surprising to find that one of d'Aviler's drawings, devoted to the design of mouldings, featured both the technical terms employed by the author and those in use among the workmen (Fig. 2).[23] For the latter, the torus became an ovolo, the echinus a quarter-round, the Doric cima a dado. What this juxtaposition of two series expressed, was a desire for coordination founded upon compatible systems of nomenclature (Fig. 3).[24] The architects and the engineers would subsequently seek to reorganise production while making use of the traditional skills of the workmen.[25] Yet, in the meantime, their day-to-day routines were governed by a division of the edifice in terms of a prescriptive logic. Within the continual process of building, theory allowed one to identify labelling strategies. The architects' taxinomia in reality provided a framework for negotiation. Theory represented such an ideal framework, which had to be endlessly adapted to fit actual conditions of work. Rigorous though the rules might seem, the actual practice of architecture was organised in terms of a set of variations and compromises. Caught between the town and detail, two terms which might almost bear comparison with the two infinities of Pascal, theory offered practitioners guidelines rather than fixed rules.

Detail thus represented a limit internal to theory, which both threatened the conceptual edifice and served to open it out on to the world of production. Around the question of detail there unfolded, as we have seen, a negotiation which was crucial to the effective operation of the theory, and which worked at two different levels, one involving debate among the architects, the other entailing discussions between architects and workmen.

Subsequent developments should therefore not be a cause for surprise. In the course of the eighteenth century, utilitarian determinations became more rigorously enforced, and less leeway was permitted in the actual execution. In such a context, architectural detail was progressively impoverished, until, in the work of Boullée and Ledoux, it became a kind of commentary upon decisions of an essentially collective nature, which were treated as the basis of architecture. At the same time, the notion of decoration lost the many resonances which it still had in the work of an architect such as Jacques-François Blondel. Durand reduced it to the elements of the buildings and his engineer pupils to questions of rational organisation. The demise of classical theory may also be understood in these terms.[26]

Nevertheless, there had been something contradictory in the status of architectural detail from the very beginning. For architects had always sought to be expressive, while at the same time being the bearers, long before the engineers, of a desire for rationalisation. The sensual nature of the materials

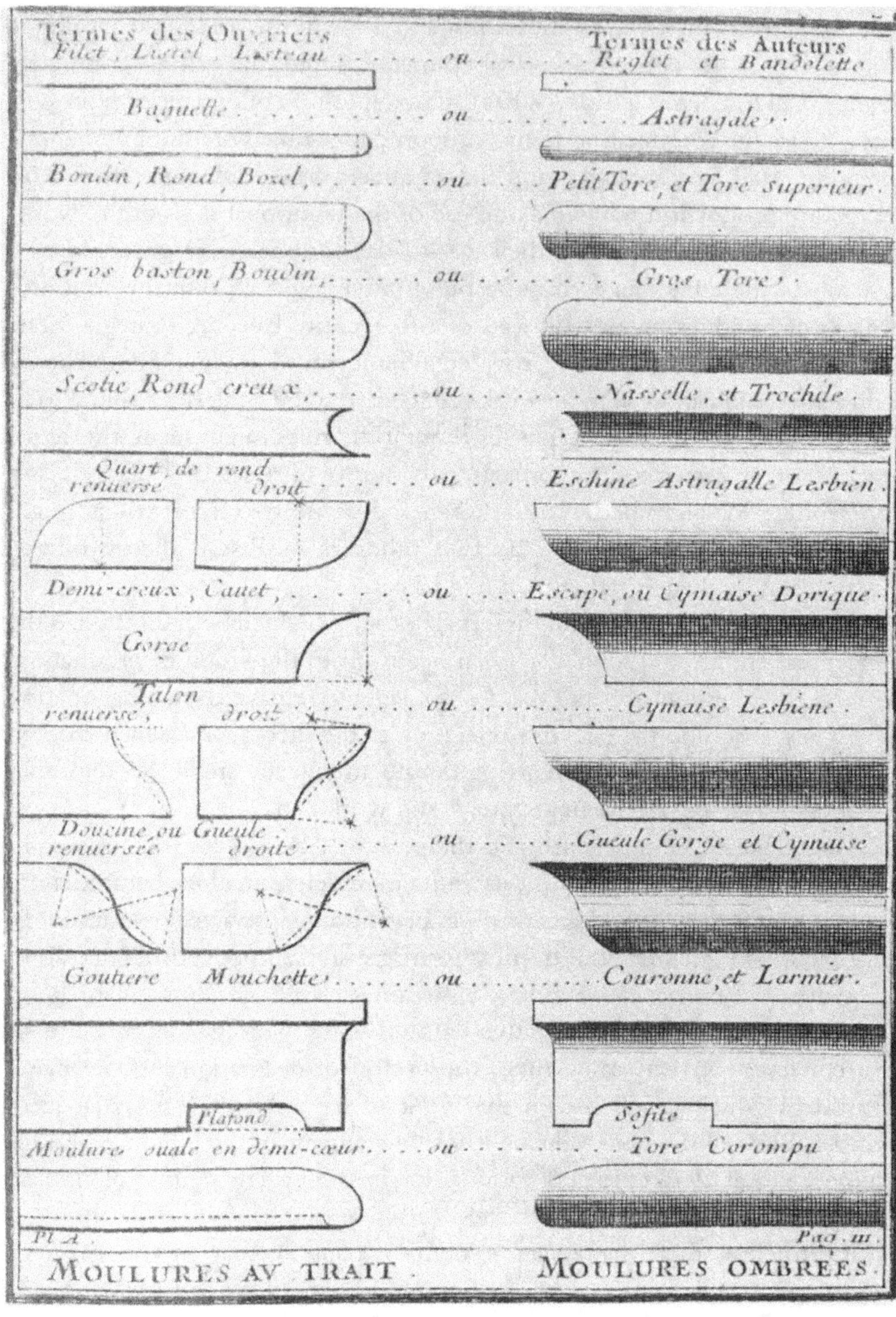

FIG. 2 C.-A. d'Aviler, 'Mouldings'. Masons' terms and those of the authors are on opposite sides of the page.

FIG. 3 C.-A. d'Aviler, frontispiece of the *Cours d'architecture*, which contained 'a full explanation in alphabetic order of all the terms' used in the art of building. The quest for a terminology common to both workmen and scholars gave rise to a dictionary, which d'Aviler placed at the end of his *Cours*. Much appreciated by professionals, this work was reprinted several times in the course of the eighteenth century.

and/or good execution of the architectural sections were the two criteria which served to define the approach of someone like Perrault and, by the same token, betrayed its contradictory character, while at the same time seeming to anticipate in more than one respect the concepts of the Enlightenment.

DESGODETS, PERRAULT AND THE FIRST ATTEMPTS AT RATIONALISATION

The founding in 1671 of the Academy of Architecture was an event of the utmost importance. Placed under the protection of the Surintendant des Bâtiments du Roi, Colbert, the Academy had three main tasks, namely, the elaboration of an official doctrine, the development of modes of instruction that might propagate that doctrine – both of these tasks being undertaken by François Blondel – and, finally, technical control.[27]

François Blondel, who was both an engineer and an architect, typifies the theoretical aims of the Academy, and he became its first director.[28] His *Cours d'architecture*, published between 1675 and 1683, is characterised by a determination to establish architectural practice upon fixed rules, based upon geometry and harmonic proportion. The contributions of Antoine Desgodets and Claude Perrault look slight in comparison to Blondel's monumental opus. Nevertheless, Desgodets and Perrault set out to interpret the fundamental tenets of rational theory.

In 1682, Antoine Desgodets' *Edifices antiques de Rome, dessinés et mesurés très exactement* was published. Although it was of little theoretical importance, this work, commissioned by Colbert, nevertheless provided Perrault with the basis he needed for his own work. In exposing the inaccuracies of the surveys of Palladio, Serlio and Labacco, and in pointing out the often disconcerting boldness of the proportions adopted by the Ancients,[29] Desgodets placed theoreticians on the horns of a dilemma. Recent authors had tended to claim that the proportions they gave were in fact derived from the example of Antiquity, and yet the latter now turned out to be marked by an oddity opposed to all *bienséance*. Because of the precision of his measurements, Desgodets gave his readers the impression that the dilemma was a real one, borne out by the evidence.

Wolfgang Herrmann has given a detailed analysis of the controversy between Perrault and Blondel.[30] As is well known, the translator of Vitruvius had caused the Academicians, and Blondel in particular, some surprise. In his

Ordonnance des cinq espèces de colonnes selon la méthode des anciens (1683), Perrault clarified his position, expressing more trenchantly what had previously been merely a statement of opinion.

Taking as his starting-point the observation that Desgodets had made regarding the range of different proportions given by the Ancients to buildings and the confusion of the Moderns in this respect, Perrault compared the members of architecture with those of the human body. If, Perrault pondered, there were immutable rules, the members should at any rate be 'sufficiently large in size to allow architects the freedom to increase or diminish the dimensions of the parts, depending upon the needs that may on various occasions arise'.[31] All that Perrault was doing here, in responding to the problem posed by the range of different proportions used, was taking his inspiration from the daily practice of architects.

Yet the gap between theory and practice was ultimately to undermine architecture's pretensions to being an immutable science, modelled upon mathematics. Once the absolute nature of proportion was questioned, the status of the architect, as guardian of rules deemed to be fixed and certain, was under threat also. In order to answer the objections which had been raised against him, objections using the traditional analogy between architecture and music, Perrault posits two sorts of beauty, one that was 'positive', 'founded upon reasons that carry conviction', such as the richness of the material used or the appropriateness of the execution, the other deriving from custom and the arbitrariness of conventions.[32]

Although the notion of 'arbitrary' beauty had been all the rage at the end of the seventeenth century, it was ultimately a somewhat secondary aspect of the doctrine developed by Perrault.[33] On the other hand, the problem of the relationship between the making of architecture and the structure of society constituted one of the key themes of the *Ordonnance*. In emphasising the role played by habit, Perrault was drawing upon a venerable tradition of argument respecting society and the values which it conveys. 'Whatever is a truth on this side of the Pyrenees is an error on the other', the adage went, as if the structuring nature of habit meant that an architect had to adapt what was built to the customs of his time.

Someone living at the end of the seventeenth century would not have found this line of argument surprising. The Dispute of the Ancients and the Moderns reached exactly the same conclusions,[34] with the Moderns expressing astonishment at the uncouthness of Greek poetry, as strange to their eyes as the proportions of the classical temples. Claude Perrault's doctrine may be compared with the positions expressed by his brother Charles in his *Parallèle des Anciens et des Modernes*,[35] in which he rallied to the defence of his contemporaries, advocating a beauty which was measured and subject to rules, a position which Boileau later tried to oppose in his *Traité du sublime*, a work based upon Longinus.[36] Whether it was architecture or literature that was involved, the origin of the rules had still to be specified. In the definition of such rules, taste was involved, which was a product of habit and of education, and such a definition already presupposed some notion of progress. Even though Perrault made some quite explicit references to taste, it did not yet have the importance for him that it was to assume subsequently; the *Ordonnance* had nevertheless pioneered a line of argument that the Age of Enlightenment would take much further.[37]

Perrault may well have been less interested in this question than in the ultimate purpose of his doctrine. It was this alone that made it possible to resolve the contradiction involved in an approach which claimed for itself both 'the freedom to increase or to diminish the dimensions of the parts' and a normative will generating new rules. Notwithstanding its own criticisms of the fanciful proportions given by the authors of earlier treatises, the *Ordonnance* itself gave rise to a new system of proportions. In order to justify his own position, Perrault began by contrasting the simplicity of his own propositions with the complicated approach of his precursors. According to Perrault, the architects who had preceded him 'had effectively said that, in order to give the proportions for a beautiful face, one should allow nineteen degrees from the roots of the hair to the bridge of the nose; twenty and three-quarter degrees from the bridge of the nose to its tip, and nineteen and three-quarter degrees from the tip of the nose to the tip of the chin', whereas he himself proposed 'to make these three spaces of equal length, allotting each of them twenty degrees'.[38]

In more concrete terms, Perrault's *Ordonnance* introduced a new module, which was equivalent to one third of the diameter of the column,[39] and he gave the height of the latter in terms of whole multiples of the module, according to an arithmetical progression of two (Fig. 4). Thus, a Tuscan column would measure 22 modules, a Doric column 24, an Ionic column 26 modules, and Corinthian and composite Orders would measure 38 and 30 respectively.[40] With the exception of the pedestal for the Tuscan Order, which would measure 6.11 modules, the pedestals would follow an arithmetical progression. On the other hand, Perrault fixed the height of the entablature at 6 modules for all of the Orders.[41]

FIG. 4 C. Perrault, *Ordonnance* (1683), the five orders of architecture in arithmetical progression.

This system was plainly in conflict with the positions advanced in Ouvrard's *Architecture harmonique* (1679), upon which François Blondel's own work had relied.[42] The latter had advocated harmonic progression,[43] and here there was the simplicity of an arithmetical sequence. Whereas harmonic proportion favours the inner coherence of an Order, the use of an arithmetical sequence serves above all to simplify the overall organisation which the various Orders constitute. The purpose of such simplification, according to Perrault, was to make the proportions easier to memorise and, by the same token, easier to combine with each other.[44]

The dispute between Perrault and Blondel thus turned upon a different perspective upon the basic problems which architecture poses. The *Ordonnance* tended to favour the architectural conception over the architectural object, whereas the *Cours d'architecture* was chiefly preoccupied with the latter.

The *Ordonnance* was thus the occasion for a wholly new theme, namely, a rationalisation of the conception, which was effected in two separate stages. First of all, insofar as allowance was made for differing customs, some leeway was permitted in the interpretation of the rules. At this stage, the only distinction between Perrault and his precursors lay in the fact that the former sought to theorise the gap between principles and actual architectural production. But, no sooner was this freedom announced than it led to the elaboration of a system of norms which the architect might well be satisfied to follow to the letter in order to live up to society's expectations. What had at first appeared approximate was therefore condensed into a series of precise measures.

The doctrine expounded in the *Ordonnance* was thus a mixture of freedom and constraint, both of which proceeded from reason, a mixture which seemed to clear a path for the attempts at rationalisation pursued by the engineers of the Age of Enlightenment, who denounced the lack of fit between physico-mathematical theory and the empirical realities in order the better to exploit it subsequently.[45] Like the engineers, and long before them, Perrault tried to assign exact values to what had formerly been the object of more or less protracted negotiation. On the eve of the eighteenth century, reason already sought to respond to society's expectations by providing it with laws.

The empirical aspect of Perrault's approach ought also to be emphasised. The structuring aspect of custom gave rise, in his view, to the need for observation, and consequently for experiment, the importance of which he had stressed in his *Essais de physique* (1680–8).[46]

Finally, imprecision constituted one of the major shortcomings of the classical age which, from Descartes to Leibniz, had been far more concerned with deductive relations than with quantitative aspects.[47] Perrault, on the other hand, clearly had an ambition to create an architecture which was codified with precision, and which would thus enable one to begin a process of assimilation of theory to practice, an amalgam which represented yet another of the aims of the engineers of the Enlightenment.[48] In their very simplicity, the arithmetic sequences of the *Ordonnance* could well be seen to prefigure the behavioural rules of the emergent science of engineering.

However, one should not exaggerate the scope of Perrault's conceptions. Their radical character very probably derived to a great extent from the eccentricities of the man himself. His quarrels with Blondel should not allow one to forget the profound complicity between them, which survived all their various disagreements. Blondel anyway appreciated the advantages of Perrault's system, writing that the *Ordonnance* 'contains an infinitely easier manner than all the others, as far as the use of the five Orders of architecture is concerned'.[49] Handling with tact the teachings of the tradition, Perrault did not draw all possible conclusions from the 'freedom to increase or to diminish the sizes of the parts' upon which his system was in fact based. Referring to rules that were still very classical, he did not appreciate the importance of the sublime either, although Boileau had already had intimations of it.[50] Nevertheless, it cannot be gainsaid that the *Ordonnance* uncovered a certain number of tensions, and restated the problem of the articulations between theory, architectural production and the social order. On the other hand, the question as to whether the beautiful should be regarded as ideal or as arbitrary was quickly settled in favour of the former, and in the course of the eighteenth century this debate itself was to yield to others.

THE STATE, THEORY AND THE ARCHITECTURAL PROFESSION

The complicity between Perrault and Blondel was now linked to their status as important agents of a state whose role was becoming ever more significant.[51] The *Ordonnance*, like the *Cours d'architecture*, had anyway been commissioned by the state, which, from Colbert to Louvois, had striven to control production.

In the domain of architecture and planning, this ambition found expression first of all in the shape of the Academy of Architecture, whose importance was emphasised above. As a part of the process by which manufactures were established on a new footing, the recasting of the customs system in 1664, the regulating of the guilds in 1673 and the creation of the Academy belonged to a series of measures designed to provide a framework and positive direction for production and exchange. Serving in an advisory capacity as far as royal buildings were concerned, the Academy also intervened in problems involving the organisation of trades.[52]

This is the context in which the work of Desgodets and Perrault, and also that of Blondel, should be understood. The theoretical work of these three figures was intimately related to the state's preoccupations at the end of the seventeenth century.

With the publication of the *Edifices antiques de Rome*, the theme of measure, as we have seen, became current.[53] As professor at the Academy between 1719 and 1728, Desgodets also drafted a course of lectures which was devoted in part to the ways and customs of the building trade.[54] Desgodets' approach, and the care he took to describe and order the realities of everyday production, ought to be considered in relation to these sorts of enquiry, which became current at the end of the seventeenth century. Colbert, for example, interrogated his Intendants in 1664 with a view to compiling an inventory of the infrastructures of transport;[55] he also requested the Academy in 1678 to survey the churches of the realm, noting the source of the stones with which they were made.[56] In the course of studying Palladio's treatise, the Academy also enquired into the techniques involved in the building of bridges, setting up several committees to look into the matter.[57] Reviews of published books, enquiries into the conditions of production, and studies of customs all reflected the same concern with exhaustive description, a preoccupation which the engineers were to share, albeit in a different guise, during the Enlightenment. It was felt that, before undertaking any form of reorganisation whatsoever, one had to describe. In its early years, therefore, the Academy took an interest in a wide range of technical questions; Desgodets' course represents one of many attempts at an inventory.

On a more theoretical level, the work of Blondel also has the appearance of an inventory, since formal and mathematical considerations, architecture and the engineer's art were inextricably mixed up in it. The works of the first director of the Academy seemed to involve a project for an all-embracing science.

To judge by this measuring and describing, unifying and ordering of knowledge, experimenting and recommending, one might almost suppose oneself in the next century, so closely and deceptively does this programme resemble that subsequently adopted by engineers. Yet the same words may well refer to different phenomena, for at the end of Louis XIV's reign it was merely a question of pioneering ventures, possible lines of development, and of the bare outlines of investigation. These investigations are anyway obscured by the complex play of short-term strategies, such as the sort of clan politics which so drastically restricted the field of action of men like Colbert or Louvois.[58] The seventeenth century saw the assertion of grand principles, all of which conflicted with a highly fragmented reality. Desgodets, Blondel and Perrault very probably did not appreciate all the implications of their statements, or at any rate did not pay full attention to them. One would have to wait until the eighteenth century for any real awareness of what was involved in the rationalisation of production. For the leading agents of Louis XIV, these issues were not yet crucial.

As far as architecture is concerned, this awareness was also delayed by the fundamental ambiguity of the academic institution. Conceived as a part of the machinery of the state, the Academy of Architecture very quickly transcended this narrowly administrative role, so as to become the place where the profession of architect was defined:

> With regard to what has been represented to the King . . . namely, that several master masons, entrepreneurs and other persons involved with buildings now without any right soever make so bold as to assume the title of architects of the King, in order to win credit for themselves and, by means of this title, present plans, construct every sort of building, both public and private, which for the most part prove to be very defective because of the shortcomings of the abovementioned masons and entrepreneurs . . . His Majesty the King has expressly forbad all entrepreneurs . . . to assume the title of architect of the King, apart from those whom His Majesty has chosen to form a part of his Academy of Architecture.[59]

On 7 March 1767, Colbert announced in the above terms the decision to restrict the title of architect of the King to those who were Academicians.

The King and his chief minister had intended that the architects honoured in this way should constitute a corps, much as the engineers of the Génie or later those of the Ponts et Chaussées did. It did turn out like this in practice, for the protection accorded to the title of architect gave rise in fact to the emergence of an independent profession, aligned with the administrative authorities, but nevertheless distinct from them, a profession in which one can already recognise certain features of the liberal architect of modern times. This development caused a number of tensions between the Academy and the Surintendance des Bâtiments, which were to culminate in the dissolution of 1767, following a conflict in which the architects contested its authority over them.[60] Admittedly, almost a century was to pass before this extreme measure was taken; the Academy was anyway immediately reconstituted. Nevertheless, from the end of the seventeenth century, ambiguities respecting the academic institution had repercussions at the level of theory, and it was never wholly clear whether the latter addressed itself for preference to the state or to professional practice. As far as Blondel and Perrault were concerned, the emphasis was probably more on the public authorities; one would have to undertake further research before deciding what the precise target of Desgodets' course of lectures was. The architects were in fact faced with an insoluble contradiction, for while the state distinguished them from masons and entrepreneurs, conferring upon them the status of protected artist and upholding their claims against the workforce, it also proved to be a hard taskmaster, and its basic concerns put theoreticians in an increasingly difficult position. Caught between the need to win recognition from the prince and that of maintaining one's independence, the Age of Enlightenment was to seek in vain for a middle way. The engineers, on the other hand, were to present themselves as architect-bureaucrats, and they owed their success to their submission.

3

THE IMPACT OF THE ENLIGHTENMENT ON ARCHITECTURE

THE NEWTONIAN MODEL

The dispute between Perrault and Blondel could well have sounded the death-knell of a theory whose foundations were insecure. However, this was not to be, even though Briseux devoted a good part of his *Traité du beau essentiel* to a refutation of Perrault's arguments.[1] What happened was that the problems, and the issues associated with them, changed. Although the eighteenth century was the heir of the *Grand Siècle*, it was also characterised by new tendencies which account in part for the subsequent development of architectural theory.

The increased influence of Newton, as distinct from Descartes, was one of the most crucial factors.[2] The *Meditations* were not republished between 1724 and 1824.[3] During this same period, Voltaire declared himself a fervent disciple of Newton, and Madame du Châtelet translated the *Mathematical principles of natural philosophy*.[4] Descartes had come to seem a distant and somewhat forbidding ancestor, whereas Newton's thought now appeared to be at the very heart of the century's preoccupations. His account of the bond between mathematics and the world seemed to provide scientists with the answer they had been seeking, while his distinction between laws and causes seemed to prefigure Kantian rationalism. Voltaire's contemporaries regarded first causes as irrecoverable, and preferred to draw conclusions from experiment, a word epitomising a curiosity that ranged from an analysis of the workings of the soul to the properties of matter. Everything became the object of experiment. The Abbé Nollet studied electricity (Fig. 5), the Abbé Bossut the mechanics of fluids, while Lavoisier concentrated upon the phenomena of combustion.[5] A literary genre emerged which dealt simply with theories and systems, empirical data and hypotheses, instead of searching for ultimate truth. Otherwise, Diderot's *Traité du beau*, or Merian's treatise devoted to the topic of blindness, could never have been written. With the advent of Newton, a provisional ethics could be discerned, poised midway between an uncertainty as regards first causes and the myopia characteristic of the systems. Reason came to be regarded as a middle way, whose limits Kant was to investigate towards the end of the century. On another level the *Principia* saw the uniqueness of nature as affecting even its own operations, and this notion justified the vertiginous deployment of diversity. Describing became a priority in the course of the eighteenth century, even if one had to relinquish the great classical systems and settle for an alphabetical list. Such attempts at an inventory, although they may sometimes have been chaotic, were commonplace, and ranged from D'Alembert and Diderot's *Encyclopédie*[6] to Panckoucke's[7] more 'methodical' and ambitious venture.[8]

As the idea of nature emerged, a new humanism – whose chief protagonists were Locke, Condillac and Rousseau – gained currency. Sensation on the one hand, mathematics on the other, marked the limits of human understanding.[9] Immediacy and pure reason were destined to converge once again, like the two Pascalian infinities which they had replaced.

Finally, the eighteenth century was characterised by modes of social and political thought that are too familiar to require comment.

Nature, reason and experiment were held to constitute the poles of human activity, while sensation and mathematics represented its limits. Apart from problems specific to itself, architecture also had to be defined by this framework. On the eve of the Revolution, the sorts of investigation advanced by architects were those of society as a whole, as it uncovered a series of contrasts, both within nature, between the beautiful and the sublime, and within reason itself, which seemed vacillating, shadowed and threatened with actual dissolution. If this gloom was to be dispelled, one would have to set up even more experiments, reorder knowledge and rediscover the great explanatory systems.

SOCIETY AND TASTE

Although the notion of taste had been of only secondary importance in classical doctrine, it was to become central in the course of the eighteenth

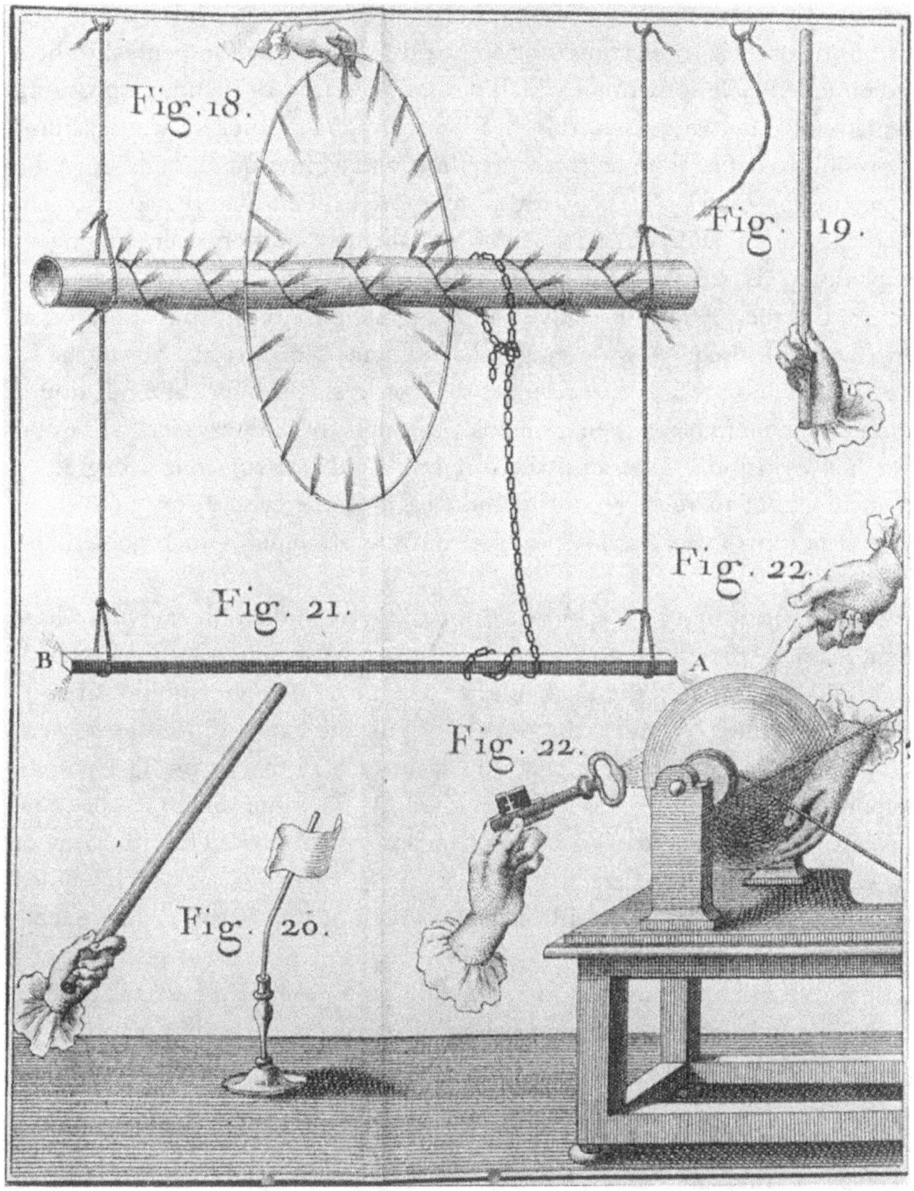

FIG. 5 J.-A. Nollet, *Lettres sur l'électricité* (1753–67), experiments in electrostatics. The feathers emanating from the electrified bodies represent the flow of an 'electrical matter' identical to that of light and fire. The Abbé Nollet's mechanistic theory was subsequently to be destroyed by Franklin's researches.

century.[10] Architecture was not immune in this respect, and we find the various theoreticians referring to it constantly:'Taste is said in general to be a certain indefinable something which pleases', declared Boffrand, emphasising immediately afterwards that 'this idea is very vague'.[11] Indeed, it was almost impossible to define taste with any precision, and Montesquieu's article on the subject in the *Encyclopédie* was hardly any more explicit than the assertions of Marivaux or of Diderot had been.[12] Yet all these authors were at one in emphasising its relative nature, dependent on climate and other individual factors. But the vague and relative nature of taste did not much matter, for, as Briseux emphasised: 'A wide range of sentiments works to the advantage of the sciences and the arts: it accelerates their progress through the emulation it produces: it illuminates their principles through the enquiry and discussion which it occasions'.[13] Besides, such diversity did not rule out a degree of standardisation; there were still boundaries, and one would not cross them with impunity. Taste was inseparable from the principles which guided and framed it.

This emphasis upon the standardisation of taste is linked to a new development, namely, the concern to relate architecture to society. On his deathbed, Louis XIV had advised the Dauphin not to imitate his own ambitious building programmes.[14] Under the Regency, major state commissions were virtually unknown, and were taken up again only in the 1750s. The Louvre colonnade, for example, was left unfinished. At the same time, private commissions were undertaken, to higher standards of comfort. The first years of the eighteenth century were characterised by the rococo style, which seemed to express an aristocracy freed from the burdens of the *Grand Siècle*.[15] Architects had to fashion new tools to suit this clientele. The art of planning was perfected;[16] and buildings had to express their purpose, in the details of their ornamentation. The architect's taste and that of his client had perforce to converge. Enthusiasm for this taste had faded by the 1750s,[17] as the demands of rationality increasingly opposed the unfettered fantasy of the rococo style, but a crucial step had been taken. For Jacques-François Blondel, who played a key part in the call to order, advocacy of classicism was in no sense construed as a retrograde step. Thus, in his *Cours d'architecture*, the spirit of *convenance* represented the agreement that must prevail between the architect and his client.[18] Based on a nature which was to be both imitated and perfected, architecture sought to reproduce in stone all the grades of the social hierarchy from prince to tradesman. The house of a prince would therefore be graced by columns and pediments, while a private person would have to be satisfied with a more modest articulation. Yet this clear hierarchy of functions and effects was to remain largely an architect's dream; on the eve of the Revolution, parvenus of all sorts, such as the actresses and tax farmers for whom Ledoux worked, sought to give a princely character to their residences. Theory had thus to deal with this subversion of traditional categories.

THE SPREAD OF KNOWLEDGE

We are concerned with a context in which problems became more generalised. Faced with commissions in which requirements were ever more precisely defined, sometimes taking opposing turns, such as the need to relate the external form of a building to its internal operations,[19] architects tried to provide a coherent solution. In the end the focus at issue shifted, the most striking instance being the developments in the theory of the architectural Orders.

Eighteenth-century treatises refer frequently to the dispute between Perrault and Blondel, siding as a rule with the latter. Briseux's determination to refute Perrault's arguments one by one was, however, an isolated instance. Boffrand, Blondel and Soufflot no doubt accepted the notion of the *beau idéal*, but without undue emphasis. Soufflot and Boullée clearly regarded the entrenched positions of Perrault and Blondel as the outcome of an argument which had 'degenerated into a dispute, passion and party spirit having got the better of them'.[20]

At the same time, the matter of proportion lost its importance. Although François Blondel was preferred to his rival, he was still relatively little read. As Patte observed in his *Mémoires sur les objets les plus importants de l'architecture*, Blondel's *Cours*, 'which contains so many excellents precepts, tends to offer discussions that are scholarly rather than appropriate for ordinary use'.[21] Ordinary use comes close to eclecticism here. Supporters of Perrault, like Patte, argued in vain for establishing 'average' proportions, and for sticking to them.[22] In treatises the systems of Italian and French authors were merely set alongside, much in the manner in which Diderot, in his writings, enumerated highly diverse philosophical positions. Dupuis' *Nouveau traité d'architecture comprenant les cinq ordres des anciens, établis dans une juste proportion entr'eux, avec un sixième ordre nommé ordre français* was published in 1768, but the notion of the correct proportions to be given to the orders of the Ancients had been well and truly undermined by Leroy's earlier discoveries regarding the Greek

Doric,[23] while the invention of a French Order, which had been one of the main preoccupations of the architects of the *Grand Siècle*, had become an anachronism. Indeed, the preceding year, the Academy of Architecture, as official a body as could be imagined, had declared, after its examination of Potain's *Traité des ordres d'architecture*: 'One must therefore not regard treatises on the proportions of the Orders of architecture as anything more than elementary works serving to clarify the ideas of those wishing to learn about these Orders'.[24]

The limited interest shown in the theory of Orders masked a more profound change. If the question of proportion gradually lost its urgency, it was because architectural theory was undergoing a transformation, and was addressing more general problems. An attempt was now made to think inclusively; a building was now regarded as a whole composed of parts, but its coherence was no longer achieved solely through modular coordination, as was evident in the discussion on the arrangement of churches, which, from Cordemoy to Laugier, was more concerned with the structure of space than with details of the ordinance.[25] Even in the instance of Blondel, who regarded decoration as the crucial aspect of the architect's work, planning and structure assumed a new importance. Whether one referred, as Laugier did, to primary needs or, as Blondel did, to the classical tradition, architecture was now to be expressed logically, even if one aspect or another might elude the rules.

One may therefore speak of a rationalist current, which may perhaps constitute the real backdrop, before neo-classicism flourished, to the eighteenth century. Cordemoy, Laugier, Blondel and Patte were to be its representatives in the realm of theory, alongside Gabriel, Perronet and Soufflot, architects whose work celebrated the same principles of clarity and legibility. Admittedly, differences of opinion, even actual disputes, sometimes divided these figures. Whereas Blondel and Gabriel were particularly interested in the ordinance and planning of buildings, Perronet and Soufflot addressed problems of structure and construction. As is well known, Patte clashed with Soufflot over the question of structure.[26] But within a framework that was still largely traditional, Enlightenment rationalism sought above all else to provide an answer to the questions of the day; its failure to do so marked the demise of classical theory.

Before this failure was to become fully apparent, numerous problems were to grow more pressing. Architecture had to adapt itself to contemporary needs, but it was still not clear how this was to be done. Even if one advocated a large-scale approach, it was no longer possible to disregard the new tensions affecting architectural theory, which compounded those studied in chapter 2, and which seemed now to threaten its very existence.

LANGUAGE, NATURE AND EXPERIMENT

The essential concern of architecture was its relation to society. If an architect was to respond effectively to the full range of concerns of the social order, the means at his disposal, when compared with the size of the field he had to cover, seemed singularly limited. In a human collectivity, almost all forms are a priori possible, but architecture could avail itself only of the five Graeco-Roman Orders to express the diversity of social conditions and customs.

Architects met this challenge in a number of different ways. J.-F. Blondel represents one approach. As we shall see in chapter 4, he tried in his *Cours* to enhance the science of architecture by exploiting all its possibilities. The procedure was thus to isolate a range of effects, from the most striking to the most discreet, each of which corresponded to a rank or social position. Architectural space was organised in terms of the pyramid of status, but it might also, through its character – nuances that were 'suave, interesting, firm or vigorous',[27] but also terrible or threatening – reproduce the whole gamut of sensations known to man. In its desire to be exhaustive, Blondel's *Cours* is reminiscent of the *Encyclopédie*.

If he adopted such an approach, the architect encountered two major difficulties. The first concerns the discrepancy between language and space. Blondel's desire to describe committed him to using language to separate, to classify and to describe. Yet architecture is irreducible to any descriptions of it; the crisis of classical theory was to some degree that of a language which John Summerson has called the 'Latin' of architecture.[28]

A second drawback arose out of the proliferation of architectural knowledge. In the course of the eighteenth century, voyages, whether to China or to Greece, became more and more frequent;[29] both the imperial gardens of Peking[30] and the baseless Doric of Paestum and of the Peloponnese were discovered (Fig. 6). At the same time, the art of construction was being perfected, and planning was becoming more complex. Blondel's *Cours* was a prisoner of its still classical structure, and it seemed to exhaust itself in the quest for a unity that was forever eluding it.

Disturbing questions now loomed on the horizon, concerning the necessity and the unity of architecture. The mediation of nature meant that necessity and unity were anyway becoming confused, as is apparent from Laugier or Patte's treatment of the theme of the primitive hut.[31] The reimposition of theory was gradually becoming synonymous with the quest for its natural foundations.

Although Blondel was still relatively little concerned with nature,[32] the theme was subsequently to become a crucial aspect of the architectural debate, along with sensation, which allowed one direct access to nature. 'No one has yet written on the analogy of the proportions of architecture with our sensations', declared Le Camus de Mézières in the introduction to his *Génie de l'architecture* (1780), and he explained a little further on that: 'Concerned with these observations since my youth, my enthusiasm has been sustained through fixing my attention upon the works of nature. The more I examined, the more I recognised that each object must have its own particular character, and that often a single line or contour is sufficient to express it'.[33] In referring to a single line or contour, his choice of words is revealing, for it suggests that nature should enable one to be economical with language, thus avoiding over-loading or an anarchic profusion. It was in the same spirit that, a few years later, Boullée wrote his *Essai sur l'art*, in which simple forms and light, the stimulators of feelings, constitute the only possible response to the excesses of an architecture defined in terms of details.[34]

Blondel thus marks the limits of the classical heritage. His opposition to Boullée and Ledoux was based on principles similar to those invoked by Rousseau at about the same time regarding the relationship between language and onomatopoeia, between articulation and the cry.[35] At the end of the eighteenth century, architectural rationalism encountered this cry without being able to reduce it to the serene articulation of a discourse. In the case of the engineers, as we shall see below, a series of factors actually enabled them to play on these two registers. Language and sensation, the beautiful and the sublime, were integrated within one and the same approach.

So far we have merely raised the question of the relationship between architecture and a given society. We need now to investigate the way in which such a relationship evolved historically. As Ernst Cassirer has emphasised, 'the common opinion that the eighteenth century was an "unhistorical" century is not and cannot be historically justified'.[36] Blondel himself had already posed the problem of history, in referring to a tradition whose sometimes outdated character he himself recognised. The relativity of taste was not only geographical but diachronic also. We thus learn in his *Architecture françoise* or in the *Cours de l'architecture* that the Egyptians favoured solidity in their architecture,

FIG. 6 D. Leroy, *Les ruines des plus beaux monuments de la Grèce* (1758), a hypothetical reconstruction of the Propylaea. Stealing a march on the Englishmen, Stuart and Revett, who had visited Greece at the same time as himself, Leroy was the first to reveal the principal features of the architecture of the Peloponnese.

the Greeks elegance and the Romans majesty.[37] Condillac and Rousseau were preoccupied, in another context, with the origins of society. If, moreover, society itself was subject to unceasing progress, one had to try and specify the conditions of such progress in the domain of architecture.

Sensation and mathematics constituted, as we have said, the limits of human activity; experiment, on the other hand, represented its trajectory, whether it was the acquisition of knowledge or the development of exchanges between men that was involved. Architecture's progress, along with the means employed to measure it, therefore presupposed the need for experimentation. One had therefore to opt for sensation or for calculation.

While the engineers decided in favour of calculation, though at the same time positing its intimate connection with sensation,[38] the architects made the opposite choice. The case of David Leroy's *Histoire de la disposition et des formes différentes que les chrétiens ont données à leurs temples, depuis le règne de Constantin le Grand jusqu'à nous* is particularly telling here. Published in 1764, the *Histoire* was one of the first books explicitly devoted to the evolution of an architectural type, that of the church. Alongside his history of the various kinds of temple, summarised in a comparative engraving (Fig. 7),[39] Leroy expounded his ideas regarding the sensations to which architecture gives rise: 'whatever the cause of the sensations which architecture makes us experience in general, one may rest assured that it is from the nature, the force or the number of these sensations that the judgement which we bring to bear on the various buildings which present themselves to our gaze arises'.[40]

Now, it is on the basis of this judgement that the historical view proceeds. From the outset, history as architects have conceived it has therefore been committed to an exploration of the bases of sensation, and consequently of its relation to nature. Boullée's Utopia was to involve an attempt to see this history crystallised in the unique instance of revealed design. Once this radical approach was set aside, the initial question, from which all else had stemmed, was nevertheless reiterated: how does one experiment in architecture?

In the course of the eighteenth century, architectural knowledge became more and more detached from the sciences. While the latter enjoyed some spectacular successes, to which engineers might refer, architectural debate seemed already to be marked by a degree of circularity.

Thus, to the problems addressed towards the end of the *Grand Siècle* were added a number of others – the limits of language, the combinatory of sensation, the circularity of experience and of history – which reflected new issues. Enlightenment rationalism, in spite of its desire for coherence and its will to promote an overall approach to architecture, was never to succeed in establishing a lasting compromise between exigencies that proved more and more difficult to reconcile.

We need now to examine this situation, and more especially to investigate what it conceals. We must also take into account the professional strategies and the ambiguous attitude of architects towards the state. In the course of the eighteenth century, ties between the Academy of Architecture and the authorities were relaxed.

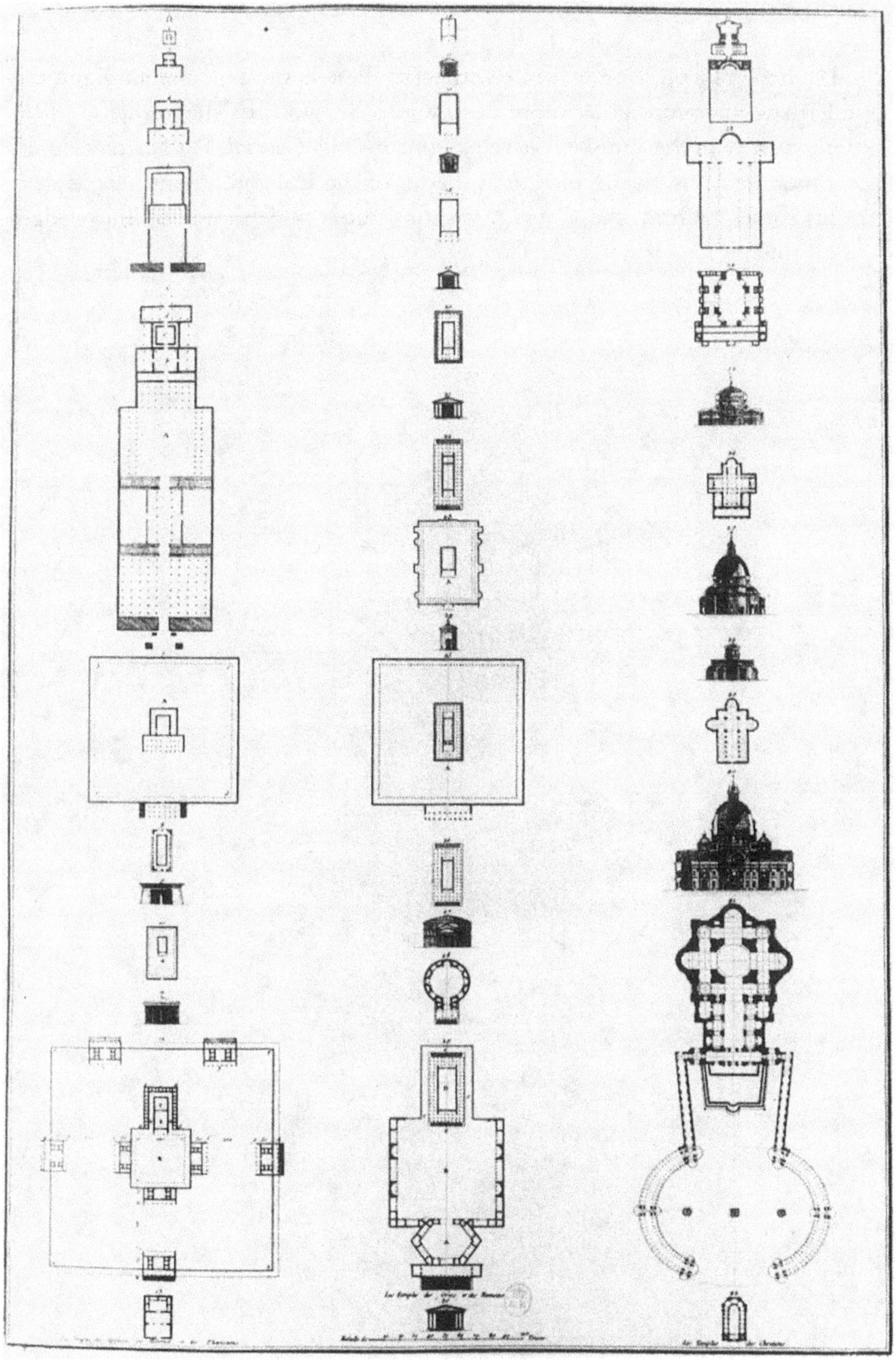

FIG.7 D Leroy, *Histoire de la disposition et des formes différentes que les chrétiens ont données à leurs temples* (1764, second edition 1770), parallel between ancient and modern temples.

Jacques-François Blondel bears witness to these hesitations and ambiguities, and I have therefore gone more deeply into his work in chapter 4, so as to compare it with the approach of the engineers. His *Cours de l'architecture* stands as a monument to the architectural theory of the Enlightenment, just as that of his namesake from the *Grand Siècle* constituted the epitome of knowledge of his time.

4

'TOWARDS A CLASSICAL ARCHITECTURE': JACQUES-FRANÇOIS BLONDEL AND THE 'COURS D'ARCHITECTURE'

THE TASTE FOR KNOWLEDGE AND ITS DIFFUSION

As a man who was more concerned with theory and teaching than with professional practice, Jacques-François Blondel held a very particular place among eighteenth-century architects. His building at Metz and at Strasburg was paralleled by projects, the majority of which are mentioned in his *Cours*,[1] but, for his contemporaries, Blondel was first and foremost a teacher committed to the development and diffusion of knowledge.

Blondel's work as an engraver reflects these preoccupations. Like Patte, he returned again and again to engraving, providing the plates, for example, for the 1738 edition of d'Aviler, his own *Architecture françoise* and his *Cours d'architecture*. From science to the mechanical arts, from the treatises of Abbé Bossut to the plates for the *Encyclopédie*, engraving was in the eighteenth century inseparable from the movement of ideas. Yet Blondel was interested not only in its expressive possibilities; he plainly had a pronounced taste for matters of lay-out and typography. As Blondel remarked of his first book, 'Men of art will see from the vignettes, the *culs-de-lampe*, the decorated letters with which these two volumes are embellished, combined with the beauty of the newly cast letters, the reglets, the choice of paper and the printing, that this venture has been carried to the highest possible point of perfection'.[2] Even in a piece of publicity of this kind, Blondel, like Patte, displays an awareness of the sensual nature of the printed object, as if knowledge and its supports, the book and the engraving, were inseparable.

Yet Blondel described himself as an architect. A passage in his *Cours* reveals the indefatigable curiosity that stirred him. The architect must, Blondel believed, 'move from calculation to design, from drawing to perspective, returning then to ornament, and finally to the landscape', but he should also turn himself into a traveller, visiting everything and inspecting everything, from the most famous monuments of the past to the workshops where the joinery and mirrors for the decoration of rooms are made.[3] Blondel also insists that the architect should meditate upon the principles and the application of art, and he returns thus to books and engraving. In his *Discours sur la nécessité de l'étude de l'architecture,* Blondel lists those works an architect should know. A reading of Palladio, Scamozzi and Vignola was of course recommended, but mention was also made of Alberti's *Treatise*, 'a work that is too little known, but study of which is indispensable', and of Fischer von Erlach's *Architecture historique*. Alongside the classics, Derand's *Architecture des voûtes* and Frézier's *Treatise* are also mentioned.[4] From the theory of the Orders to matters of construction, the culture of architecture constituted a whole world, and one with which one should familiarise oneself patiently.

Blondel's patience may well have been his most characteristic quality. Beyond his taste for study and his love of books, he seems to have been indefatigable in his search for knowledge. The dogmatic character of his teaching was often belied by remarks that bear witness to a high degree of receptivity to the problems of his age.

BLONDEL AND THE ENLIGHTENMENT

As a contributor to the *Encyclopédie*, and friend of Perronet (whose talent as an engineer he repeatedly praised),[5] Blondel naturally shared most of the convictions of his age. Like Ange-Jacques Gabriel, whose counterpart in the field of theory he may be said to have been, he was beset by conflicting impulses. Beneath the apparently conservative surface, one might find in Blondel an echo of the problems that preoccupied the most progressive architects of the day.

He had, for example, a relatively complex attitude towards the subject of Gothic architecture. Furthermore, his position clearly shifted, between the *Architecture françoise* (1752–6) and the *Cours*. In the former, Gothic was still deemed ridiculous, loaded down with 'frivolous'[6] ornaments. In the *Cours*, on the other hand, he praised the proportions of those cathedrals 'which reveal that rules have been followed'.[7] When presenting a plan for a cathedral, for example, Blondel even went so far as to adopt the dimensions of Notre-Dame.[8] Though expressing his approval of the façade of the Cathedral of Orléans,[9] he gave to the cathedral of Metz a Doric portal, designed to harmonise with the rest of the building.[10] 'Imitation for imitation, why should we not take as a model whatever in our finest Gothic works is most interesting, such as great height and the lightness of the vaulting, that continuity of form which continues from one pier to the next',[11] wrote Blondel in the *Cours*, adopting an attitude that Soufflot himself would not have disavowed.

On numerous points, Blondel was thus aligned with the opinions of the century. He saw that it was necessary to rationalise the town, arguing, like Patte, that hospitals and cemeteries should not be concentrated in the centres.[12] He had a fine understanding of the tendencies of his time, so long as he was not asked to commend the taking of architectural liberties. In the latter case, his attacks were forceful; his judgement of Ledoux's hôtel d'Uzès in *L'homme du monde éclairé par les arts*, is well known.[13]

One cannot help but wonder why a man whose knowledge was in other respects broad should have suffered so curious a mental block, evident enough in his writings. It may well have been that his loyalty to tradition involved a well-defined aim, opposed to the experiments of a generation which he himself had nonetheless helped to form.

'PATIENT RESEARCH'

'Having been disregarded and buried as it were in my study for several years', declared Blondel at the start of his first book, 'it must come as a surprise to see me make my debut with a subject that is so important and demands so much experience'.[14] This book, *De la distribution des maisons de plaisance et de la décoration des édifices en général*, appeared in 1737. Its author, although thirty-two years old, had not yet published anything and, the descriptions of projects contained in the work notwithstanding,[15] seems to have built nothing either.

His major treatises were to appear much later. The first four volumes of *Architecture françoise* were published between 1752 and 1756, while the *Cours* appeared between 1771 and 1774, its last volumes being completed by Patte. As this slow progress implies, Blondel's ideas long remained hesitant. His engravings for Mariette's republication of d'Aviler reveal him to have been akin to the sophisticated graces of the rococo. In the *Distribution des maisons de plaisance*, he allowed himself liberties that he would later reject. The *trumeaux* in a country house, for instance, were adorned with busts on the corbels,[16] while the plan of a garden pavilion was based on polygonal forms,[17] the pointless complexity of which he would later criticise in the *Architecture françoise*, on the grounds that they gave to the interiors of buildings the qualities of a piece of indented lawn.[18]

By comparison with these early works, the *Cours d'architecture* is much more sober in spirit. Pointless decoration was condemned, and planning was more carefully analysed. The articulation of façades became more repetitive, to the point of tedium. Such transformations expressed an aspiration towards a 'classicism', the interest of which was not always quite clear.

This 'classicism' in fact poses a number of problems, such that its formal coherence is open to question. The designs upon which the *Cours* is based were anyway inferior to those built at Metz or Strasburg, for they serve a theoretical programme the constraints of which are continually in evidence. The *Cours* represents the culmination of an approach comparable in some respects to the *Encyclopédie*, and therefore subject to the same uncertainties. Whether it was a question of decoration or of the planning of buildings, Blondel endeavoured to review and to order the architectural knowledge of his time in terms of the theoretical imperatives analysed below.

DIVISIONS AND CLASSIFICATIONS

The first volume of the *Cours* begins with a series of essays on the 'sources of the art'. We learn, among other things, that architecture is the daughter of necessity. In order to find shelter, men had to build huts, and, in order to defend themselves, fortifications; when prosperity favoured them, they needed to travel and so designed boats.[19] Blondel used the myth of origin to explain civil, military and naval architecture. As with Condillac and Rousseau, the origin could explain everything, a concept that was held also by engineers such as Prony, who employed them in order to divide the sciences in relation to their various lines of evolution.[20]

As Rousseau implicitly admitted, the origin might never have existed, but the notion of a first, undivided state enabled one to impose crucial divisions. The architecture with which Blondel was concerned was predominantly civil, even if the architect might also design fortifications or ships.

This initial division was succeeded by a set of distinctions and listings, an overwhelming catalogue containing more entries than one could readily identify. Blondel multiplied lists of objects and qualities. There were twenty-four kinds of staircase and forty or so architectural characters. A column, for example, might be classified in terms of its material, its form and its position. In material terms, it might be hydraulic, metallic, precious, *rocaille* or lattice-work.[21] In terms of form, one had to distinguish between banded columns, annular columns and those which were fluted, smooth, massive, pastoral or rusticated.[22]

The *Cours* thus proceeded by means of endless, often heterogeneous listings. Its aspirations towards an encyclopaedic status and exhaustive itemisation betray the shortcomings of any analysis based upon labelling alone. The abstruse and, in the last analysis, relatively chaotic quality of the *Cours* was further reinforced by the instability of its basic principles. Proportion was no longer, as it had been in Delorme, 'divine'; *convenance*, character and taste were not powerful enough to effect the hierarchical reorganisation of knowledge. The structure of the *Cours* seems limited to the juxtaposition of parts which were tending to become autonomous. Notwithstanding the listing of the parts of architecture and of their gradations of character, the edifice of knowledge seemed to be disintegrating.

This anarchic proliferation was matched by a fairly slack structure. Discourses, dissertations or sequences of maxims were interposed between chapters, in such a way as to compromise at times the clarity of the exposition. In fact, the real guiding thread of the *Cours* was the pedagogic aim which it everywhere affirmed. The author justified the countless repetitions in terms of the need to accustom pupils to the basic truths of the art.[23] The *Cours* thus resembled an itinerary in the course of which the patient apprenticeship of reason was applied to art.

Nevertheless, there was one division which dominated the others. For, like most Enlightenment theorists, Blondel referred to the distinctions employed by d'Aviler in his *Cours d'architecture* of 1691 (see Fig. 8), where architecture was defined by three parts, decoration, distribution and construction. For Blondel, decoration and distribution were the essence of architecture. Since he showed no real enthusiasm or aptitude for construction,[24] this chapter will be mainly devoted to the first two aspects of an architect's operation.

DECORATION

Decoration, as noted, represented one of the main parts of architecture. Far from being restricted to the deployment of the Orders or to the register of form, it used *convenance* to regulate a whole series of relationships between architecture and social requirements. When, for example, Blondel surveyed the various kinds of building, from palaces to private houses, from theatres to hospitals, it was in terms of the different kinds of decoration used. The *Cours* was thus trapped within a rule of function which remained strictly dependent upon questions of ordinance. The decoration of a building was meant to express directly its use and the rank of its owner. But this desire for legibility was to become more equivocal at the end of the eighteenth century, when the transparent house invoked by Rousseau would relinquish the social display required by Blondel.[25] If it was to appear necessary at all, architectural decoration had then to evolve.

Dominant modes of thought served, nevertheless, to obscure the fact that a breach had occurred. *Convenance* continued to be the crucial criterion, as is testified by Diderot's astonishment when confronted with the pavilion of Mlle Guimard, which led him to describe Paris to Catherine the Great as a town where a courtesan might return to her house, passing through a screen of columns.[26] In spite of the contortions to which the young architects subjected it, decorative order was still seen as a reflection of social structure.

NATURE AND IMITATION

Nevertheless, nature remained a crucial anchor. In the Enlightenment, no reasonable operation failed to refer to it. Architecture had therefore to proceed from nature, from a nature whose involvement was not restricted to the framework of primary needs, but which also enabled one to interpret the subsequent developments of art. From a theoretical point of view, Blondel remained fairly vague. His thought was much less radical than that of the Abbé Laugier,[27] while Le Camus and, above all, Boullée pushed their analyses much further. There were echoes, however, in the *Cours* of the dominant ideas of the period, foremost among them the doctrine of imitation. 'The taste of nature in its simplicity may well have presided over the birth of architecture', but 'the latter, reduced to art, has managed to perfect nature',[28] wrote Blondel, referring to that 'perfection through imitation' which constituted one of the central themes of the aesthetics of mimesis. It is also worth noting the

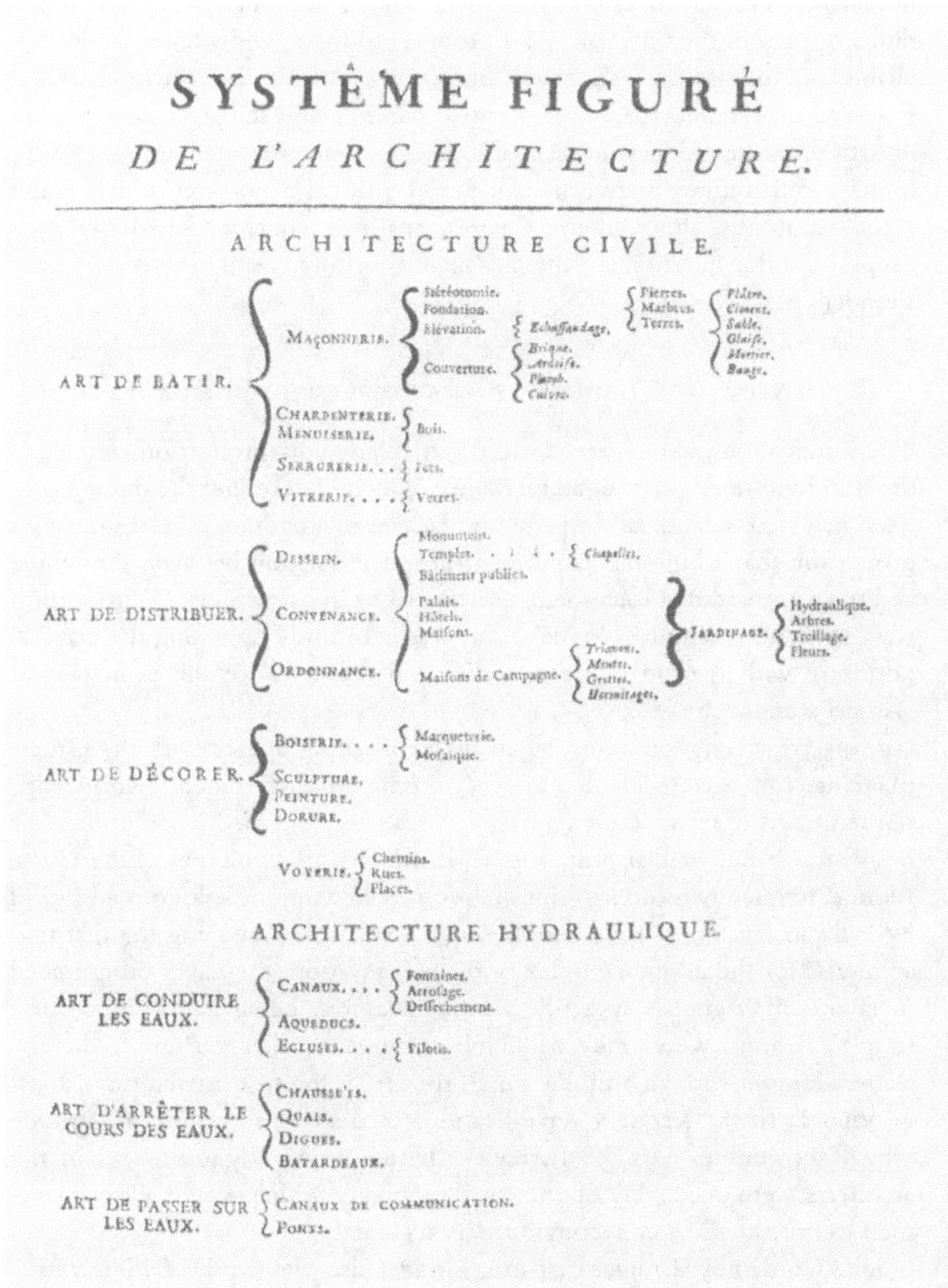

SYSTÈME FIGURÉ
DE L'ARCHITECTURE.

ARCHITECTURE CIVILE.

ART DE BATIR.
- MAÇONNERIE.
  - Stéréotomie.
  - Fondation.
  - Élévation. — *Echaffaudage.*
  - Couverture. — *Brique. Ardoise. Plomb. Cuivre.*
  - Pierres. Marbres. Terres. — *Plâtre. Ciment. Sable. Glaise. Mortier. Bauge.*
- CHARPENTERIE. MENUISERIE. — Bois.
- SERRURERIE. . . — Fers.
- VITRERIE. . . . — Verres.

ART DE DISTRIBUER.
- DESSEIN. CONVENANCE. ORDONNANCE.
  - Monumens.
  - Temples. . . . . — *Chapelles.*
  - Bâtimens publics.
  - Palais.
  - Hôtels.
  - Maisons.
  - Maisons de Campagne. — *Trianons. Menées. Grottes. Hermitages.*
  - JARDINAGE. — Hydraulique. Arbres. Treillage. Fleurs.

ART DE DÉCORER.
- BOISERIE. . . . — Marqueterie. Mosaïque.
- SCULPTURE.
- PEINTURE.
- DORURE.

VOYERIE. — Chemins. Rues. Places.

ARCHITECTURE HYDRAULIQUE.

ART DE CONDUIRE LES EAUX.
- CANAUX. . . . — Fontaines. Arrosage. Dessechement.
- AQUEDUCS.
- ECLUSES. . . . — Pilotis.

ART D'ARRÊTER LE COURS DES EAUX.
- CHAUSSÉES.
- QUAIS.
- DIGUES.
- BATARDEAUX.

ART DE PASSER SUR LES EAUX.
- CANAUX DE COMMUNICATION.
- PONTS.

FIG.8 C.-A. d'Aviler, 'Tabulated architectural system'.

numerous allusions to essential beauty, which, 'although scattered in the different parts of the universe',[29] art proposes to focus, and which would thus enable one to establish architecture upon solid foundations. Yet Blondel also observed that architecture, in contrast to painting and sculpture, arts of pure imitation, created objects instead of imitating them. Architecture was therefore located midway between nature and artifice, a position which could prove hazardous, although the *Cours* seems little concerned to look more deeply into this question. In the last analysis, nature merely served to justify Blondel's basic principles.

PROPORTION, ORDER AND THE 'EXPRESSION' OF ORDER

This justification was concerned more particularly with proportion, defined as the 'most interesting part of architecture', which 'determines the dimensions, sizes, height and depth of the plans and façades of a building'; 'it was through proportion that a building acquired an intimate relation between the whole and the parts, and that each member came to be in its own place'. Proportion was learnt, according to Blondel, who was here simply restating the classical position, 'with the aid of mathematics and the study of the principles of classical architecture'.[30]

More concretely, one must begin by familiarising oneself with the Orders of architecture and by learning to apply them with discernment: 'Knowledge of the Orders of architecture consists in a study of, and meditation upon their origin, kind and peculiar properties'.[31] In the first two volumes of the *Cours*, Blondel frequently returns to this question, reviewing the solutions advanced by Palladio, Scamozzi and Vignola (Figs. 9 and 10). Rebuking the first two architects for the undue complexity of their position, the *Cours* pronounced in favour of Vignola's system,[32] whose measures, being based upon fairly simple relations, were easily deducible from each other.[33] Blondel thereby aligned himself with an attitude which runs from the first French translations of Vignola to the frequent republications of d'Aviler's version, an attitude which accounts in part for Perrault's. In espousing Vignola's system, the architects were effectively opting for a simplicity which would allow workmen to execute the architectural Orders accurately.

Like Cordemoy, Laugier or Patte, Blondel also viewed the Orders with a degree of pragmatism. Although he did not hold that one could adopt a mean proportion as Patte, for example, had suggested, in his *Mémoires sur les objets les plus importants de l'architecture*, he nevertheless felt an instinctive distrust for all

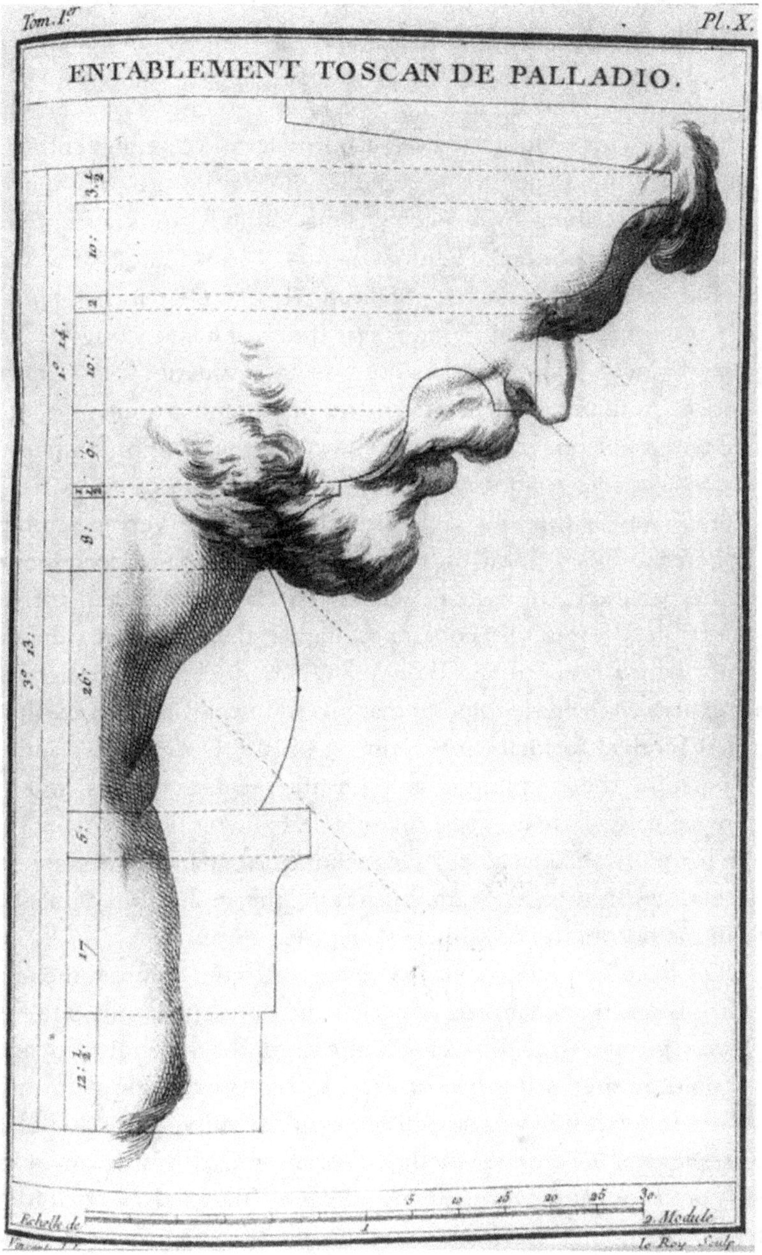

FIG. 9 J.-F. Blondel, *Cours d'architecture* (1771–7), 'Palladio's Tuscan entablature'. The anthropomorphic representation corresponding to Palladio's profiles.

excessively dogmatic justifications, which for him ran counter to all reason. Nothing could be more opposed to Briseux's beliefs than this observation in the *Cours*: 'If there is no satisfactory demonstration which actually proves the need for proportions in architecture, there is certainly none proving the reverse either'.[34] This sophism is a typical product of the eighteenth century, which tended as a rule to condemn Perrault's more wayward pronouncements while at heart subscribing to his views. Thus, Blondel took exception to all notions of beauty as 'arbitrary', while straightway conceding that proportions might be employed with a degree of latitude. In the *Discours sur la nécessité de l'étude de l'architecture*, Claude Perrault was indeed ranked alongside François Mansart and François Blondel as one possessed of a superior knowledge of the architecture of Antiquity.[35] Such assessments of Perrault are fairly symptomatic of the ambiguities of Enlightenment architectural theory. For, by renouncing the ideal character of proportion, theory in fact lost one of its crucial justifications. What was in the long run threatened was the very possibility of a formal coherence based upon the system of measures and correspondences between the parts and the whole, which the classical age had increasingly sought to refine. It was this coherence, rather than the question of ideal proportions, which preoccupied Blondel. Where decoration was concerned, the *Cours* explained the series of links that united the parts one to another. This concern led Blondel to focus, for example, on the Doric Order and on the problem posed by the coupling of columns in re-entrant angles, and thereby to venture an answer to a question which François Mansart and Libéral Bruant, in particular, had tackled.[36] He reckoned as a part of the same order of ideas the 'licences and abuses of art, concerning the application of pilasters and columns in the decoration of buildings' (Fig. 11).[37]

But architecture's repertoire was not restricted to the column, the incorrect use of which Blondel condemned. The column represented only one element, although very probably the most eloquent one, in the vocabulary of architecture. It should be used with discernment, by always relating it to the other means available for producing an impression of solidity, grace and elegance. The 'expression' of an Order, whether Tuscan or Composite, might convey its effects, often operating even in its absence. Intended to strengthen the character of a building, the mere coordination of dimensions was enriched with an erudite play of inflexions and of nuances. One would thus distinguish between combinations of mouldings, forms of balustrades (Fig. 12), windows or niches, which varied from one architectural Order to another.

As in the classical age, the 'expression' of an Order did not really do more

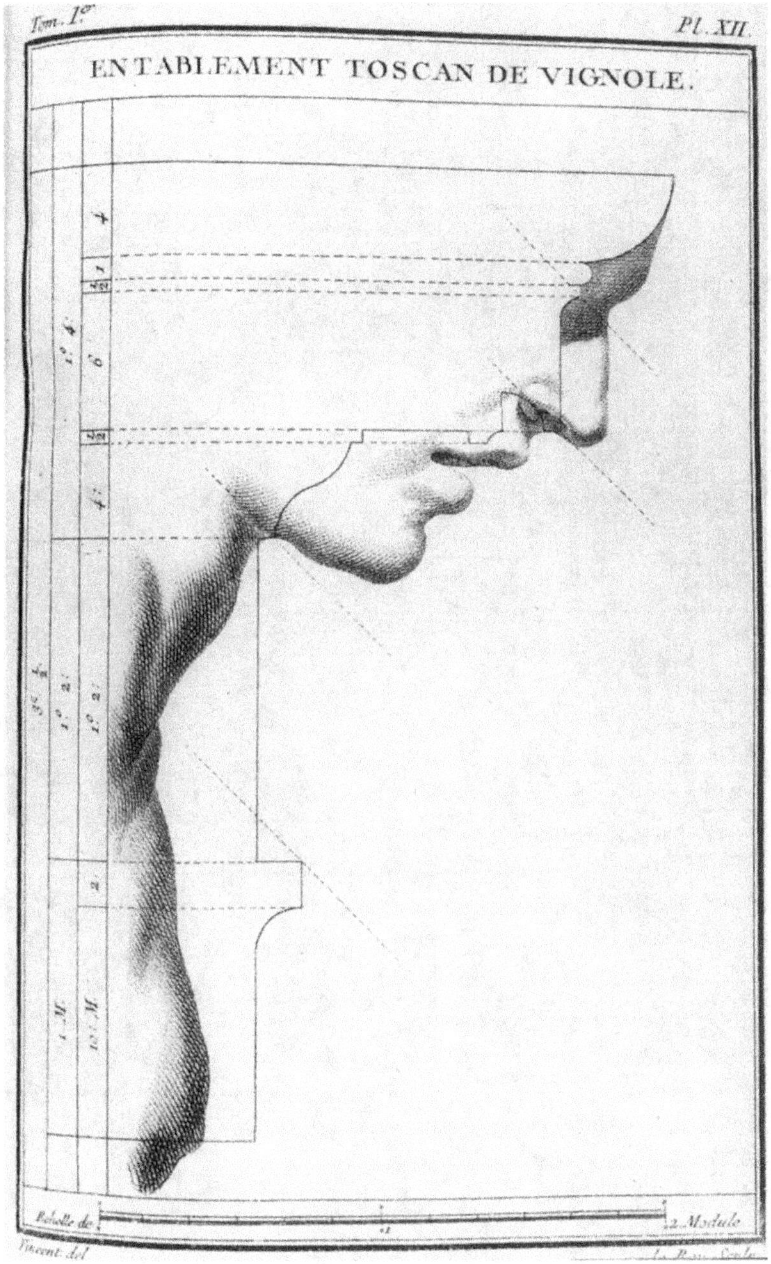

FIG. 10 J.-F. Blondel, 'Vignola's Tuscan entablature'. The simplicity of Vignola's entablature is contrasted to the complexity of the Palladian profiles.

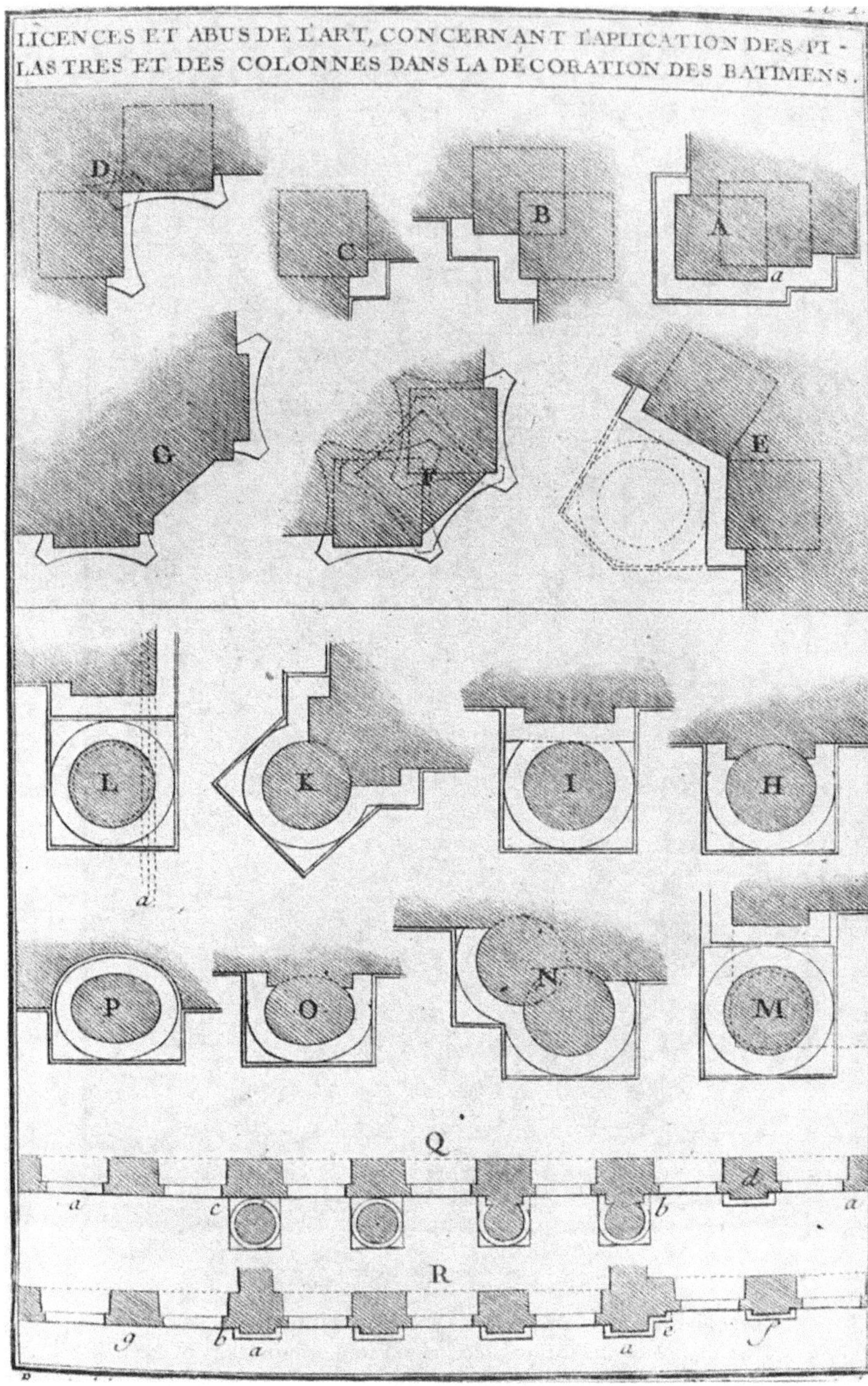

FIG. 11 J.-F. Blondel, 'Liberties and abuses in architecture relating to the use of pilasters and columns in the decoration of buildings'.

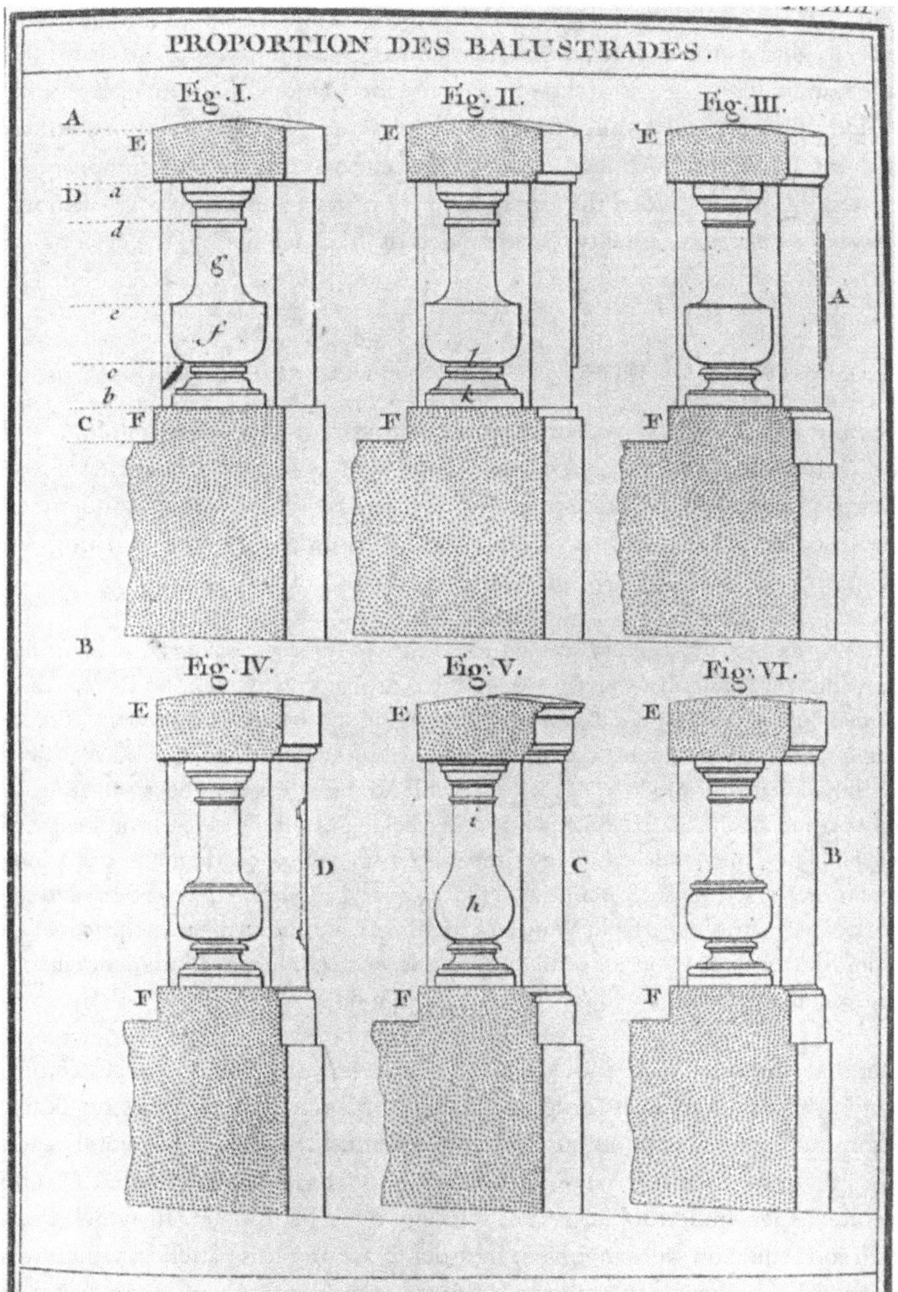

FIG. 12 J.-F. Blondel, 'Proportion of balustrades', I and II: Tuscan order, III: Doric order, IV: Ionic order, V: Corinthian order, VI: Composite order.

than provide a framework for operations, within which looser rules, linked to taste, might be articulated. One of the main aims of the *Cours* was to form this taste, upon which good architecture depended, and which enabled one to avoid both the sterile application of principles and any too obvious a betrayal of them. There thus arose an ethics of decoration, an ethics of compromise, located midway between the harsh austerity of rule and a free play of motif, between an excessive fidelity to the models of the Ancients and the eclecticism of the Moderns.

SOBRIETY OF EFFECT AND ARCHITECTURAL RATIONALISM

Reserve and propriety were the main qualities required in an architect, and the cultivation of these demanded a degree of self-denial. Good taste was always supposed to prevail over unbridled fantasy, and Blondel condemned the distractions of a century which had set aside the wise teachings of tradition:

> A few years ago, our century seemed to be that of the rococo; today, without our knowing why, things have taken another turn. At that time, Greek and Roman taste seemed cold and monotonous; now, we are taking up the civilised product of those peoples; and, without giving the matter too much thought, we ask that other nations slavishly imitate our mode of decoration, while we ourselves have been imitating, in our apartments, the bizarre ornamentation of Peking or, in the exterior ordinance of our buildings, the ponderous taste of the earliest inventions of Memphis. For a long time now, the style of such architects as Cuvilier and La Joue has been preferred to the work of Mansard and Perrault; Watteau's paintings have ousted the masterpieces of Le Brun. All that there is left for us to do is to introduce Gothic taste into our architecture, and the signs are that this may well happen soon.[38]

Blondel contrasted this thoughtless eclecticism with genuine architecture, which should, even in its boldest flights, be measured. One must no doubt begin each project with enthusiasm, but one must then 'study separately each of its main parts, as if it were a matter only of those to be executed'.[39] The architect then undertook a patient refinement of his project, in which both taste and reflection were involved. In order to acquire this patience, which was so crucial a quality in an architect, Blondel went on to recommend that one spend time with sculptors: 'We wanted the young artists who are devoted to architecture to learn drawing from our skilled statue-makers rather than from our famous painters; for experience has taught us that they would thereby become more truly knowledgeable in the beauty of forms, less extravagant in their handling of contours, and more cautious in their compositions'.[40]

Thus sculpture, more even than painting, constituted a model for those young artists wishing to embark upon an architectural career. In spite of frequent allusions to the wide range of emotions to which painting gave rise, Blondel distrusted its illusionist character. Sculpture, on the other hand, was more moral than the art of painting; presenting more practical difficulties from the start, it was more akin to the work of an architect. The effect achieved matched the effort invested, so that the aesthetic was the outcome of a discipline, which thus took on moral connotations.

This morality was very different from that underlying Ledoux's well-known counsel: 'If you would become an architect, begin by being a painter'.[41] Like Ledoux, Blondel aimed to seduce, but rejected effects which he judged too facile. In the last analysis, Blondel's morality was close to that of Laugier, who demanded a degree of intellectual rigour. The *Cours* is therefore linked to that rationalist current which, from Cordemoy to Patte, unceasingly called for a return to simplicity. Architecture should be content to perfect nature instead of being distracted by an endless series of whimsical inventions. Blondel thus excoriated the 'Greek taste' in furnishing, whose forms offended the eye, and which was no more than the arbitrary transposition of elements borrowed from the external ordinance of buildings.[42] His argument here is plainly close to that of Laugier. Since its allegiance is to the same order of ideas, the *Cours* condemned the abuse of ornamentation, along with all forms of unreasoned seduction.

THEORY AND DESIGN: THE ARCHITECTURAL CRITIQUE

This quest for an authentic simplicity did, however, pose problems regarding the relation between theory and practice. Insofar as theory acknowledged the relativity of proportions and the existence of transitional modes, it recognised the need for a means of control of built form that might be adapted to this set of variations. The moral character of architectural aesthetics, and the emphasis which was laid upon the architect's intellectual rigour and self-control, allowed no disregard of the wide range of variation in principles, or the difficulty of reconciling rules and invention.

Like the majority of architectural treatises of the period, the *Cours* is characterised by a mixture of genres. The system, with its principles and rules, was perpetually at odds with the freedom the architect enjoyed in elaborating his design. It was not sufficient merely to imitate; one had to choose.[43] Blondel was clearly aware of this freedom of choice; the strict application of rules could only give rise to a cold, soulless building.[44] Yet theory was chiefly concerned with rules or doctrines, for they alone could be raised to the level of a science. Genius could not be taught, and its 'successful inventions' were the outcome of experience rather than instruction. In order to regain its effectiveness in relation to design, theory, whose precepts were 'limited',[45] had to be linked with 'the reasoning of art', an architectural critique enabling it to overcome its deficiencies. More precisely, theory and critique constituted two aspects of a single approach. Blondel's *Cours* switched endlessly between the two registers, shifting from one to the other when argument proved fruitless or, conversely, when the accumulation of examples threatened to lead the reader astray. The alliance of theory and critique enabled one to discover a middle way, which respected both the simplicity of principles and the proliferating nature of experience.

THE EXAMPLE OF MANSART, LEMERCIER AND LE VAU

Numerous buildings are illustrated in the *Cours*. Blondel also refers readers to the *Architecture françoise*, and to his four volumes devoted to the principal buildings of Paris. The merits and the shortcomings of the Palais du Luxembourg, the Chapel of Val-de-Grâce, the dôme des Invalides or of the pavilions of the Louvre are reviewed in turn, with a wealth of detail which often proves disconcerting to a modern observer. In the case of Lemercier's pavilion on the *cour carré* at the Louvre (Fig. 13), for example, Blondel begins by criticising the excessive height both of the caryatids and of the balustrade which crowns the attic. The four-edged dome also seemed to him to be too heavy, given the multiplicity of architectural members of which the ordinance was composed. The latter was also judged to be overarticulated, although the purity of the mouldings and the perfection of the ornaments was duly noted. Blondel's inspection proceeded by endlessly shifting from the whole to one detail or another. The arcades on the ground floor seemed to him too large, annihilating, so to speak, the girth of the columns; the three interlocked pediments at the top were an unacceptable abuse, but no praise was too high for the refinement of the figures in the frieze of the entablature of the second order, even if the lintels that formed part of this frieze were a little too bold.[46]

This complex survey of the eye of a series of apparently trivial details, returning to take in the overall balance of the forms, in order the better to

FIG. 13 J.-F. Blondel. 'Central features of one of the façades of the interior of the courtyard of the old Louvre, on the rue Fromenteau side'. The somewhat overloaded ordinance was not wholly without qualities, in Blondel's estimation.

divide its attention again subsequently, was linked to an aesthetic pleasure which we find hard to comprehend today. Yet Blondel's critique is more ambitious in scope than this simple framework would suggest: 'The reader should bear in mind that judicious criticism, as we understand it, is the work of reason; but, to enjoy it, and to convince whoever it is aimed at, two equally important conditions should hold, namely, one should be able to improve upon the work one is criticising, and one should know how to make one's account agreeable to the reader.'[47]

In spite of this bold claim, Blondel's writings seem to have contributed to architecture no more than his conservative stance, where he displays his loyalty to the classical tradition. In the praise he bestowed upon Mansart, Lemercier and Le Vau, Blondel presented himself as the staunch opponent of the most innovatory currents of his time.

While this judgement requires some amendment, Blondel's praise of the achievements of the *Grand Siècle* was not without nuance, and in the *Cours* he even went so far as to criticise certain details of Mansart's work.[48] Only the château of Maisons (Fig. 14) escaped unscathed, but one may well wonder at the fascination it held for Blondel. The verticality of the building, the complex interplay of its volumes, and the superimposition of the Orders all contravened principles expounded elsewhere.[49] Furthermore, the buildings described in the *Cours* were all very different from each other, and it is noteworthy that a fair number of the projects of Lescot or Delorme were judged of interest in spite of their manifest incorrectitudes. Blondel's tastes were undoubtedly broader than his trenchant pronouncements would lead one to suppose. Alongside these declarations, one sometimes comes across startling remarks, such as the following: 'in the last analysis, our French architecture is perhaps more convenient than genuinely interesting',[50] or again, 'less biased than one would imagine against the varied products of Italy'.[51] In spite of the latter, Blondel showed no compunction in slating Maderno, Borromini and Bernini.[52] In the end, however, such contradictions are of little real import. Architectural criticism presupposes that one is 'able to improve upon the work one is criticising'. If Blondel's polemic was so trenchant, it was because he aimed at some future design.

RATIONALISM AND TRADITION

This intention is perceptible, first of all, in the architects chosen from the tradition to represent a range of crucial attitudes one might adopt when

FIG. 14 J.-F. Blondel, 'Central feature of the château of Maisons on the garden front'. The château of Maisons was Blondel's favourite building, and he arranged for his pupils to visit it every year. 'Certainly not everything in its ordinance is to be imitated', he wrote, 'but it is composed in such a way that nothing in it can be changed' (*Cours*, vol. 3, p. 88).

undertaking a particular design. 'Men of talent', such as François Mansart, were thus distinguished from 'men of taste' such as François Blondel or Claude Perrault, both of which categories were opposed to 'men of genius', the most complete example of which was Jules Hardouin-Mansart.[53] Talent and taste consisted, above all, in one's mastery of the profession of architect; genius, on the other hand, could overturn all the rules and sometimes even corrupt them. In contrasting François Mansart with Jules Hardouin-Mansart, one was therefore stressing one or the other of the ever-present options in architecture. One's approach to a project involved a confrontation between knowledge and imagination, rules and invention. Blondel undoubtedly preferred François Mansart to Jules Hardouin. Praise of tradition here in fact concealed a fairly strong conviction. Like Laugier or Patte, Blondel saw architecture as an inclusive problem. Unity and simplicity were the qualities most prized in a design, and Blondel was to reproach the 'artists of the new Italy' for having 'preferred to unity, to beautiful simplicity, the intersection of forms, the distortion of the architectural elements, the contrast of forms'.[54] 'Symmetry must be regarded as one of the chief beauties of architecture',[55] but it was also desirable to treat buildings as 'pyramids' by skilful positioning of frontispieces, pediments and roofing and, more generally, by aiming at a hierarchical composition, which would be clearly arranged around a central feature.

Designed to promote a rationalist approach, the *Cours* was also opposed to anything that might inhibit the movement in the composition of the building and the free play of its parts.[56] Blondel therefore criticised overlappings and interpenetrations, along with those pediments which encroached upon the entablature of an upper floor or those ornamental keystones which break the arches. Blondel rejected the contorted mouldings and truncated members introduced by Palladio or Borromini, and recommended that one avoid all superfluous ornamentation. He also emphasised the disadvantage of superimposing the Orders, since the variation in the module from one level to the other created countless difficulties. The ideal building would consist, if possible, of a basement and a ground floor; attic stories were generally proscribed, because they also involved virtually insoluble dimensional problems. Architecture should be as simple and as legible as a noble speech.

Unity should not, however, result in an excessively impoverished expression. Aside from the liberties taken by Borromini or Bernini, what Blondel found fundamentally irritating about Italian architecture was the massive quality of the buildings and the bareness of the façades (see Fig. 15). To the 'affected and inappropriate simplicity of the Roman palaces', one was

FIG. 15 J.-F. Blondel, 'Façade of the Palace of Duke Mattei, built according to the plans of Charles Maderne'. To Blondel's mind, this palace was an example of an 'affected and none too decorous simplicity'.

supposed to prefer a play of breaks, insets and set-backs, which served to 'uphold the style' by 'movements which the graces of art allow upon certain occasions'.[57] These movements could be achieved by a judicious placing of decorative elements, by means of a contrast between the frontispiece and background or by effects of 'pyramidal' composition. Blondel was thus campaigning for an architecture that was hierarchical and articulated 'in the French manner'.[58]

This condemnation of Italian architecture enables us to understand what separates Blondel from Boullée or from Ledoux. Their differences hinged upon the way in which they apprehended space rather than upon any abstract principles. Blondel still had an eye for the nuances of the architecture of the *Grand Siècle*, its fluidity and a learned play of variations on its equilibrium. If he was opposed to Orders that were colossal, it was not for reasons of *convenance* but because, to his mind, the eye tires itself in attempting to take in columns that are too high.[59] It was very probably for similar reasons that he found a profusion of ornament irritating. Blondel's aesthetic was founded upon the visual pleasure to which slight discrepancies between the built form and rule gave rise. It was therefore bound to reject forms which were based upon internal violation, like the so-called Greek revival, the excessively bold lines of which were an offence to the eye. The 'cubism' of someone like Ledoux unbalanced a system favouring the expression of nuances over that of contrast.

It might perhaps be worth undertaking a study of the link between the development of optical values and the transformation of architectural sensibility in the course of the eighteenth century. From Le Camus de Mézières to Leroy, from Boullée to Ledoux, the problem of light in fact held a central place in architectural thought. We know that Newtonian optics displaced that of Descartes during this period; Voltaire, for example, developed a theory of light resting upon a principle of universal attraction extended to the laws of optics. In his *Eléments de la philosophie de Newton* (1738), he therefore supposed that solid bodies exercised a power of attraction over the emission of light.[60] At the same time, a preoccupation with the subjective phenomena of vision is evident, as in the ocular keyboard of Father Castel,[61] or the works of Buffon or Lambert. It is as if problems in optics were being transposed to other domains. The geometrical and mechanical perspective of the *Grand Siècle* gave way to lines of enquiry which were better equipped to take on the results of experimentation. Seventeenth-century optics reasoned in terms of impact, pressures and shocks, with matter being thought to bathe in a kind of luminous fluid which reacted to its contact as much as a liquid coming up against the walls of a vessel. With the diffusion of the Newtonian model, another relation, based upon more dynamic forms of interaction, was very probably established. Bodies were now thought to generate fields of force; in capturing the light, they could modify it, just as a prism enables one to break it down into separate colours.

Experimentation with prisms was rightly regarded as crucial in the eighteenth century,[62] and it is tempting to transpose such lessons to the domain of architecture. Whereas classical architects had been content with the play of light upon the surface, buildings were now designed that reflected other investigations. The day had now to be decomposed, with its counterpart revealed. Architects opted for smooth and massive forms, cubes or spheres, which created as much light as shade and provided a basic palette. Boullée was very probably the architect who pushed this approach the furthest. His design of 1784 for a cenotaph for Newton demonstrates how day could be turned into an eternal night; his design for a cathedral enables one to enumerate all the effects that arouse impressions of happiness or sadness in the soul:[63]

> It is light which produces the effects. These cause diverse and contrary sensations in us depending upon whether they are brilliant or sombre. Could I succeed in spreading in my temple magnificent effects of light, I would instill the sentiment of happiness in the spectator; on the other hand, I would instill only sadness if the temple were to present only sombre effects. If I could prevent the light from arriving directly, and therefore have it penetrate without the spectator perceiving where it came from, the resulting effects of a mysterious light would produce inconceivable effects and in some way a truly enchanting kind of magic.[64]

Nothing could be further removed from the 'sombre effects' of this 'mysterious light' than the religious buildings 'to the author's design' described in the *Cours d'architecture*. Blondel presented plans for a cathedral,[65] for a parish church,[66] together with the plans and cross-sections for an abbey and monastery church (Figs. 16 and 17).[67] His rationality led him to a simple, almost severe ordinance. In the wake of Laugier's writings, the *Cours* denounced the heaviness of piers, opting for isolated columns instead.[68] However, to make construction easier, Blondel kept arcades above the columns, in spite of the recommendations of the author of the *Essai sur l'architecture*.[69] Blondel's designs are in fact reminiscent of the churches of Flanders or of the eastern regions of France, as, for example, the Church of Prémontrés at Pont-à-Mousson.[70] They are therefore related to Gothic construction and to the German *Hallenkirchen*. Blondel was thus inspired in a fairly direct manner by the Gothic.

In those churches detailed in the *Cours*, the light penetrates directly, without interruption, through large windows. It reinforces the architecture rather than refashioning it.[71] Blondel evinced a desire for unity and legibility which the neo-classical architects were soon to dispense with. In the case of Boullée, for instance, unity would be based on other principles; the building, the spectator and the light would be indissociable, architecture being defined by their interaction. Blondel, on the other hand, believed that built form existed independently, as it were, of any effect it might produce on a spectator. The unity of architecture derived from its articulation alone, as revealed through the qualities of line and the repetition of structural and decorative elements.

A RIGID CLASSICISM

At the end of the eighteenth century, however, this unity became uncertain. The specialisation of functions tended to multiply spaces, and even Blondel was not wholly unaffected by this process. His design for the Royal Abbey of Saint-Louis at Metz (Figs. 18 and 19) was significant in this respect.[72] Three almost independent spaces – nave, sanctuary and choir – are juxtaposed, with a gallery encircling them, designed to control this multiplicity of volumes. This explosion of spaces seems to derive from the complexity of the programme, and from a concern for a hierarchical relation in the interior, not easily reconciled with the principles of the external composition.

Other designs exemplify this same problem, for instance, the 'Palace with a sixty-six *toise* frontage' (Figs. 20, 21), for which Blondel provided plans, sections and elevations.[73] Apart from a central feature and the four outlying pavilions, each of which has an added storey, the building consists of no more than an arcaded ground floor. The central feature, together with the pavilions framing it, seems to rest upon this ground floor, in the manner of those neo-classical elements which Emil Kaufmann described as 'autonomous'.[74] The side elevation and the section of this palace are even more revealing, above all if one recalls Blondel's professed admiration for the château of Maisons. At the same time as it declares its allegiance to classicism, the *Cours d'architecture* marks its limits.

Blondel was nonetheless aware of such difficulties, and he tried to resolve them through the repetition of the same elements. The porch and the gallery surrounding the church of Saint-Louis at Metz, and the Corinthian arcades of the Palace with a sixty-six *toise* frontage, were designed to re-establish the unity of the composition. Blondel saw repetition as a means of restoring architectural coherence.

Against the background of a strengthening of forms, evident from the 1750s on, in reaction to rococo, one observes an explosion of the elements of composition, often in response to the constraints of the programme, together

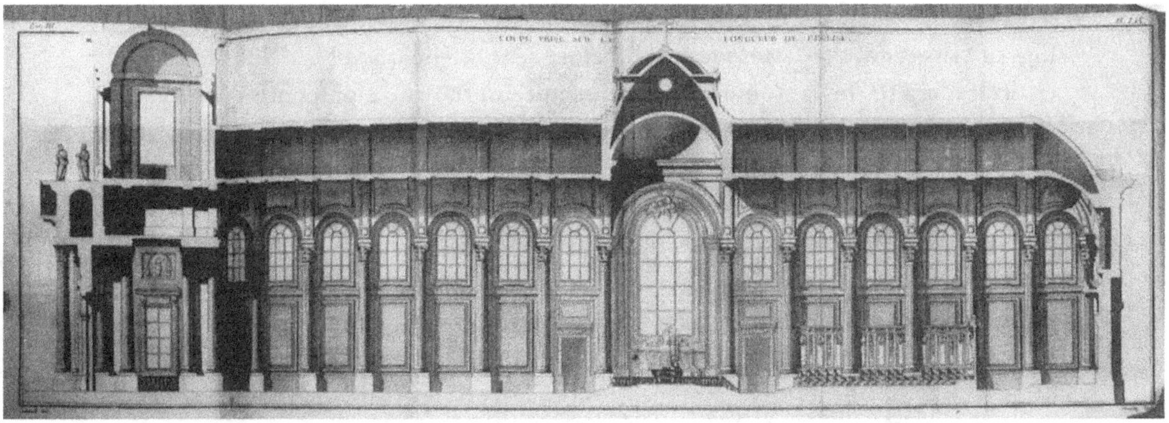

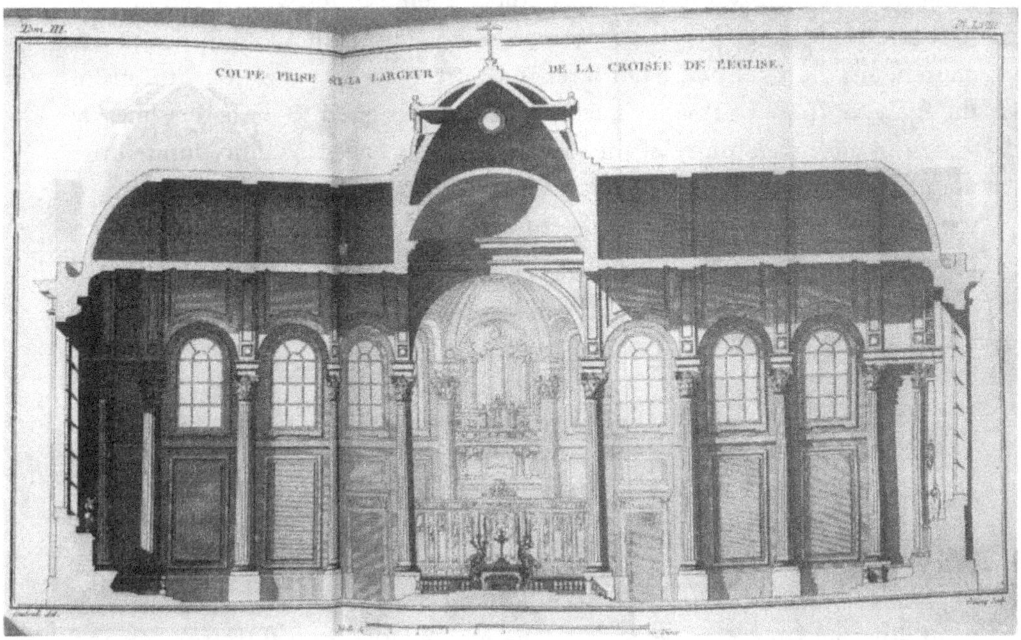

FIGS. 16, 17 J.-F. Blondel, sections of a monastic abbey church, from the author's own drawings.

with repetition and monotony. Despite Blondel's desire to maintain the teachings of classicism, the *Cours* would therefore seem to be one of the links in the chain leading to the 'autonomous' architecture of the late eighteenth century.

Although Blondel was still able to appreciate all the richness of the tradition of Mansart, Lemercier and Le Vau, he was ineluctably distancing himself from it. It is tempting to regard his desire to re-establish links with the purity of a tradition as an anticipation of some of the preoccupations of neo-classicism. One would merely have to alter the reference-points of this tradition, to measure the gap which separates it from contemporaries, and to acknowledge that a complete return to it was impossible, in order for one of the key themes of neo-classical aesthetics to emerge.[75] Blondel very probably had intimations that such a radical return was impossible, but he refused to analyse it. Likewise, repetition – which was to become a crucial theme in his work – did not yet have the significance it was to assume subsequently. The colonnades of the churches in the *Cours* were not intended to recreate the infinite by means of the finite, as those of Boullée's cathedral were designed to do;[76] they merely served to affirm the unity of the design. Yet this quest for unity limited the undulating flow of classical decoration; subject to the heightened demands of sobriety and of rationalism, architectural detail lost its richness, shedding in particular its structural principles. Blondel's Utopia consisted perhaps in wishing to restore condemned tradition.

PLANNING

Guadet saw Blondel as a key figure in the history of planning, going so far as to write that it was after him that 'architecture becomes human, that is to say, practical'.[77] The history of planning is, however, a fairly long one, and one might try to establish a line of descent leading from Le Muet's *Manière de bien bastir pour toutes sortes de personnes* to the eighteenth-century treatises on the subject. By the end of Louis XIV's reign, ideas on the internal arrangement of the dwelling had been clearly formulated. In Blondel's view, Jules Hardouin-Mansart was the real inventor of the art of distribution, his château at Clagny regarded as a model in this respect.[78] It was, however, during the Enlightenment that planning really took off, inspiring works as various as Briseux's *L'art de bâtir des maisons de campagne* or Le Camus de Mézière's *Génie de l'architecture*. The rooms of a dwelling are specialised; Blondel, for example, distinguished between six different sorts of bedroom, and the first attempts were made to

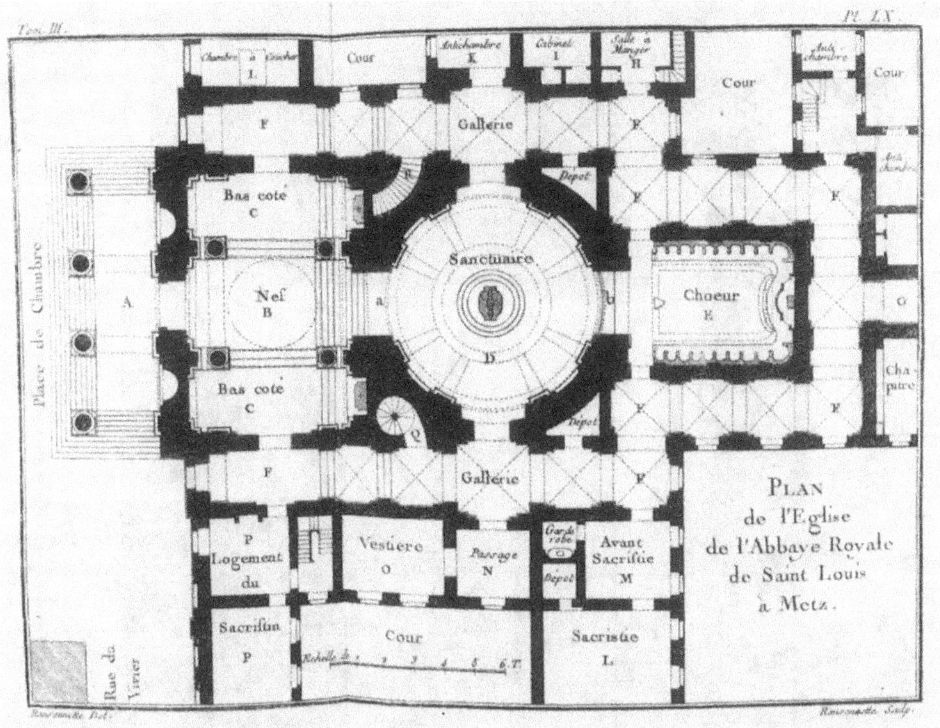

FIGS. 18, 19 J.-F. Blondel, plan and cross-section of the royal abbey of Saint-Louis at Metz.

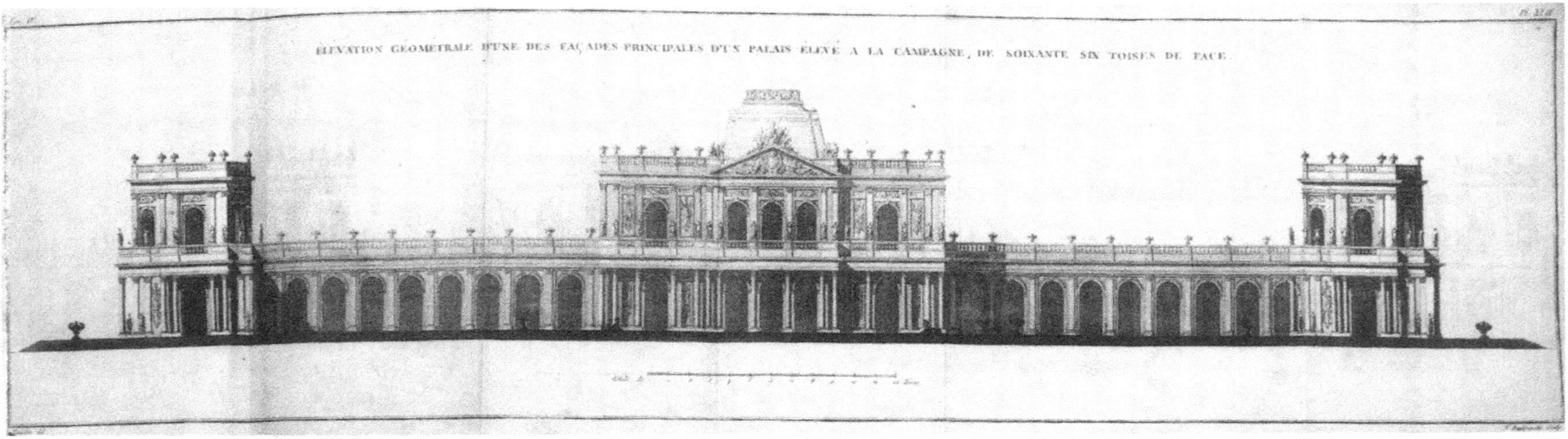

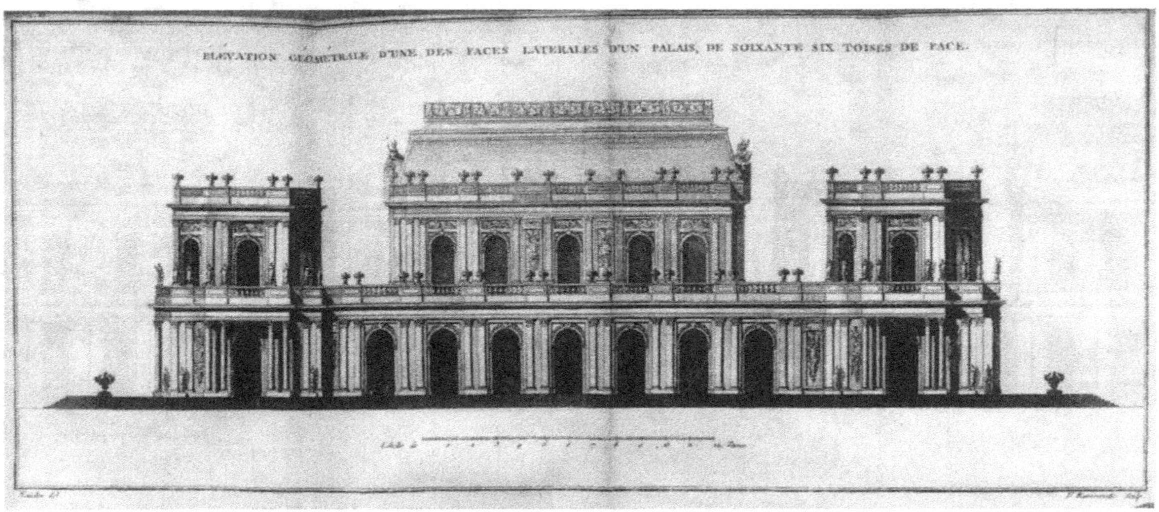

FIGS. 20, 21 J.-F. Blondel, elevations of a 'Palace with a frontage of sixty-six *toises*'. This project was probably inspired by Le Blond's 'great building with a frontage of forty *toises*', which featured in the 1720 edition of d'Aviler's *Cours d'architecture*, serving to illustrate 'the new manner of distributing plans' (plate 63N, O). But Blondel expanded the composition, which became less 'pyramidical' and more repetitive.

consider the relationship between rooms, with the corridors making an initiatory appearance, related to 'conveniences', which were increasingly used.[79]

This development now featured in a much more far-reaching transformation of social life, whereby increasing importance was attached to notions of intimacy and comfort, alongside the traditional imperatives of pomp and display.[80] Architectural coherence was thereby threatened. Indeed, one should not disregard the 'technical' character of planning, which was linked from the time of Le Muet on with the practice of measuring and with concerns of a practical nature.[81] Planning therefore presented theory with something of a dilemma. For, once a certain stage was passed, formal articulation and use necessarily came into conflict. Experiments in the rococo period had seen undeniable progress in the internal arrangement of dwellings, yet, as time passed, the relationship between decoration and planning proved conflicting. The matter of the connection between the exterior and the interior became a major issue.

This problem preoccupied Blondel throughout the *Cours*, prompting him, for example, to suggest a curious 'correction' to the Louvre colonnade (Figs. 22 and 23).[82] Observing that Perrault's façade turned it back on the rest of the building and, more especially, on the square courtyard behind, whose very existence it seemed to deny, Blondel proposed opening up the casement windows and emphasising the main door of the colonnade by means of a monumental staircase, which would give it a more striking character. His other proposals were inspired by practical concerns, such as the replacement of the oval medallions on the attic with square window openings, the substitution of semicircular arches for the flattened arches of the windows in the basement, and the use of rustication designed to reinforce the character of rustic strength in the latter. However, one should not be misled by the problematic nature of these changes into forgetting his greater aims for the nature of architecture; they served merely to emphasise the difficulty in attempting to relate the exterior of buildings to the internal arrangement.

This difficulty is not evident in Blondel's early works. His *Distribution des maisons de plaisance* presents a series of designs ranked in descending order of size from fifty to twenty *toises*, from châteaux to pavilions in the Italian style, in the great tradition of Perret and Le Muet. The exterior/interior relation is not yet seen as the conflict it would subsequently become. At the most, the need to reconcile the registers of decoration and of distribution was emphasised. The same is true of the *Architecture françoise*, which provided Blondel with the opportunity to clarify his position through a critique of examples inherited from the past. In his *Discours sur la nécessité de l'étude de l'architecture*, distribution was still defined, fairly narrowly, as: 'the division of the rooms which constitute the plan of a building, whose situation depends upon the various uses to which state or formal, social and intimate apartments are put'.[83] The *Cours*, on the other hand, testifies to a broadening of Blondel's perspective, for there he was to write that 'distribution, as we understand the term here, is something completely different: it is not sufficient to think of one's arrangement in terms of formal rooms, those of social gathering or intimacy; it is not sufficient to establish their true diameter, to regulate their height and to vary their forms'.[84] Planning was therefore no longer concerned solely with the arrangement of rooms in a dwelling, but also with the positioning of the main features of a building, with the design of the gardens and with the handling of access.

This change of perspective was evident in the arrangement of the *Cours*. It was intended at first that the distribution of apartments would be considered before that of the service buildings and of the gardens.[85] Blondel later reversed the order of contents, and spoke first of the distribution of the external elements, then of that of the rooms of the dwelling. This reorganisation implied a notion of general principles for planning related to the landscape as a whole, a science of planning overtaking the symmetry of the façade, constituted by decoration. The verticality of the wall would be counterbalanced by the horizontality of the plan and, at the intersection of these two planes, there would be a conflict at once spatial and theoretical, with planning, the daughter of necessity, being opposed to decoration, which aimed to perfect nature by means of art. The theory of the origins of architecture was thus extended by a number of practical points:

> If we go back to the source, we find that the naturally industrious bees gave men their first notions of the art of dividing and sharing out the interior of their dwellings . . . Cultivation of the fields, which subsequently led to the sharing out of lands, introduced men to the idea of property. They then found themselves obliged to mark boundaries, to plant hedges, to dig ditches, which would distinguish their own holdings from those of their neighbours.[86]

The use of terms such as 'division', 'sharing' and 'property' is reminiscent of Rousseau's *Discourse on the origins of human inequality* and it is therefore tempting to compare Blondel's preoccupation with the exterior/interior relation with the Genevan philosopher's obsession with transparency.[87] It is as well to

FIG. 22 J.-F. Blondel, 'Central feature of the façade of the Louvre colonnade'.

FIG. 23 J.-F. Blondel, 'Design for the great central feature of the Louvre colonnade'.

be cautious, however, for notwithstanding the numerous references in the *Cours* to a state of lost innocence, Blondel was in fact aligned with Voltaire: 'We would perhaps have more virtues, were we to live as the first peoples in the world [had done]. On the other hand, let us think of the times and consider, with Plutarch, how happy we are, and how unhappy our fathers were'.[88] As a lover of luxury, and also of mirrors, which served to extend rooms and to link the exterior and interior by means of reflection,[89] Blondel was a man who accepted dividing and sharing; the art of planning guided him in his choices.

This art, whose relative novelty and specifically French character[90] was again and again emphasised in the *Cours*, was divided into two main sections, reproducing in a minor way the conflict between planning and decoration. A distinction was made between external arrangements and the planning of apartments.

EXTERNAL PLANNING

External planning concerned the arrangement of the main building, of the courtyards and outbuildings, together with the lay-out of the gardens. Factors such as general salubriousness and favourable aspect enabled one to decide upon the siting of the dwelling. Around it, kitchens, stables, courts and fore-courts were to be arranged in a graduated composition as 'pyramidal' as a façade. After these general considerations, Blondel turned to the art of gardens. As was to be expected, Le Nôtre's creations were superior, in his view, to the 'new productions of England'.

The *Cours* was nevertheless not dogmatic on this point, restricting itself simply to observing that the fashion for rustic gardens took off 'more through fickleness than through reason'.[91] At a time when the ha-ha was all the rage in both France and England, Blondel applauded those views which gave out on to the countryside, at the end of artfully contrived parterres and thickets.[92]

The *Cours* goes on to mention 'the various objects which may be exposed to view in well-ordered gardens', namely parterres, lawns, amphitheatres (Fig. 24), fountains, vases, staircases and terraces. There then follows a list of enclosed elements, for example thickets and greenhouses. Blondel illustrates general principles by means of examples, presenting the plans of a number of gardens of his own invention: a park designed for the Archbishop of Cambrai,[93] and two others for German princes (Figs. 25 and 26).[94]

What strikes us now when we read the principles espoused by Blondel is their declared intention of subordinating the garden totally to the dwelling. The buildings and the park surrounding them should constitute a single entity, a spatial entity without break or opposition. 'Those who admire contrasted forms will undoubtedly disapprove of the regularity we have favoured in the distribution of this plan', Blondel wrote with respect to his design for the Archbishop of Cambrai at Chateau-Cambrésis, but he declared a little later, 'I have merely followed my own natural taste in this respect.'[95] This natural taste ruled out any breach in tone. Everything was set in place as if architecture had been projected on to the landscape; nothing was more telling in this respect than the 'green rooms' formed by clipped hedges which mimicked the arrangement of interior spaces (Fig. 27). Ballrooms, trellis rooms and open-air theatres seemed to make up an apartment all of box-woods and arbours. Here again, through its rejection of violent contrasts, the *Cours* ended up with a relatively stable aesthetic. The park in the French manner, which had still been experimental in Le Nôtre's time,[96] had become in Blondel's case the petrified reflection of an architecture that was itself under threat. At a time when the writings of William Chambers were being translated and lovers of gardens were composing irregularities and picturesque accidents,[97] the *Cours d'architecture* was trying to preserve an impossible unity, with external space being defined as a simple extension of architecture. Just as the garden was modelled upon the latter, so too the sequence of courtyards and terraces allowed controlled access to the body of the main dwelling, while the out-buildings were arranged like a miniature town centred upon the residence of the master. This overtly hierarchical arrangement stands out in sharp contrast to the engineers' conception of a domain which would be overrun with lines and organised as a network.[98]

INTERNAL PLANNING

Blondel may have permitted himself a few liberties in the *Distribution des maisons de plaisance*, but the *Cours* offered a much more rigid system. The *enfilades* constituted the chief beauty of a room, and an architect's task was first and foremost to establish them. One had to observe 'a gradation relative to the intended purpose, and to the use of each room',[99] from soberly decorated vestibules to those sitting-rooms in which all the resources of art were deployed. As far as internal distribution is concerned, we find all the major principles of decoration, transposed to the articulation of the rooms:

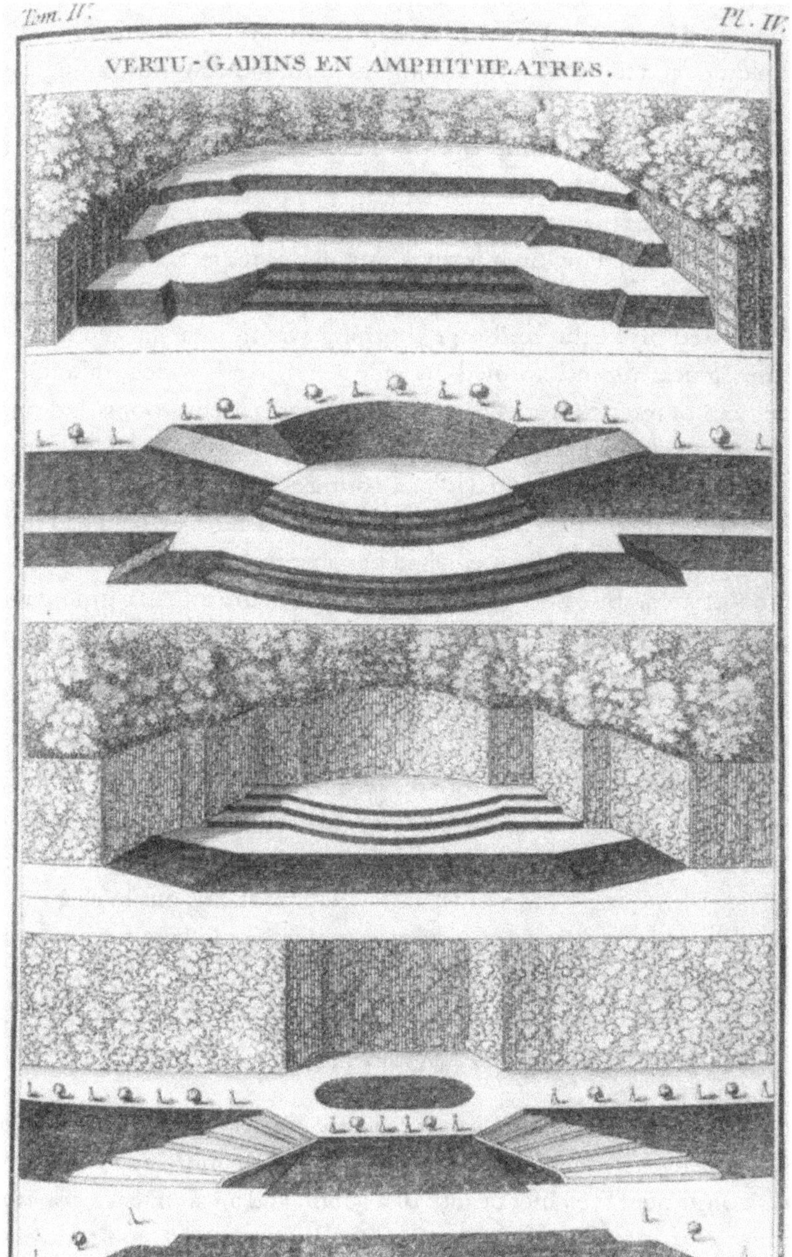

FIG.24 J.-F. Blondel, 'Amphitheatres'.

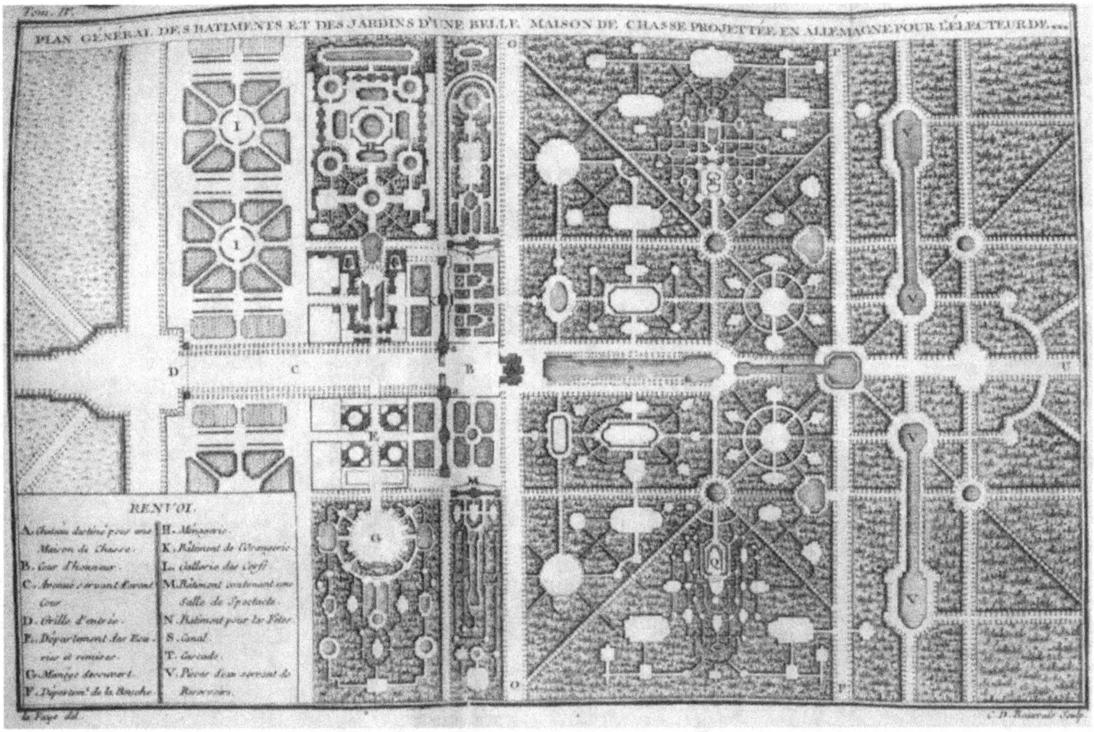

FIG. 25 J.-F. Blondel, 'Site plan for the buildings and gardens of a grand hunting-lodge projected in Germany for the Elector of . . . '.

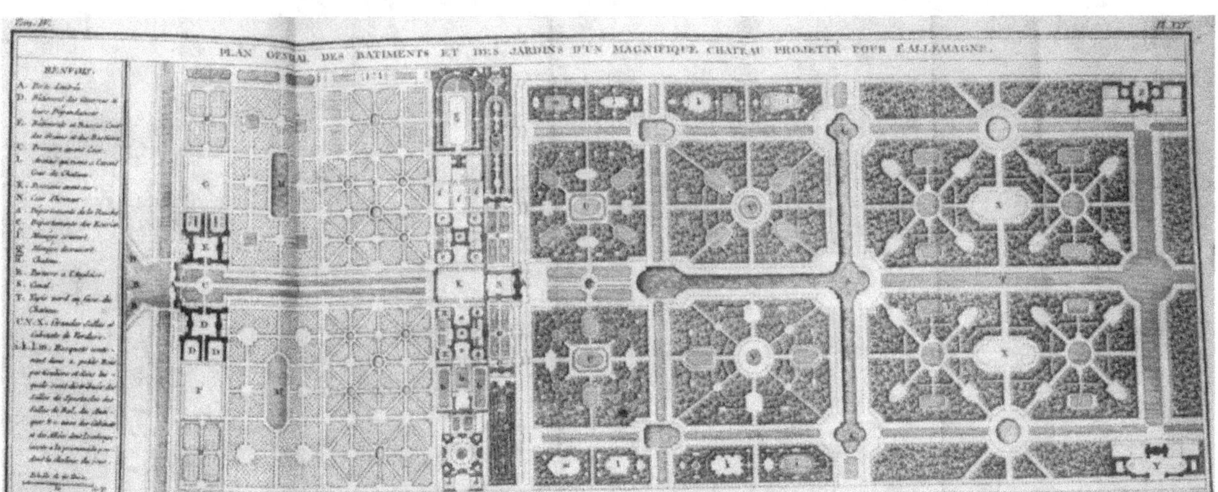

FIG. 26 J.-F. Blondel, 'Site plan for the buildings and gardens of a magnificent château projected for Germany'.

FIG.27 J.-F. Blondel, 'Various designs for complex palisades, serving to decorate ceremonial gardens'. The ordered universe of the gardens directly echoes that of the apartments of the dwelling.

By the spirit of *convenance* relative to planning, we understand the varying degrees of magnificence which one should seek to impart to the interior of the apartments, depending upon the rank of the owner who was intending to live there, as well as the attention the architect should pay to making the main rooms more or less large, varied in form, with good views, well-lit and separated, depending on their respective uses.[100]

By symmetry, we understand the regularity of the elements opposed, one to the other; the need to place chimneys and piers in the middle of the length and width of rooms. Symmetry also involves the relation which should hold between the different rooms of an apartment.[101]

By proportion, one should understand how crucial it is to give the different rooms in a building heights that are appropriate to their various widths and to their particular use, as well as to establish a direct relation between their length and breadth.[102]

The analogy between decoration and distribution could be pursued further, since there was a correspondence between the pyramidal character of the façade and the gradation of the *enfilades*. If buildings were classified in terms of the number of their floors, a distinction had also to be made between single, double or semi-double building forms, depending upon whether there was a single *enfilade* over the whole extent of the building, a principal *enfilade* with another, less important one running parallel, or two main *enfilades* in parallel. One can discern here an attempt to resolve the conflict between exterior and interior by reducing it to a problem of geometry involving the combination of a whole series of horizontals and verticals.[103] In contrast to Ledoux, Blondel sought to avoid the paradoxical effects which might insinuate themselves between plan and elevation. He saw it as a matter of reconciling volumes rather than contrasting them in varied projections. One might enlarge the windows, one might tinker with the height of the rooms or with the external ordinance, but never should a cylinder start on the ground as a square.

Blondel favoured *enfilades*, symmetrical rooms, reviving the kinds of gallery which the seventeenth century loved to place in the wings of private town-houses; in this respect, the plans given in the *Cours* are, in the last analysis, fairly conventional. In order to illustrate his principles, he used the Palace with a sixty-six *toise* frontage, already mentioned. The plan was organised in a fairly simple manner; in the centre there was a large salon in the Italian style, preceded by a vestibule, two large rooms and the grand staircase, followed by apartments on either side, while bedrooms, studies and galleries occupied the wings of the building. In order to emphasise yet again the difficulty of attuning decoration to distribution, Blondel gives two versions of the palace (Figs. 28 and 29). The first version, which has an H-shaped plan, is chiefly determined by the exterior arrangement;[104] the U-shaped plan, on the other hand, gives more importance to the distribution of the apartments.[105] When it was a question of choosing between these two options, Blondel opted for decoration, declaring that one should sometimes be prepared to sacrifice the interior to the exterior.[106] In the hierarchy of architectural theory, as well as in architectural practice, decoration therefore took precedence over distribution. Blondel also listed the various rooms of which apartments were composed, namely, vestibules, salons, antechambers, *salles de compagnie* and *salles d'assemblée*, studies, bedrooms and galleries, providing in addition many details with regard to their use and the conveniences one might derive from them. In this respect, the *Cours d'architecture* represents a transitional stage between the classical age and the discovery of domestic intimacy. In the Palace with a sixty-six *toise* frontage, this intimacy was still fairly restricted, being reduced in most cases to angled passages or dressing rooms of limited size positioned between two *enfilades*. For Blondel, service rooms were spaces whose status was still unresolved.

By the term 'apartment' Blondel understood 'the linking of several rooms, having for their object the same general purpose; each of which, however, being liable to have a specific use'.[107] One then had to distinguish between the two different kinds of apartment, those of state or display, those of society, and those which were private.[108] State rooms were still more numerous than the little rooms where one might relax, the dressing rooms, or the small recessed rooms, but the *Cours* also described the 'S-trap places', brought from England, use of which was becoming more and more common.[109] At the end of volume 4, there are also a number of plans drawn up by Franque, an architect friend of Blondel,[110] which, through their more modest dimensions, provide intimations of a more everyday sort of architecture than that of the Palace with a sixty-six *toise* frontage. While the palace enabled one to list the various rooms of which a dwelling consisted in a sort of hieratic catalogue, the plan of an abbot's house (Fig. 30)[111] or the ground floor of a private house[112] indicated other possibilities. The detail of the drawing seemed at last to temper the solemnity of the grand principles. A geometry involving arcs of a circle, ellipses and triangles, and artifices ranging from sunken recesses to the use of nooks through the imaginative use of panelling, marked the possibility of another approach than that of a *bienséance* based on noble and royal references;[113] 'If it is a question, for example, of compensating for angles, irregularities and slants, one has recourse to circular, elliptical, triangular or irregular forms which, when employed properly, invest the whole work with a kind of charm, altogether preferable to a strict and almost always monotonous symmetry.'[114]

Nevertheless, the Palace with a sixty-six *toise* frontage held a privileged place, and the fact that this place was accorded it marks one of the limits of Blondel's approach. In providing the framework for a listing of the rooms of a dwelling, the palace served, as I have noted, like a catalogue. This usage was linked to the hypothesis that programmes remained valid, that a range of subtractions might enable one to shift from the palace to the simplest of dwellings. Use and function were thus thought to be established on a continuous scale. But what the eighteenth century may perhaps have invented was a discontinuity of scale, the substitution of a heterogeneous array of social practices for the notion of a single hierarchy. The state and its institutions were no longer reducible to private life, the dwelling-place of a poor person no longer had anything to do with some fragment or other of a palace. No matter how many primitive huts one added together, one would never arrive at the residence of a prince. Ledoux foreswore, perhaps unconsciously, Laugier's dream, when he left the poor man sleeping beneath a rainbow instead of building him a Doric shelter.[115]

Where planning is concerned, we therefore find the same preoccupation with unity and coherence, and the same obsession with the continuity of effect in Blondel's architecture already apparent in the field of decoration. On the eve of the Revolution, this position would seem already to have become a blind alley, if the success of the engineers, who based their work on a wholly different set of premises, is anything to go by.[116]

'CONVENANCE' AND CHARACTER

The unity and coherence of architecture also found expression in terms of principles. Beyond geometry, these principles were organised into a system which I shall now try to describe in more detail. Principles and the practice of building were in fact intrinsically connected; *convenance*, for example, determined each and every one of an architect's decisions.

Blondel was undoubtedly the supreme theoretician of *convenance*, and he gave a new importance to this term of the art, which had long been subject to

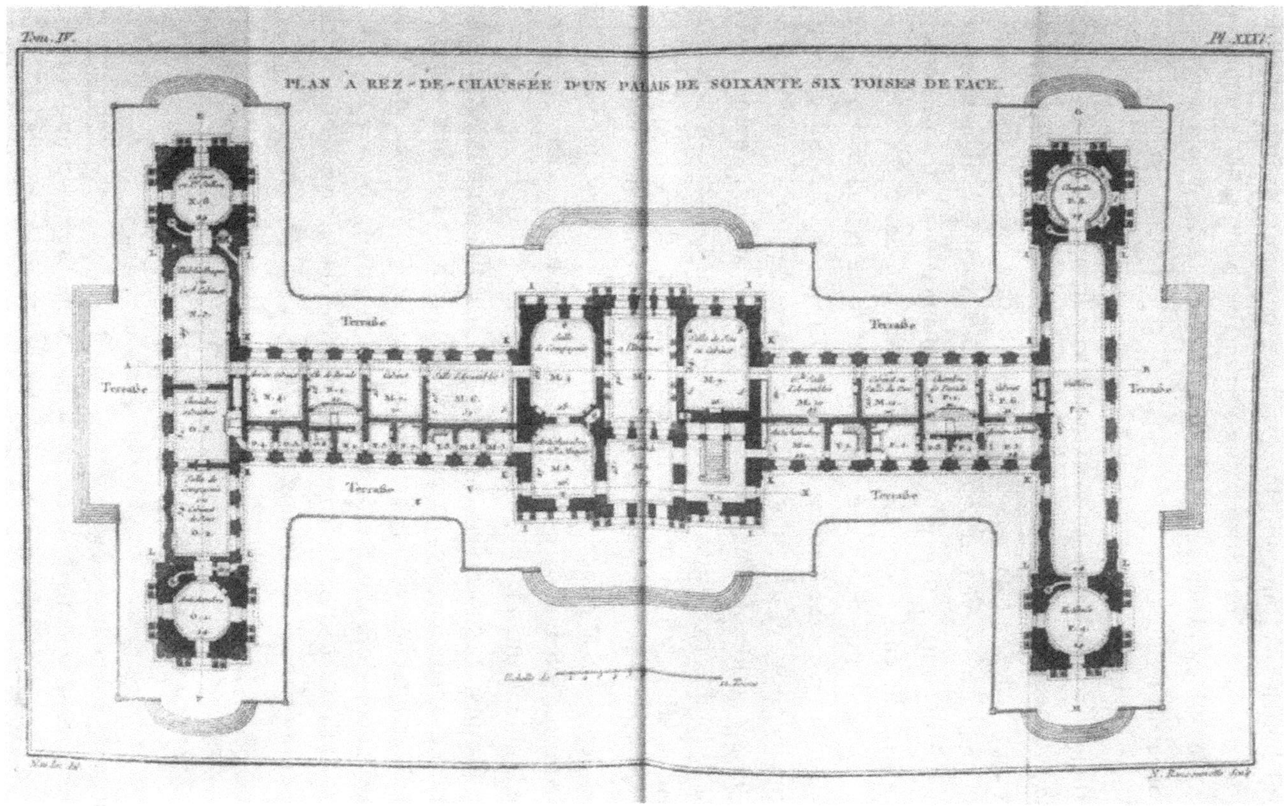

FIGS. 28, 29 J.-F. Blondel, plans for the ground floor of a 'Palace with a frontage of sixty-six *toises*'.

Tom. IV. Pl. XXXVI.

AUTRE PLAN A REZ DE CHAUSSÉE DU MÊME PALAIS DE SOIXANTE SIX TOISES DE FAÇE, COMME LE PRÉCÉDENT.

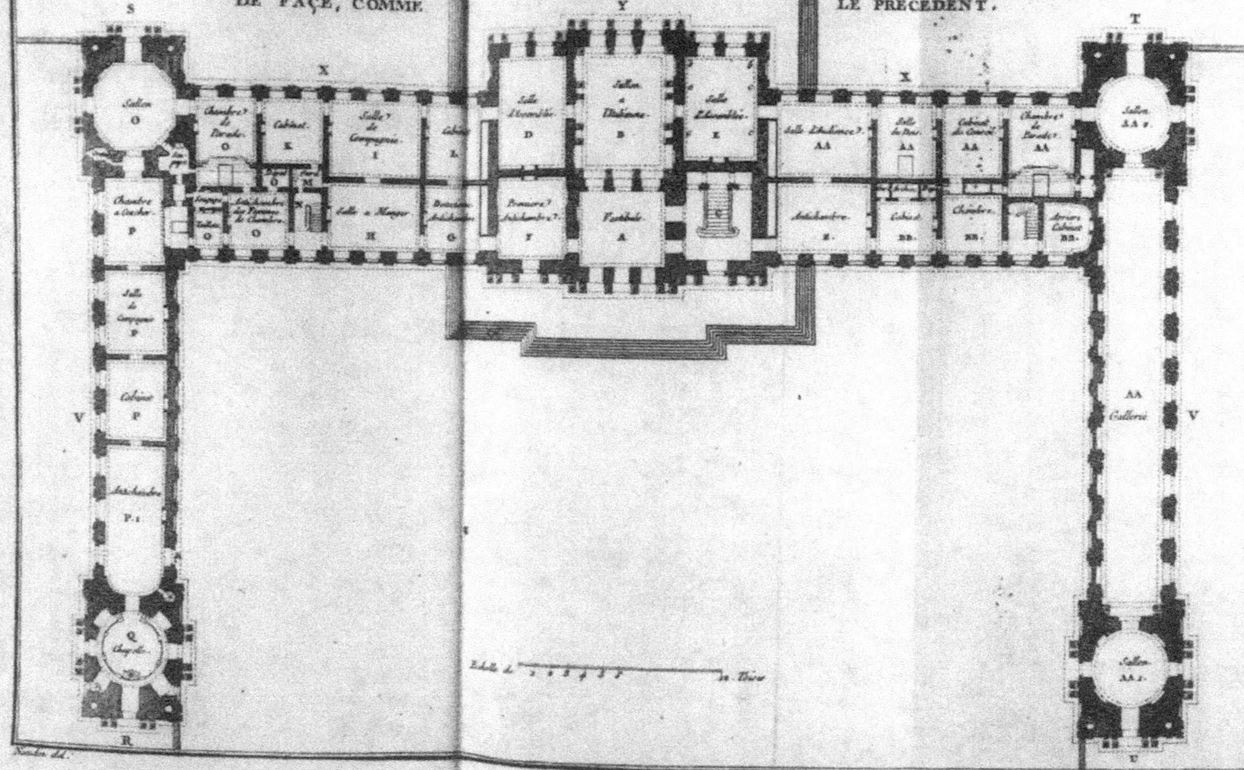

Neufm del. N. Ransonnette Sculp.

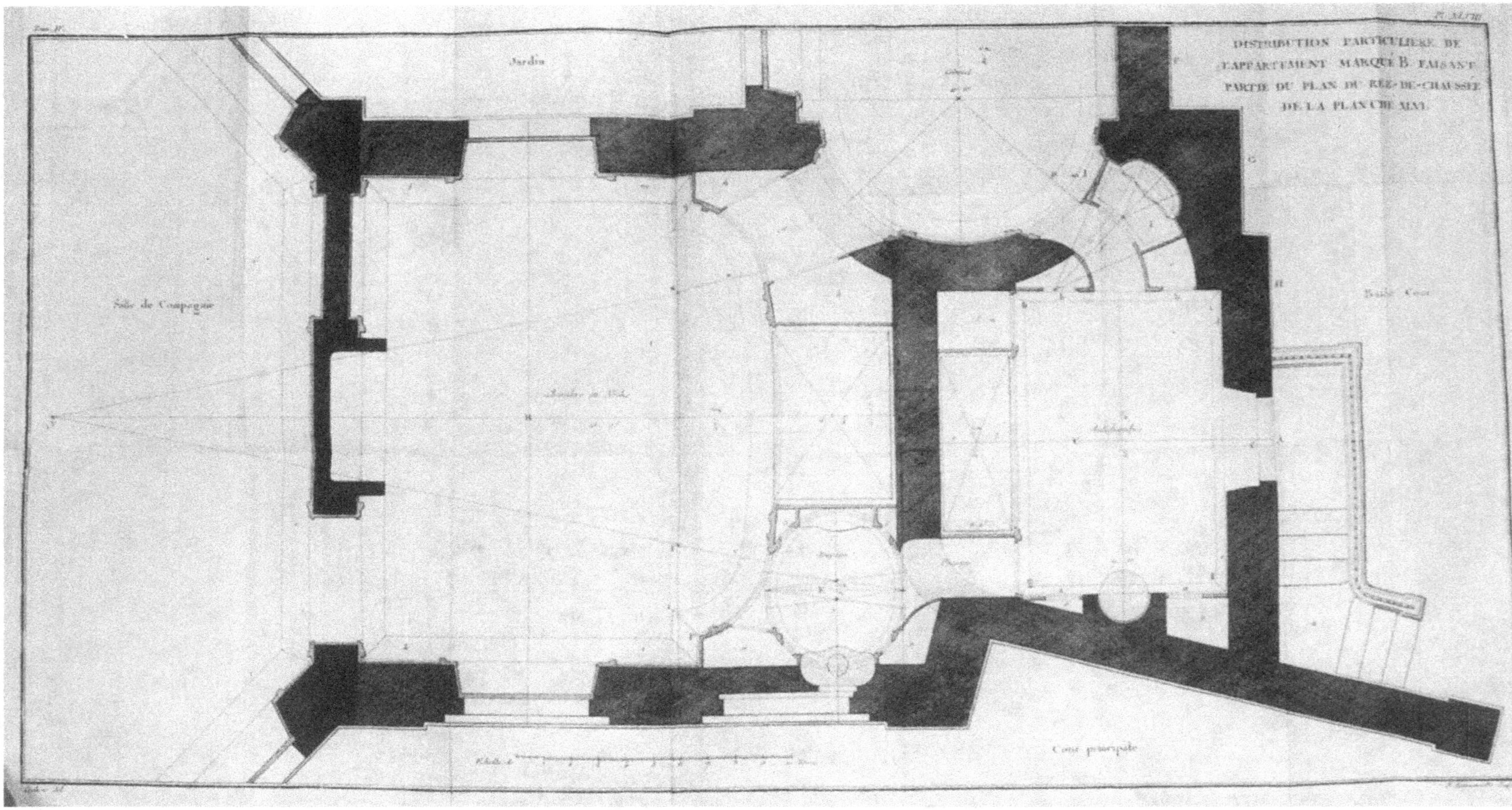

FIG. 30 J.-F. Blondel, detailed arrangement of an apartment in the abbot's house of the Abbey des Prémontrés at Villers-Cotterets, from a drawing by the architect Franque.

the rules of *bienséance*.[117] In his writings, *convenance* seems to work on several different planes. It concerns, first of all, the spatial definition of the design, as the *Distribution des maisons de plaisance* emphasises:

> The spirit of *convenance* instructs one as to the choice of position, the correctness of proportion, and the convenience of arrangements; it helps one to discern which materials are appropriate to the site upon which one is building; it guides one in the task of assembling or working such materials; it keeps the ultimate aim of one's enterprise ever in view and, in a word, it is through *convenance* that a building may have all of its perfection and that one may find in it a pleasing correspondence of the parts with the whole.[118]

Yet *convenance* was still more far-reaching, since it also covered the domain of *bienséance*, a quality defined as 'the decorum one should observe with respect to each type of building, depending upon the rank of the persons for whom one is building it'.[119] If seemliness is especially appropriate for sacred buildings, public squares and royal palaces,[120] *convenance* was supposed to generalise this 'decorum' so as to suit the rank of the particular patron. *Convenance* was the principle underlying the coherence of the whole design, and in the end it therefore referred both to the set of problems one encountered as it proceeded, and to the quest for a perfect fit between architecture and its commission.[121]

Convenance therefore represented a 'spirit' as much as a fixed concept. Since the architect reflected upon this principle even when elaborating the details of a building, it served to guarantee the unity of the outcome. This unity was especially concerned with the relation between the exterior of the buildings and the proposed shape of their interiors. It is worth emphasising the double nature of *convenance* in this respect, since it extends from questions of ordinance through to economy and expenditure. *Convenance* is located precisely at the point of intersection between the beautiful and the useful, and between decoration and distribution. One can therefore understand why Blondel should have made so much of this principle, for, by enabling him to reconcile apparently contradictory demands, it in fact constituted the most synthetic expression of his rationalism. 'Architecture', wrote Blondel, 'is one',[122] and *convenance* was the name bestowed upon this unity marrying taste and reason.

The need to 'give different forms to buildings', 'depending upon the rank, dignity or wealth of the owners', was not without its difficulties, especially if one bears in mind that one had at the same time to abide by 'constant principles', and that this constancy consisted in the first instance in fidelity to the classical vocabulary. There were only five Orders of architecture, whereas society distinguished between a host of different conditions. In order to resolve the problems occasioned by this disparity, Blondel tried to reduce the register of architectural effects; the systematic inventories of the *Cours* took their inspiration directly from this.

But it was not sufficient merely to list the constitutive elements of architecture; one had also to indicate what the legitimate combinations were, and to compile a catalogue of them, prior to specifying the kinds of correspondence which might unite the range of architectural effects to the various uses of society. The theory of character could then be assimilated to this set of prescriptions:

> Since all the different products based in architecture should bear the mark of the particular purpose of each building, all should have a character which determined their general form, and which makes known what the building actually is.[123]

In his *Cours*, Blondel tried thus to transpose to architecture the notion of character, a concept which had emerged towards the end of the seventeenth century with theoreticians such as Boileau or Dubos.[124] Architectural character consisted first of all in the five Orders of architecture, whose 'type', which had been fixed since the Renaissance, assumed a new prominence. The Tuscan Order represented a somewhat rustic solidity, the Doric was robust and masculine, the Ionic more feminine and the Corinthian rich, elegant and light, whereas the Composite Order evoked an almost adolescent grace. The character of the Orders was then transmitted to the whole building through their 'expression', through the design of the entablatures, doors, casement windows and balustrades. The contrast between smooth and hard, the interplay of different members in the decoration and the pyramidal effects were all intended to transform architecture into a poem which was by turns eloquent or reticent. Blondel revived for this purpose Horace's *ut pictura poesis*, which he transformed, as Boffrand had done, into *ut architectura poesis*:[125]

> Does not painting, in the variations it offers in the ordinance of its canvases, inspire in the eyes of its viewers the image of strong or tempered passions, of action or of rest? In poetry, do not the heroic, lyrical or pastoral genres express in different forms the various affections of the soul. In music, with a fairly limited number of signs, does one not succeed in expressing by turns terror or desire, heroism or voluptuousness? Why then should not architecture, which is more wide-reaching in its precepts, and no less susceptible to taste than the other arts, supply architects with the resources to invest their different kinds of production with infinite variety?[126]

At the end of the first volume of his *Cours*, Blondel gave a list of some of the characters to be encountered in architecture. One could distinguish between sublime, admirable, original or pleasing productions, between architectures which were varied, free, exact, symmetrical, feminine, mysterious, terrible or barbarous, and so on through forty or so different characters.[127]

This listing prompts one to enquire as to the exact scope of the theory of character in the *Cours*. It is apparent, first of all, that the list is a heterogeneous one, for it presents as being on the same plane purely formal notions, such as symmetry, and others, taken from the realm of sensation, such as the genre of the terrible. Indeed, Blondel never defines what he understands by character; this imprecision very probably reflects an ambiguity which he never managed to resolve.

The ambiguity lay in the relations between *convenance* and character: 'the spirit of *convenance* buttressed by a taste for art, leads the architect to choose the character of one of the Orders in order to designate the kind, the importance and the use of the building which he wishes to decorate'.[128] The whole problem consisted, as we have seen, in establishing what the rules of correspondence between order and use were. However, through the intermediary of *ut architectura poesis*, the theory of characters was tending to shift the debate on to the terrain of sensation, or, to be more precise, it was bringing under the same heading form and its effects upon the sensibility of the spectator. It was nevertheless the case that this sensibility differed in its structure from that of society, and that character was not easily defined. It was easy enough to grasp that a palace should be majestic and a prison terrible, that a triumphal arch should evoke the idea of grandeur, and that a temple should awaken a contemplative feeling, but how should one describe a bourgeois dwelling or programmes of everyday utility, whose existence the *Cours* did in fact acknowledge? In spite of assertions regarding the prominent role of *convenance*, architecture and society displayed their effects along two separate lines which would not readily converge.

There is nothing surprising in this situation if one thinks of the primary function of art, namely, to imitate the most universal qualities of nature. Since the theory of characters was designed to reconcile the general with observation of the particular, it came up against the question of types. Architecture could not imitate society directly, nor could it even reproduce all the articulations, unless it had recourse to a typology. But the *Cours* rejected this approach. There would be an alternative, which would entail referring to the universality of a nature which determined both sensation and the chief features of the social edifice; Blondel's failure to arrive at a rigorous definition of character may then be explained in terms of his lack of interest in the natural bases of architecture. In the eighteenth century, typology and the exploration of nature were anyway linked, as is plain from the example of someone like Boullée, who was planning a *Receuil d'architecture privée* at the same time as he was drafting his *Essai sur l'art.*[129]

Although Blondel never looked more deeply into the ubiquitous contradiction between the imitation of nature and the representation of buildings as social, the question of the representation of the latter had nevertheless been raised, since character by itself did not serve to resolve the problem.

A kind of linguistic overbidding seemed then to occur. It was as if, by multiplying causes and effects, and by piling up references and distinctions, some unity might emerge from the proliferation of terms. I have frequently emphasised the encyclopaedic aspirations of the *Cours*, which may have reflected a wish to endow architecture with a structure as complicated as that of society. Having failed to arrive at simple rules of correspondence between architecture and society, Blondel sought to give both domains the same degree of complexity, leaving it to the architect to define the terms for a comparison between them.

Yet a definition of this sort refers to language and to the unintended consequences of the accumulation of signs. There were already too many characters in the *Cours*; too many or too few characters, it was fundamentally the same thing. Endless divisions were as impenetrable as chaos. Theory, in its aim to be exhaustive, became suffocating. This charged atmosphere was diffused through the use of violent combinations and contrasts. A purified syntax was soon to supplant the catalogue of characters compiled by Blondel.

TALENT, TASTE AND GENIUS

Without the finesse of an architect, *convenance* and character became a dead letter. Such qualities concern design directly, and together with them one might look also to talent, taste and genius:

> The man of talent, as we understand the term, is someone who, being steeped in the knowledge of the theory of architecture, and the practice of building, produces nothing without having taken into account the proportions established by the Ancients, and followed by the best of the Moderns; he is someone who, through the science of design, knows how to observe exact relations between the masses and the

main parts of his building: someone who, enamoured of the rules of the art and of the laws of symmetry, knows how to combine in his design external and internal planning, external with internal decoration and, finally, solidity with the economy which should be observed in all manner of enterprises.[130]

As we have seen, talent consisted of a mastery of the rules. It was plainly reckoned the most indispensable quality in an architect, since François Mansart was considered by Blondel to be the quintessential 'man of talent'. Pierre Lescot, Philibert Delorme and Libéral Bruant[131] were also held to possess this attribute. Nevertheless, if one proved to be lacking in taste, one's talent would only produce compositions which, though correct, were cold and monotonous:

The man of taste, as we conceive him to be, is someone who knows how to blend the style of architecture with the *convenance* of the building; who knows how to decrease or increase the sizes as defined by the precepts of his art, in accordance with the point in the distance from which the building is meant to be viewed; it is someone who knows how to select ornaments, decide upon their attributes, their allegories and their relief. The man of taste knows how to ascertain the secrets of his art, to deploy the forms, and to grasp when he should make them symmetrical and when he should contrast them; he alone is therefore capable of crossing the boundaries prescribed by the rules, and of transcending precepts, without however allowing himself liberties which taste itself might repudiate.[132]

If, therefore, talent was concerned with the general organisation of design and with the correct application of principles, taste, which 'is very difficult to explain' and which 'can scarcely be more than felt',[133] has far more to do with the elaboration of detail, a play of variations and inflections in which the classical architects were past masters. The two qualities which were essential in an architect were talent and taste, obedience to rules and their nuanced interpretation. Genius, on the other hand, was in Blondel's view a fairly marginal factor. 'An artist's genius cannot be too much praised',[134] and yet one should beware of his inaccuracies. The positive aspect of genius consisted in the felicitous deviations which it authorised, which intensified the pleasure of the eye, and which were more far-reaching than those prescribed by taste. Jules Hardouin-Mansart and Boffrand were the most complete examples of what genius, for good or ill, could produce.[135]

The distinction between talent, taste and genius in fact masks a particular stance towards design. Blondel declared, in a lyrical tone which seems to anticipate Ledoux, that 'architecture is a creative art, concerned with buildings of every kind, whether their purpose be that of utility, security or magnificence, from thatched huts to monuments with coverings with metals that shine everywhere'.[136] Yet while Ledoux would invoke the call of genius freed from the shackles of tradition, Blondel still saw architecture as a matter of talent rather than boldness of conception. Through the great examples inherited from the past, and the criticism which put them in their proper perspective, what architecture was basically concerned to imitate was not nature but itself.

Given this framework, design involved a compromise between imperatives which were necessarily contradictory, since society in no sense thought of imitating architecture. The stiffening of forms in Blondel's work corresponded to a wish to resolve these contradictions by conforming to the established principles, by means of a geometry that was wholly lacking in artifice. One had therefore to decompose and analyse each of the parts that were to be treated, before proceeding with a definitive synthesis of them. This synthesis became increasingly difficult to effect; the *Cours* itself was not immune from this fragmentation. The talent of Mansart, who wished to resuscitate Blondel, was in fact lacking in that a priori judgement which the neo-classicists were to find in a genius to whom nature dictated its laws.[137]

'TOWARDS A CLASSICAL ARCHITECTURE'

ARCHITECTURE AND SOCIETY

Through *convenance,* talent and taste we are the better able to grasp the perspective from which Blondel viewed the relations between architecture and society, and between the architect and his patron.

Architecture was designed to express all of the gradations of the social order. An architect was an artist who sculpted in stone the image of the society which commissioned him. The ideal patron would therefore be characterised in terms of his acceptance of architectural language and of its constraints. A prince would have a richly decorated palace built, whereas a tradesman was supposed to be satisfied with a more modest façade. The bond between the architect and society was in essence a contractual one; it presupposed a shared point of view. Endowed with a legitimacy which his talent alone conferred upon him, the architect had to seduce and convince.

However, an architect's patrons would sometimes prove difficult to persuade. Vanity often led a tradesman to usurp the attributes of birth and of rank: 'all proprietors in fact have pretensions to being architects, to the extent that someone who has a passion for building would prefer to give it up rather than follow humbly the counsels of an enlightened man, who would curb him in his tastes, and prevent him from announcing himself to the world as what he believes himself to be'.[138]

Education constituted the sole means, as we shall see again below, to which one might have recourse in order to avoid such mutual incomprehension. Architects dreamed of a civilised society, in which each person would be acquainted with the principles of architecture. The title of a posthumous work by Blondel, *L'homme du monde éclairé par les arts*, presents a programme in itself. A count and a countess exchange letters and, against a background of amorous intrigue resembling a very insipid Laclos, the theory of Orders, distribution and even construction are treated in turn. The count is well informed about architectural matters, while the countess is eager to gain instruction; one cannot help but applaud this comedy, whose themes are of basic importance to architects. More prosaically, Le Camus de Mézières a little later published his *Guide de ceux qui veulent bâtir*, which contains both advice on financial matters and reflections on the beauty of art.

A similar attitude was adopted towards institutions. There, too, it was a question of training the grand agents of the state in the understanding of architectural language. In his *Discours sur la nécessité de l'étude de l'architecture*, Blondel sang the praises of those Intendants who had improved the great provincial towns;[139] Ledoux, however, was later to complain about the ignorance of the bureaux respecting everything having to do with architecture.[140]

If I have entitled this section 'towards a classical architecture', it is because it was this same preoccupation with communication between architect and society which, in the last analysis, justified Blondel's attachment to the classical tradition. The claim that there was a specifically French manner in fact betrayed a desire to see architecture constitute the privileged expression of the collective consciousness, of that 'spirit' which Montesquieu attributed to the various nations. In rebelling against the fashions of the day, whether Italian or English, the *Cours* also invoked the influence of climate and of customs.[141] Architecture was supposed to reflect the reasonable bases of society rather than its decorative caprices.

We are now in a position to ask just how pertinent this approach was at a time when the engineers were establishing a new set of power strategies. The question which in fact arose was that of the need for an architecture which reflected society, sending back to it its own image as embellished with the help of art. Utility, however, was not *convenance.* Did one therefore have to transcend masonry, at whatever cost, in order to fix a simple image for all eternity? Was the 'perfected' imitation of nature a worthy terminus for one's art?

At the end of the eighteenth century, economic debate was concerned with the source of wealth. The physiocrats and the utilitarians were both unstinting in their praise of the soil and of trade. The engineers' role was legitimised, as we shall see, in terms of its contribution to the general prosperity. By enclosing itself within the cycle of consumption, even if it was only a question of social appearances, architecture had already condemned itself to an irredeemable gratuitousness.

THE LIMITS OF THE TRADITION

Although I do not wish to survey all of the elements considered in the course of this chapter, I feel that I should conclude by emphasising the crucial shortcomings of Blondel's approach.

The first of these shortcomings stems from the actual nature of architectural knowledge, which was becoming increasingly out of touch with the sciences. The mathematics of the architect was based entirely on the geometry of the *Grand Siècle*, evincing a total disregard for recent advances in analytical enquiry or for the crucial contribution of physicists and natural scientists to an understanding of the world. An explosion of archaeological knowledge also served to undermine a body of theory whose explanatory powers were becoming increasingly irrelevant. Though Leroy taught alongside Blondel at the Ecole de l'Academie,[142] the baseless Doric of the Greeks was not permitted to intrude on a theory centred on the classical tradition.

This same tradition was responsible for the severest handicaps from which architects suffered, for, as I have already pointed out, the *raisonnement de l'art* was continually threatening to reduce architecture to an imitation of what had already been produced, so that imitating nature became synonymous with imitating architecture. If this problem was never addressed directly in the *Cours*, it was very probably because it could have destroyed the theoretical edifice built up by Blondel, which was based upon the doctrine of imitation. For the underlying principles of such a doctrine could not in fact accommodate themselves to an architecture that was self-referential. In order to reconcile the artificial with the natural, one would have had to develop a new approach. If the neo-classical architects were to change the grounds of the debate, it was by devoting themselves to an exploration of the rules of invention, which they held to be the equal of the universal laws of nature. Boullée was to turn his back on tradition, and to subsume history within the initial concept of design; Ledoux, on the other hand, was to study a system of composition based upon oppositions and contrasts. Given this framework, architecture became no more than a succession of bold presentations, its manner of projection being nonetheless based upon the geometrical embodiment of reason.

But Blondel's referral to tradition resulted in a kind of blockage. Among the new themes discernible in his work one can identify that of the independent expression of parts and even, sometimes, that of a break separating the centre from the outlying part. It was certainly not by chance that the author of the *Cours* trained the majority of the revolutionary architects at the Ecole des Arts or at the Academy of Architecture. For all Ledoux's ironies about the master, 'circumscribed by the five Orders',[143] Blondel very probably addressed the same questions as he was to do.

But the failure of Blondel's attempt to restore the classical ideal may perhaps have had a deeper cause, stemming from an argument for continuity that had ceased to be appropriate. In claiming kinship with a tradition based upon the subordination of the parts to the whole and upon the regulated interplay of proportions, the author of the *Cours* thought of architecture and its relation to society as a continuous and homogeneous space. Just as the module enabled one to account for all of a building's measurements, so a single scale could be applied to the different programmes of architecture, from a palace to a bourgeois residence. If Blondel rejected a typology of the kind which had been developed by Neufforge, it was because of his reluctance to put his architectural solutions into different compartments. When it was applied to a specific design, Blondel's reticence in this respect produced some strange results. By a curious paradox, the *Cours* only rarely represented complete buildings, preferring instead to concentrate upon their frontispieces or upon such-and-such a detail of their ordinance. Similarly, the curricula which Blondel proposed to students at the Ecole des Arts were concerned with such subjects as a triumphal arch, a monumental fountain, the frontispiece of a palace or the decoration of a belvedere,[144] or with a set of isolated objects or fragments of architecture. The argument from continuity thus involved treating architecture as a composition arrived at by means of parts conceived in isolation. The Palace with a frontage of sixty-six *toises* shows what the outcome of such a venture would be. By the end of the eighteenth century, the very principles underlying Blondel's approach in the *Cours* had come to seem anachronistic.

5

THE ENGINEERS' 'SYSTEM'

A NEW ACTOR

While Blondel was trying to reorganise architectural knowledge, a new figure made his appearance, the state engineer, a proponent of general development policies. Through the construction and maintenance of thoroughfares, the engineers of the Ponts et Chaussées developed planning and supervision policies which reflected a change of attitude towards the built world. Although there was no conflict as such between Enlightenment architects and engineers, the emergence of a technological state apparatus would eventually provide a challenge to the assumptions of architectural theory and practice.

The eighteenth century was not wholly unacquainted with the figure of the engineer, for, from the time of the Renaissance, he had been involved in the construction of machines, fortifications, bridges and, finally, hydraulic projects.[1] However, up to the end of the eighteenth century, the art of the engineer was still regarded as a branch of architecture.[2] Before the creation of the Ponts et Chaussées, there was only one properly organised corps of engineers, that of the officers of the Génie, who were the heirs of Vauban and who were responsible for fortifications.[3]

During the Enlightenment, the two professions of architect and engineer, which had previously been quite close, began to separate. Engineers of the classical age thought of themselves as architects, whereas the engineers of the Ponts et Chaussées, who succeeded them after the Regency rejected this comparison. Aware that they embodied a new ideal, these men identified with a specific knowledge and practice.

These changes came slowly at first, since the corps of the Ponts et Chaussées around 1720 still included architects such as Boffrand or Gabriel, but they gained momentum later, with the introduction of courses of instruction reserved for engineers. Alongside the Ecole des Ponts et Chaussées, which was founded in 1747, one should also mention the Mézières School of Engineering founded in 1748, the School of Mining founded in 1783, and finally the Polytechnique, which appeared in 1794.[4] Although it is the engineers of the Ponts et Chaussées who have been most studied, one should bear in mind that they were a part of a fairly complex organisation, which also involved officers from the Génie, those responsible for the Mines, together with a group of 'private' engineers working in the provinces, for towns or private individuals.[5] One cannot understand such an arrangement unless one refers to the world of scholarship and of the academies, whose role need no longer be emphasised.[6] One can thus pass in review a whole series of institutions through which a homogeneous technical and scientific milieu was constituted, and one which was more powerful than in the past. The engineers of the Ponts et Chaussées played a pioneering role here, and their organisation and work are altogether symptomatic of the changes that were under way.

THE BIRTH OF THE MODERN LANDSCAPE

The importance of the engineers of the Ponts et Chaussées is directly linked to the birth of the modern concept of landscape and to the definition of a roads policy, which was becoming one of the state's chief priorities.[7] Before discussing the corps and the Ecole des Ponts et Chaussées we should perhaps undertake a brief retrospective survey, in order the better to understand this phenomenon.

In the eighteenth century, landscape was still a relatively recent factor, which can be traced very schematically to the end of Louis XIV's reign, at a time when the need for coherent development policies became apparent, and when the scientific world was preoccupied with fixing once and for all the image of geographical reality.

In the context of his development of trade and manufactures, Colbert felt the need to attain to an overall concept of domestic circulation within the realm and also of exchange. Roads were then classified in terms of their economic importance,[8] and the first attempts were made to calculate profitability.

During this same period, 'the' map of France was begun in the hope that it would provide an accurate picture of the realm. The prominent names which recur in the history of 'the' map are those of the Abbé Picard and the Cassini.[9] The first accurate outline of the coast of France was published in 1693. From 1670 to 1718, the meridian was triangulated; the venture was revived and completed during the Revolution, under the supervision of Borda, Delambre and Méchain. The far-reaching attempt at pinpointing and exploring places, which was begun at this time, was only completed at the very beginning of the nineteenth century, when the topography of the mountain areas was also established.

Owing to this series of transformations, territorial reality was conceived in a different fashion. In a world that was still under the sway of custom, the idea arose of space described by the circulation of men, ideas and commodities, and covered by services which respond at a distance. In unifying cartographic representation, the first campaigns of triangulation paved the way, albeit indirectly, for the disappearance of local particularities. Their anarchic juxtaposition was to yield to a concept of landscape in the modern sense, measurable and comprehensible in terms which were as much technological as economic.[10]

The engineers were soon to be entrusted with the task of equipping and managing this territory. In differentiating themselves from the architects, men whose projects were limited in space and in time, the engineers were then able to develop overall planning policies; through being formulated in terms of a perspective involving the conquest of nature and of social progress, their activities were to become ever more extensive.

THE CORPS OF THE PONTS ET CHAUSSÉES

At the end of the seventeenth century, there was, as we have seen, no corps of civil engineers that could launch such schemes. This lacuna was filled by the Duke of Orléans' creation, in 1716, of the corps of the Ponts et Chaussées,[11] which was entrusted with the maintenance of highways.

The Ponts et Chaussées had been modelled from the beginning on the Génie, and its establishment of equal status for architects and engineers appointed by the King to oversee the larger sites[12] placed it fairly and squarely within the framework of rationalisation which Colbert had pioneered. It was under the direct control of one of the *intendants des finances* and, under the authority of the Controller-General of Finances, a guardianship which would later prove, through the dynamism of the Trudaines,[13] and the active support of Turgot and Necker, extremely profitable. To begin with, there was an architect serving as chief engineer, three inspectors and twenty-one engineers of the Ponts et Chaussées, or one for each region of the kingdom.[14] These engineers were assisted by sub-engineers,[15] together with a technical personnel consisting mainly of draughtsmen.

In the early years, the corps functioned in a decentralised fashion. Recruitment proceeded piecemeal, with each engineer enjoying substantial independence within his own region. He would define and then carry out his own manning policies, and merely had to keep the inspectors of the Ponts et Chaussées informed.

The shortcomings of this system soon became apparent. Planning policies suffered from a lack of coordination which rendered them highly inefficient. Modes of recruitment were also responsible for disparities between the engineers, some of whom proved to be utterly incompetent.

No sooner was Trudaine appointed 'intendant with special responsibility for Bridges and Highways' in 1734, than he set about rectifying the situation. He began by reorganising the administration of the Ponts et Chaussées, and then, in 1744, he set up the Drawing Office in Paris, the precursor of the Ecole des Ponts et Chaussées.

THE ÉCOLE DES PONTS ET CHAUSSÉES

The creation of the Drawing Office[16] was a turning-point in planning policies, even though it followed logically from that of the corps of the Ponts et Chaussées. For the first time, a central supervisory institution existed at managerial level. Likewise, for the first time, the recruitment and the training of engineers charged with the conception and execution of public works was combined. With the advent of Perronet (Fig. 31), in 1747, the Drawing Office was entrusted with a twofold task, involving both the coordination of the engineers' activities on the ground and the training of 'young persons intending to pursue a career in Ponts et Chaussées'.[17]

Although the emphasis was upon pedagogic questions, the professional aspect was no less important, since the pupils, who were described by

FIG. 31 D. de Carré, medallion depicting Perronet, end of the eighteenth century, ENPC wash-drawing, with lights in white gouache; 20.5 × 26 cm. This was probably a competition piece from the Ecole des Ponts et Chaussées, and it testifies to the ascendancy Perronet enjoyed over his pupils.

Trudaine and Perronet's regulation as 'employees', came to the Office to draw road maps.[18] At first, instruction took place on the job, as it were, for no specific course of teaching had been anticipated.[19] The radically innovatory nature of the Drawing Office consisted less in its organisation than in the kind of work done there. It appeared to be the first technical state structure whose competence extended throughout the whole country. The maps established by the employees of the Office featured the roads which they proposed to build. The pinpointing of geographical features was linked to concrete projects, thereby helping to fashion a new approach to planning.

The rapid completion of France's road map reinforced the pedagogic vocation of the Drawing Office, while its professional commitments virtually ceased. Furthermore, after 1756 the Office went by the grandiose title of Ecole des Ponts et Chaussées. Up until the end of the eighteenth century, Perronet's original organisation was in broad outline maintained. Lesage,[20] in his report of 20 Nivôse Year III on the 'original foundation of the Ecole Nationale des Ponts et Chaussées', gave a fairly accurate account, specifying that 'from 1747 to 1794, there were some sixty to eighty pupils or supernumeraries at the school. The better educated among them attended courses in everything from arithmetic to hydraulics, frame-building, stone-cutting and the calculation of earthworks'.[21]

The principle of mutual instruction described by Lesage, whereby the best educated pupils taught their fellows the rudiments of the various sciences, constituted one of the outstanding characteristics of the Ecole des Ponts et Chaussées in the Age of Enlightenment.[22] The instruction given evidently bore little resemblance to any modern form of teaching. In reality it was through actual projects, and especially upon the occasion of competitions and periods of probation, that pupils were trained.

The first competition to be held at the Ecole des Ponts et Chaussées was in 1747, when Trudaine and Perronet's regulations instigated two yearly competitions, with prizes of 300 and 400 *livres*. The following year, it was decided to launch an additional competition, and up to 1775 their number was limited to two or three a year. In 1775, Turgot's *Instruction* led to the reorganisation of the Ecole, whereupon further competitions were introduced.[23] The latter became a permanent means of assessing pupils and, in a context known for its fierce rivalries, they actually served to structure the institution up to the 1800s.[24] Since there was no degree course, the practice of building represented the sole point of reference in a form of training that was still relatively unstructured.[25]

Finally, probationary periods completed the pupils' training. They were usually held in the summer, thus enabling prospective engineers to familiarise themselves with technical processes used on the site. When substantial works were in progress, pupils might absent themselves from the Ecole for a fairly long period.

Although there was much that was ill-defined about the training which the Ecole des Ponts et Chaussées offered, it had become a universally admired model by the end of the eighteenth century. It made available a form of knowledge whose efficacy none at the time disputed. Since its orientation was towards actual building, this knowledge enabled engineers to adapt themselves to a wide range of different situations. The increasingly good reputation which the Ecole des Ponts et Chaussées enjoyed was due more to its readiness to respond to a swiftly changing reality than to any systematic organisation of teaching.

THE ENGINEERS' 'SYSTEM' AND ITS PRINCIPLES

Territory and planning, corps and Ecole des Ponts et Chaussées were four terms connected to each other in a series of different permutations. Territory was defined, as we have seen, by a development policy which the corps of the Ponts et Chaussées aimed to put into practice. The corps and the Ecole were in their turn indissociable, for the local placing of engineers presupposed a single system of training. Establishing the map and centralisation represented two aspects of a single approach.

This approach involved a new kind of institutional set-up which was, in contrast to the dispersed character of architecture, a genuine system. Yet this system was not only concerned with development; in addition, it was designed to reap the benefits of the state's attempts to supervise production, attempts which had already been apparent at the time of the founding of the Academy of Architecture. As the model of the liberal architect, whose descendants we are only too familiar with, slowly emerged, the engineers were engaged in reformulating the problem of the structuring of space.

What the engineers in reality sought was mastery over the structuring of space, a mastery achieved by means of a series of rationalisations and reorganisations. More particularly, they would have to rationalise their conception of design, and of the conditions under which it was executed.[26] In the eighteenth century, a programme of this sort was inevitably hampered by scientific and technological shortcomings, as well as by problems of a professional sort.

Initially, the Enlightenment saw little real progress in construction and in development. Genuinely accurate examples of the breaking stress of materials were not achieved until the advent of Navier and Cauchy.[27] Engineers' calculations of resistance were therefore very approximate. Scientific formulae were still of relatively little use to an engineer wishing to make provision for the problems he might encounter.[28]

It was anyway virtually impossible, under the *Ancien Régime*, to imagine a recasting of professional structures. The hands of the engineers, in particular, were tied by the traditional organisation of trades, those guilds which were finally to be abolished as late as 1791, by the Le Chapelier law.[29]

The attitude of the engineers towards these difficulties was somewhat different from that of the architects, although it was of course determined by their position in the service of the state, whose demands upon them were apparent at every turn. But their ambitions found expression in a form which transcended the institutional framework. To achieve their aims, the engineers developed a set of responses whose impact upon architectural building was soon felt.

In the eighteenth century, architecture and, more generally, design, were still quite central to the engineers' preoccupations. Above all else, it was a matter of rationalising the way in which works were both conceived and put into practice, a rationalisation which, as we shall see below, occurred at a number of different levels.

The first level was that of training. The engineers' claims in this period touched upon pedagogical definitions. From the rudiments of the art to complete mastery of the *métier*, the profession was defined in terms of a trajectory involving a progressive acquisition of knowledge. In the course of this trajectory, one may discern the possibility of a genuine science of engineering, and one that was to prove more effective than the disparate formulae which the engineer could call upon in his everyday work.[30]

This perspective led quite naturally to the construction of works which science allowed one to situate within a theoretical framework. Although one could not depend upon mathematical calculation, one could at least envisage an exhaustive description of the project, from its architectural form to the technologies employed, not forgetting the quantities and prices of the materials. In adopting a basically descriptive approach, the engineers were effectively formulating categories which were common to both the conception and the realisation of buildings, and they were thereby preparing the way for rationalisation. While the architects sought to base their vocabulary upon the imitation of nature, the engineers located their work within a continuous process, from the initial moment of invention to the effective realisation on the building site.[31]

Since the aim was to make this process realistic, reference was again made to the land, whose presence conferred coherence and efficacy upon technological decisions. Indeed, the existence of a development policy and a register of specific intervention alongside the drawing up of plans or construction had turned the territory into an institutional reality. Territory also concerned the engineers' desire to compile a catalogue of the 'productions of nature and of art', an inventory of natural and human resources representing to some degree the equivalent of the work of explanation undertaken on the design.[32]

The engineers' institutional context entailed the superimposition of three different planes, involving the drawing-up of plans, construction and the territory. Their system arose out of the mingling of these two different kinds of approach. Nevertheless, this system was still quite close to architecture, and the engineers resembled architects who had been organised into an administrative corps. Indeed, their knowledge barely transcended the empirical and quantifiable stage, whereas building constituted the crucial part of their work.

Engineers could not therefore be characterised in terms of any particular competence but rather by an attitude.[33] This attitude was also expressed through a doctrine which evolved in parallel with architectural knowledge, and which I wish now to discuss, in order to appreciate the implications of the engineers' system.

THE BASIS OF A DOCTRINE

The engineers' doctrine was not expounded in technical books, which were for the most part based upon architectural treatises. Being designed to unify the point of view from which both the institutions and the problems raised by training, building and territorial intervention were regarded, this doctrine was actually a form of discourse shared by all the engineers. This federative discourse was developed at the Ecole des Ponts et Chaussées, where it was learned by pupils along with design.

Alongside the major competitions in architecture, the fine arts and hydraulic projects, we find 'exercises of style' at the Ecole des Ponts et Chaussées. These competitions involved the writing of a dissertation upon a subject chosen by the board of the Ecole. We get some notion of the engineers' concepts by considering the range of subjects treated, together with the essays submitted by the pupils.[34]

NATURE AND REASON

When one reads these essays, one is struck first of all by the inclusive character of the engineers' doctrine. They derived some of their general conceptions from philosophy[35] and, first of all, from nature, from which everything seemed to stem. Nature played a far more important role here than in the case of Blondel, for the engineers saw it as possessing two crucial characteristics.

First and foremost, nature was seen as productive, as the source of knowledge and, through agriculture and mining, of wealth. Without its inexhaustible fecundity, there would have been no society of men, who derived their subsistence from it.

But the possibility of an economy of knowledge and of a social organisation referred to a principle of regularity ordering the products of nature. This principle was involved in everything from the form of crystals to the laws of astronomy, and also accounted for the progressive constitution of science and of society. Reason was simply the name given to this principle acting within nature in order to organise it. More precisely, it represented the incarnation of the tendency in man which leads him to seek to perfect his knowledge and to set society on the path of progress.

It is thus relatively easy to understand why the engineers might have wished to declare their kinship with nature and with reason. Was not the engineer working to increase his knowledge, and did he not also wish to contribute to the progress of society? Through Condillac and Rousseau, authors to whom reference was often made, the competitions in style invariably displayed a concern with legitimisation.

A SOCIETY CHARACTERISED BY REASONABLE EXCHANGES

This legitimisation also involved a recognition of the usefulness of communication. An engineer's work was justified in the name of the necessity of exchange:

Before one had overcome the obstacles which nature placed between men, by cutting through mountains in order to build roads, by rendering rivers navigable with the help of causeways, by opening up useful communications by means of bridges, each country was restricted to what it itself produced. When there was a superfluity, such products were sold cheaply; a year of dearth would cause the deaths of thousands of men. Suffering discouragement, agriculture languished.[36]

The message from the pens of the pupils of the Ecole des Ponts et Chaussées was thus clear enough. Yet the improvement of communications was also understood in symbolic terms. Obstacles on the relief map were compared to the prejudices dividing society. Nature was to be perfected by means of bridges and causeways; society had to be transformed also, so as to facilitate exchanges between men. The title of a dissertation of 1784, 'Advantages and disadvantages of equality of conditions in a large society such as a whole nation', shows how the engineers' thinking, in so far as it called for a society based upon reasonable trade, was acquiring a political dimension.

In concrete terms, the equality of men chiefly concerned production. Like the physiocrats, the engineers wished to eliminate a whole series of obstacles. *Laissez-faire* was to be the order of the day; supply and demand were to be allowed to balance each other, from the grain market to the right to work.[37] In the wake of Quesnay, Dupont de Némours or Turgot, the engineers called for the suppression of customs barriers, and also of the guilds. The essay title for 1780, 'Some inconveniences occasioned by tolls and other charges levied upon the internal navigation of the realm, with respect to the Ponts et Chaussées and the general good of the state', could just as well have referred to the drawbacks of the guild system. The desire for a reasonable society thus reflected the everyday concerns of the engineers, which had shifted to a more general level.

Yet this reasonable society remained unrealisable, unless one were to permit an institutional violence comparable to that accompanying the exercise of Rousseau's 'general will'. The engineers also took from Rousseau the idea of the 'social contract' and the necessary surrender of a part of one's individual liberty to the collectivity. More especially, one had to accept that one might entrust this liberty to an authority that one had chosen for oneself. There would then be the risk of seeing an arbitrary power arise.

THE ENGINEERS AND THE PONTS ET CHAUSSÉES

In order to avert a danger of this sort, one had to return to nature, and to the moral qualities which it conferred upon man. These qualities belonged to the engineer, whose responsibility it was 'to furnish the details according to which one should pronounce judgement upon the happiness of one's fellows'.[38] As well as having taste, genius, imagination and talent, an engineer ought to be good. Thus, in 1783, it was stressed 'how important it [was] especially for the engineers of the Ponts et Chaussées to have an enlightened mind and a good heart'. Reference was repeatedly made to the engineer's goodness, whether it was a question, as in 1779, of holding forth upon 'the qualities of a sub-engineer' or, as in 1788, of listing the moral qualities 'of the young people who are destined to enter the corps of the Ponts et Chaussées'.[39] We should now consider what these discourses on virtue might mean.

As we have noted, the style contests were concerned with legitimisation. One's handling of written expression itself played a part in this legitimisation, since the engineer was supposed to have acquired 'a noble and concise style, either in order to expound a project to government ministers or in order to be able, being inspired by a love of the public good, to transmit his experience to posterity'.[40] In this context, goodness helped to establish the reputation of the engineer, who was supposed to be without blemish. Prony also advised engineers to 'keep all their actions in the open' in order 'to avoid ambiguous approaches and ill-intentioned interpretations'.[41] Crucially, 'every man destined to have authority over his fellows should match his intellectual knowledge with a just and compassionate heart'.[42] Disinterestedness, goodness and love of one's fellows were therefore aspects of a strategy of power, since the engineer was supposed to employ these qualities in order to court the favour of society. One should also note that his goodness was primarily directed at the workmen, who were supposed to be treated with 'an unyielding authority'.[43]

Yet this goodness, along with the talents which it in some sense crowned, referred to the administration of the Ponts et Chaussées, which was a focus for the thoughts of a good engineer. An engineer and his corps were in fact indissociable; it was not enough to have many talents, for, upon joining the corps, one had to take on its 'spirit'.[44] An engineer's first duty[45] was therefore to subordinate himself, and it was this which enabled him to sublimate the still-latent conflict between the 'artist', who conceived of buildings, and the corps, which financed them, between the isolated project and the overall scale of the territory. The administration of the Ponts et Chaussées represented more than the sum of its individual members; it could be compared to an ideal society, a genuine 'architecture of reason', opposed to the prejudice surrounding it.

AN IDEOLOGY AND ITS CONTRADICTIONS

It is worth stressing the ideological character of a doctrine designed to legitimise the engineer's actions. Whether it was nature, reason or the problems of society which was under discussion, it always was a matter of demonstrating the utility of bridges and highways, as well as the need for an administration devoted to their construction. The ideological nature of this doctrine is also evident in the structure of an argument which permitted the systematic exploitation of its own contradictions.

It is an easy matter to identify the latent contradictions in this argument, beginning with the problem of an institutional violence which, in spite of the goodness which might serve to redeem it, was forever liable to erupt. There might also be some tension between the corps of the Ponts et Chaussées and the individual engineer. We shall see below that nature itself was not immune to similar conflicts. The engineers' doctrine thus presupposed a number of contradictions which it aimed to resolve.

These contradictions were constitutive of the discourse itself, since utility seemed in some way to generate the goodness of the engineer, while his ties with his organisation gave rise to his *esprit de corps*. At the cost of various transformations, substitutions or assimilations, the doctrine of the Ponts et Chaussées managed to turn these contradictions to its own account, coordinating them with its own aims. Never at a loss for something to say, such a doctrine no longer sought to define concepts, but resorted instead to transitory and perpetually shifting figures. In the last analysis, utility, goodness and *esprit de corps* were merely aspects of a single process.

ARCHITECTS AND ENGINEERS

As this chapter, which has been devoted to the noble theme of the engineers' deeds and to the doctrine reflecting them, draws to a close, it remains to attempt a preliminary comparison between the architects and the engineers. For this purpose, it is necessary to stress yet again the kinship between these actors, whose roles were narrowly complementary, since the engineers eventually took the place in the machinery of state which the architects had previously denied them. This complementarity in professional terms was matched by a shared perspective regarding a number of different problems.

Like architects, engineers took their bearings from design and the spirit of *convenance*, from character and from taste. The discourse described in broad outline above therefore developed alongside these shared concepts, which it reconstituted in a perspective somewhat different from that which architecture had assigned them.

UTILITY AND 'CONVENANCE'

This phenomenon was nowhere so clear as in the case of utility, which played a part somewhat similar to that of *convenance* for architects. Utility determined the form which the engineers' design would assume in the social domain, and the use to which it would be put. Yet utility was a far more wide-ranging concept than *convenance*, since it also prescribed the concrete conditions governing the execution of the plan, and even specified how much the work would cost. Indeed, utility and economy were indissociable.[46]

Utility thus occasioned an extension of the concept of *convenance*, and one which wholly transformed it. As the engineers understood it, utility combined technical decisions and expression of the collectivity in a manner somewhat reminiscent of Rousseau's 'general will'. Whereas the architect entered into negotiations with his patron over the image a design should have, the engineer addressed himself to a human group whose needs and laws he had himself defined. By invoking these laws,[47] he might then apply a degree of violence to a minority which had fallen prey to error.

This violence was in contrast to the persuasive character of *convenance*, which presupposed a common formal language. In place of the client whom the architect sought to convince, the engineers had borrowed from the *Philosophes* a generic notion of man, together with the paradox of a 'general will' liable to have recourse to violence. The formal language to which the architect had referred was no longer constitutive of utility. However, this latter subsumed *convenance*, that same quality which governed the decoration of bridges, and which might perhaps account for Perronet's doubts about works in iron, as if they lacked the monumental character of grand constructions in dressed stone.[48] Yet there was something essentially ambiguous about the subordination of the *convenable* to the useful. An iron bridge would soon be considered as useful as a work in masonry; utility would then be able to free itself from *convenance*, and concern itself solely with imperatives of a technical kind.

CHARACTER, TASTE AND NECESSITY

Engineers frequently referred to character and taste, but, as with *convenance*, much of the earlier meaning of these terms was lost. Once these features had been subordinated to utility, they were no longer strictly necessary. One of the architect's difficulties plainly lay in this lack of necessity, which even Blondel could not deny. Character had therefore come to seem supplementary, as far as the engineers were concerned, whereas taste continued to be upheld, albeit in a fairly vague way. Fluctuations in taste were contrasted with the evident nature of primary needs. It was true that taste might still temper these evident needs, and character might render a given project more legible in social terms, yet the crucial debate unfolded on another plane, in relation to a natural realm which one had both to combat and to exploit, and in the progress of a society which the corps of the Ponts et Chaussées had to equip.

STATE AND SOCIETY

As far as engineers were concerned, the loss in meaning of the chief concept of architectural theory was undoubtedly linked to their perception of the nature of society, which was very different from that of the architects. For the latter, as we have seen, this relation operated at two different levels, namely, that of a private clientele which had first to be won over and then educated, and that of power, which had to be celebrated, by inscribing in stone the record of its noble deeds.

Whether private individuals or the state were involved, architects invariably thought in terms of a settlement designed to resolve any conflicts that might arise. Architecture employed a common register of sensations and usages to resolve these conflicts, and, since space could reflect every nuance of the social order, a design could be adapted to the society which had commissioned it.

The engineers, however, accepted a different set of premises. A twofold approach, involving both homogenisation and differentiation, allowed them to respond to the collectivity's needs. Both the country and society and its institutions were thus considered in an overall fashion, before the attempt was made to express their oppositions. Whereas architects had argued for a form of production which was regulated by a freely agreed adherence to the same formal language, which they tried to render ever more transparent, the engineers proposed a compactness of entities, which they had first of all sought to unify, and then the abrupt revelation of the tensions thus created.

The process of unification in fact generated conflicts of every kind. The country, as the engineers saw it, would thus be traversed by lines of cleavage, just as society taken as a whole would immediately break up again into individuals and factions.[49] By combining similarities and differences, by superimposing continuities and discontinuities, the engineers aimed to reflect this state of affairs, to exploit its creative possibilities instead of trying to eliminate it. Thus, their system enabled them to take new institutional constraints into account. The modern state was both to be centralised and to have a local basis. Blondel's desire for spatial reconciliation was to be succeeded by indefinite possibilities of openness and transformation. One should, however, resist the temptation to interpret the engineers' approach as a kind of dialectic based upon the play of opposites. Their thought is undoubtedly to some degree analogous to concepts entertained at the beginning of the nineteenth century, yet its structure differed radically from that. The role of contradiction, in particular, remained a limited one, for the engineers saw it more as a question of appearance, a peripheral phenomenon, than a principle set in the very heart of being. Contradiction proceeded first of all from the will to federate formerly distinct entities; situated between the whole and its parts, it did not yet operate within each of the parts thus combined. The engineers' approach developed alongside architectural theory, although it did not really seek to overthrow it; in subjecting it to a new purpose, in placing it in a new perspective, it seemed above all to anticipate a possible future.

THEORY AND SYSTEM

Finally, the difference in attitude between the engineers and the architects led to a reformulation of the problem of theory and of its relation to professional practice, a problem to which we shall return in chapter 6. Here I want simply to note that theory did not mean the same thing to the engineers as it did to the architects. When an engineer spoke of theory, he was referring almost exclusively to science and to calculation, whereas architecture aligned itself with a practice which it showed little wish to raise to the level of abstract knowledge.

It is therefore difficult to make a term-for-term comparison between architectural theory and the system of the engineers. Although it has been possible to venture a comparison between *convenance*, taste and character, through which the architects' professional strategy found expression within architectural theory, it would be ill-advised to pursue it too far on this terrain.

FIG. 32 Dumont, church projected for the competition of 1755, first prize, sketch of the façade, ENPC, pen and ink, 43 × 32 cm. The spirit of the project was still close to Blondel's teaching.

We are not in fact concerned with a set of clear-cut oppositions between architects and engineers, but rather with a series of slippages, gradual transformations which everyday practice very often obscured.

This everyday practice was still centred on design. The designs of the Enlightenment engineers resemble those of the architects to such an extent that one might be confused with the other. It is worth noting, in this respect, the kinship between the architecture compilations of the Academy and those of the Ecole des Ponts et Chaussées (see Figs. 32–5); from Blondel's teaching

FIG. 33 B. de Seuré, project for a sepulchral chapel for the competition of 1781, elevation, ENPC, pen and ink, wash and watercolour, 77 × 220 cm. The evocative qualities of this project depend upon the use of simple forms, both curved and pyramidal, which suggest death and eternity. The funereal urns and the sarcophagi, which are set in niches, provide a still more direct illustration of the function of the monument. Seuré's chapel featured in the *Collection des prix de l'Académie* (book 20, plate 4), although it was done for the Ecole des Ponts et Chaussées.

FIG. 34 J.-L.-H. L'Escaille, arsenal projected for the competition of 1780, plan, ENPC, pen and ink, wash and watercolour, 207 × 220 cm. This composition resembles a simplified version of projects realised by pupils at the Academy of Architecture around the same period.

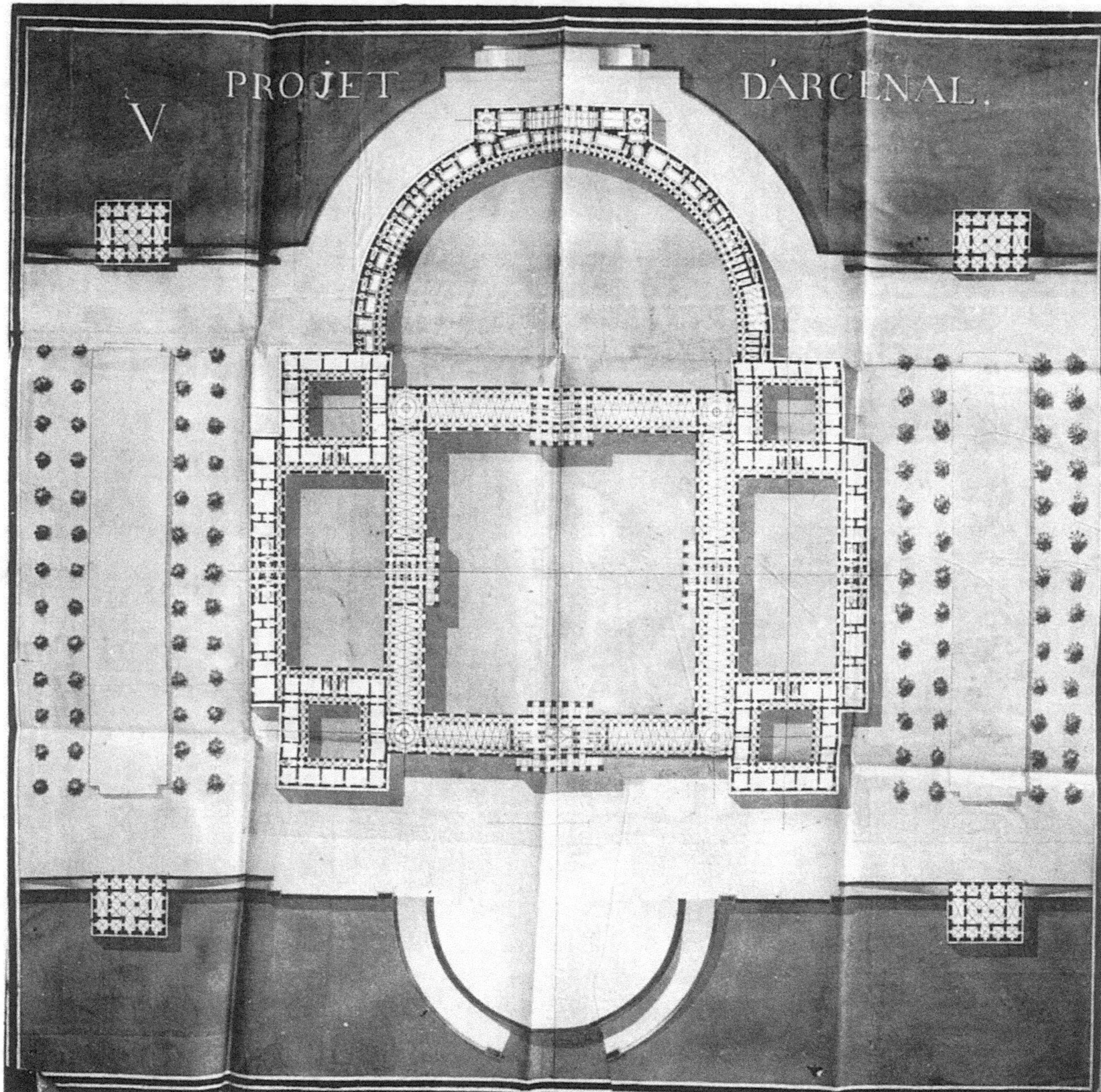
V
PROJET D'ARCENAL.

FIG. 35 J.-L.-H. L'Escaille, projected arsenal, elevation of the main façade, ENPC, pen and ink, wash and watercolour, 62.6 × 299 cm. This austere and monumental façade was designed to reconcile utility and character in such a way as to reveal the purpose of the building. The cannons at each end of the huge façade also serve to emphasise this function. The boldness of this project separates it from the tempered decorum upheld by Blondel.

to the research of the revolutionary architecture, their development was identical.[50] The differences discernible between the two professions were in reality more concerned with the status of the production of space than with its concrete modalities. At the risk of simplifying things drastically, I would be tempted to claim that design ceased to have much significance for the engineers, although it remained the basic point of reference for the architects. One should, however, be careful to note the limits of this disaffection, the consequences of which would only really begin to be felt in the nineteenth century. What occurred in the meantime was a relativisation of formal issues, and the subordination of development and of building to the demands of rationalisation and of power. The engineers' doctrine bears the mark of this change in direction of design, a change made possible by the weakness of a number of the basic assumptions of architectural theory.

I shall investigate below the increasing fragility of architectural theory, while at the same time disentangling the problems of vocabulary which we have encountered so far. We may measure the increasingly vulnerable nature of theory in terms of its capacity to tackle the major issues of the period. Among these issues, training and construction held an important place, and I shall pursue the comparison broached in this chapter in terms of these two areas.

It is not at all surprising to find that the key questions, for eighteenth-century architects and engineers, were the training of practitioners and the realisation of works. Education and its methods intrigued an age investigating the notion of progress, linked in turn to that of the indefinite perfectibility of human beings, whereas the scientific curiosity of men versed in the art of architecture had crystallised around problems of construction. These two areas of investigation, if we consider them from this angle, enable one to compare the thinking of the architects with that of the engineers.

During the Age of the Enlightenment, the writings of Laugier, Blondel and, more particularly, Patte had gradually established the town as a theoretical concept. The country was also a crucial element in the engineers' system. In much the same spirit as before, it is therefore tempting to posit a parallel between urban development, such as Patte perhaps conceived of it, and the territorial work of the engineers. The conclusions which may be drawn from this parallel serve, as we shall see, to round off the analyses touching upon training and construction.

A number of the divergences respecting theory and system dealt with in the course of the present chapter will be considered again below. As regards the whole gamut of points which we have just discussed, the architects were generally concerned to establish a coherence based upon the transparency of architectural knowledge and upon the continuity of causes and effects. The world of the engineers was gradually detaching itself from this model, and new motifs – continuity and discontinuity, transparency and opacity – were emerging. This paradox masked the negative aspects of technique as such, which were to emerge in the future.

6

THE PRINCIPLES OF TRAINING AND THE ECONOMY OF KNOWLEDGE

TWO KINDS OF COMPLEMENTARY INSTRUCTION

In the eighteenth century, questions concerning teaching assumed an increasing importance, especially in the domain of planning and of architecture. It was through teaching that the new professional models spread, as was apparent in the case of the engineers. The traditional organisation of the building trades was proving increasingly inappropriate; training enables one to identify aims being expressed in a diffuse manner in day-to-day building. Architects and engineers thus followed different principles in their teaching, yet here too it was a matter of minor differences or of rival claims. The training of the engineers of the Age of Enlightenment resembled very closely that of the architects, involving the practice of building and an apprenticeship in empirical knowledge. In this chapter, the emphasis will be upon what divided the two professions, while I refer also to the work of Laurent Pelpel for discussions of the pedagogical context common to both architects and engineers.[1]

The Ecole des Arts and the Ecole des Ponts et Chaussées were the two teaching institutions in eighteenth-century France with the highest reputations. Contemporaries had no doubt that they complemented each other, with the Ecole des Arts providing, up until the death of Blondel, the architectural training of the future engineers of the Ponts et Chaussées.[2] It is therefore something of a paradox to find that a comparison between the objectives of the two institutions enables us to measure the gap between architects and engineers.

THE ÉCOLE DES ARTS

Blondel's pedagogic preoccupations were inseparable from his thinking, since his *Cours* was primarily designed for the use of his students, and he therefore founded around 1740 the Ecole des Arts, with the aim of assembling in one place all the various sorts of talent which were useful to architecture. Alongside the teaching of theory and design, which was Blondel's own responsibility, the Ecole des Arts featured courses in mathematics, stone-cutting, wood-cutting, and even physics.[3] There was an implied criticism here of the mode of training offered by the Academy of Architecture, which was restricted to a master's course, further devalued by the professorship of De Courtonne.[4] The Academy itself had no illusions about the rival nature of the Ecole des Arts, and in 1742 refused its founder permission to deliver public lectures. Permission was finally granted in 1743, after the government had applied pressure.[5]

From the very start, in fact, the authorities had looked favourably upon Blondel's initiative. Thus, when the Ecole des Ponts et Chaussées was founded, in 1747, he was appointed professor of architecture at the new institution, and in later years the Ecole des Arts even became a kind of preliminary to the corps of the Ponts et Chaussées, the future engineers acquiring the rudiments of architecture and calculation there, before enrolling with Perronet.[6] Bounties rained down during these years: in 1750, the King granted 2,400 *livres* in scholarships for the pupils of the Ponts et Chaussées, and in 1753 there were six scholarships and a further bounty of 2,400 *livres*.[7] With the Ecole des Arts, Blondel thus set up one of the very first modern schools of architecture. The range of subjects taught was much wider than had been the case at the Ecole de l'Académie, thus taking up a tradition of openness which had virtually been lost since Desgodets.

When, in 1762,[8] Blondel was appointed professor of the Academy, he tried to bring its teaching up to date, while still running the Ecole des Arts.[9] He cut the length of the public course from three to two years, since a three-year system did not favour the admission of new pupils.[10]

Relatively traditional though his theoretical positions may seem, Blondel's role as a teacher was crucial. Boullée, De Wailly, Ledoux and Peyre were to be his pupils, as were Brongniart, Legrand and Rondelet. Blondel thus played a part in training the majority of architects of the second half of the eighteenth century. Finally, the Ecole des Arts was an original creation, and the only one capable of rivalling the Ecole des Ponts et Chaussées.

THE ÉCOLE DES PONTS ET CHAUSSÉES AND THE ACADEMIC MODEL

Although I discussed the operation of the Ecole des Ponts et Chaussées in chapter 5, I want here to consider a number of its defining features, beginning with the almost total absence, up until the Revolution, of masters' courses. From this point of view, the Ecole des Arts was more 'modern' than that of Perronet, with its regularly appointed professors.[11]

Indeed, the Ecole des Ponts bore the marks of its original purpose for a very long time. It had been designed as an instrument of technical training, and resembled in certain respects a bureau of architecture, controlled by its director's concerns. While continuing to attend complementary courses of study elsewhere, the Ecole's pupils undertook a broad range of different tasks, tidying up Perronet's major bridge, for example, or drafting plans for engineers. The Etats des Elèves, which summarised the Ecole's activities each term,[12] give us a fairly accurate impression of the various kinds of work undertaken there. One pupil, for example, received a prize for having copied out a report concerning the Burgundy canal, while another distinguished himself by drawing up a plan for the aqueduct of Arcueil, or by making road maps.

The term 'academy', which pupils at the Ecole frequently employed in their writings, may provide us with a key to these apparently disparate activities. The Ecole saw itself as a nursery of young artists practising every kind of drawing preliminary to the design. In order to perfect their technique, future engineers would thus do figure drawing and ornamentation (Figs. 36–8). The architecture competitions, which were held at frequent intervals, provided an opportunity for a display and appreciation of the pupils' talents. It is hard to imagine anything more academic than the jury for the competition of 1778 (Fig. 39), which included Gabriel, Soufflot, Moreau and Leroy alongside engineers whose task it was to examine the projects.[13] The Academy of Architecture thus looked favourably upon the work of the Ecole des Ponts et Chaussées.

This consensus might have proved illusory, if the specific aims of the institution had not been forcefully reiterated when the competitions of style were held, or if Prony had not been involved at the same time in defining a new pedagogy. Indeed, the latter had been preoccupied at a very early stage

FIG. 36 Bas-relief in the antique style for the figure-drawing competition, end of the eighteenth century, ENPC, pencil, charcoal, wash and gouache on blue paper, 38.5 × 61.5 cm. The important part drawing played in the engineers' training is evident in the sheer variety of graphic art produced. at the Ecole des Ponts et Chaussées.

FIG. 37 'Entablature which may be seen in the garden of the Palazzo Colonna at Rome', end of the eighteenth century, ENPC, pen and ink, wash, 8.15 × 59 cm. Like their equivalents at the Academy, future engineers applied themselves to the drawing of architectural detail. This diagram probably served them as a model.

FIG. 38 Acanthus leaf for the ornamental drawing competition, end of the eighteenth century, ENPC, pencil, wash and pastel on brown paper, 35 × 22 cm.

FIG. 39 L.-J. Desprez, the Ecole des Ponts et Chaussées, end of the eighteenth century, detail, Musée Carnavalet, pen and ink. This imaginary depiction of the School follows the convention of representation of academies and illustrates the range of interests, which were artistic, technical and scientific.

with the problems of teaching. While still a pupil, Prony had taught hydrodynamics (in 1777), chemistry and conic sections (in 1778), and integral calculus (around 1780), in the context of mutual teaching.[14] He also composed numerous reports on scientific subjects, but was not a man for design, although his work prefigured issues linked to the rationalisation of knowledge. Around 1790, he summarised his ideas in *Réflexions sur l'organisation d'une académie qui aurait pour objet la perfection et l'enseignement de l'art de la construction*,[15] which constituted a programme of study for the engineers, even anticipating in certain respects the teaching of the Ecole Polytechnique.[16]

Through considering Blondel and Prony, two theoreticians belonging to different generations, we are now in a position to understand what was at stake, at the end of the eighteenth century, in the disputes between architects and engineers over matters of teaching. While the architects tended by and large to remain loyal to the programme of teaching outlined by Blondel in the years after 1750,[17] the pedagogic conceptions of the engineers assumed a greater depth in the years immediately prior to the Revolution. It is therefore interesting to compare the ideas developed by Prony in his *Réflexions* with the positions elaborated by Blondel in those of his texts which were devoted to the study of architecture. Reference should certainly be made to the *Discours sur la manière d'étudier l'architecture et les arts qui sont rélatifs à celui de bastir* and the *Dissertation sur l'utilité de joindre à l'étude de l'architecture celle des sciences et des arts qui lui sont relatifs*,[18] but a special mention should also go to the *Discours sur la nécessité de l'étude de l'architecture*, intended, according to Blondel, to 'serve our pupils as a guide in the planning of their studies and of their researches'.[19]

THE DIFFUSION AND SCOPE OF KNOWLEDGE

When one reads Prony's *Réflexions* or Blondel's *Discours sur la nécessité de l'étude de l'architecture*, one is struck by the fact that they were clearly aimed at quite different audiences. Prony was addressing his peers, engineers such as himself, whereas Blondel sought to win as large an audience as possible. This difference stemmed, of course, from the object of the two texts, the first describing a programme of teaching aimed exclusively at engineers, the second being delivered at the opening of the fifth public course at the Ecole des Arts. Between 1743 and 1748, Blondel had in fact given four public courses, and in the *Discours* he reviewed them in a far from complacent manner, criticising them for having been merely 'lectures devoid of demonstrations', 'a theory too lofty for workmen' and 'dissertations [which were] too mechanical for enlightened minds'.[20] In order to correct these faults, Blondel then proposed to reorganise his course, dividing it into three series of lectures. One could then differentiate between the 'elementary course', designed for those persons merely needing to know the principles of architecture, the second, entitled 'theory course', which would be for artists and, finally, a 'course for practice', which would be aimed at workmen and contractors.[21]

There was a dramatic contrast between the wide range of circumstances and professions – men of the world, artists, workmen – at which Blondel aimed in his courses, and the limited audience addressed by Prony. This contrast served to obscure, as we shall see, a far more profound difference.

Through his public courses, the director of the Ecole des Arts sought to supply the basic elements for a dialogue between 'those who, on account of their birth, would one day hold high office', artists, architects, but also painters and sculptors, 'who all derive enormous benefits from the rules and principles of architecture', and, finally, those who 'being exclusively concerned with the construction of buildings, require a less transcendent theory'.[22] And, since a single discourse was aimed at all of them, one can comprehend the aim of forging a common language for these different individuals. Architectural knowledge might be compared to this language, which each might employ at his own level.

Prony's aims were of a wholly different nature. Thus, in one of the dissertations he wrote while still a pupil, he recommended that an engineer show himself to be 'honest' towards the great, and hide 'even the slightest traces of scientific reasoning' from the workmen, learning their language and using only those terms which they themselves used.[23]

Knowledge had two different functions here. For Blondel, knowledge constituted a language whereas, for Prony, it enabled the engineer to adapt himself to the different languages already spoken in society. In either case, the attempt was made to enter into a dialogue, but it was undoubtedly no longer the same one.

Each of these functions was as paradoxical as the other. As far as Blondel was concerned, architecture was useful, but it was supposed to serve above all else to make Paris 'the rival of ancient Rome',[24] and would thus partake of a perspective which glorified power and which reflected back to society the image of its own magnificence. For Prony, on the other hand, construction was designed to satisfy immediate needs. The resulting splendour was paradoxically enough one which was presumably the exclusive preserve of an elite but which was also supposed to be the concern of the whole of society. Conversely, the satisfaction of essential needs was achieved in mysterious ways which seemed to contradict their universality. Whereas Blondel dreamed of a shared knowledge, Prony espoused a notion which seemed to partake of the secrets of the trade. Thus, although the scientific reasoning of the engineer was hidden from the workmen, who were not equipped to understand it, it nevertheless enabled him, because its practical consequences were of so general a nature, to enter into a dialogue with all strata of society.[25]

THE ORIGIN, CONTENT AND STRUCTURE OF KNOWLEDGE

The slightest notions of the sciences and the arts suffice to allow us a glimpse of the numerous relations linking them to each other or else to our needs and our pleasures, but what may be regarded as a vague feeling, in ill-educated men, becomes a source of enlightenment and of discovery for those who have studied and analysed the operations of nature in the formation of our understanding, that is to say, the system of our affections, of our thoughts and, in general, of the habits of our soul.[26]

Prony's *Réflexions* espouse an empiricism derived from Locke and from Condillac, yet it was from the *idéologues*,[27] who kept this intellectual tradition alive during the Revolution, that he learned the most.[28] The origin of all knowledge was in sensation,[29] which derived in its turn from nature. Nature was an 'institution' from which the sciences and arts simply drew their inspiration, regarding it from various points of view, decomposing 'the groups of sensations' which it had put 'in one and the same base'.[30] For Prony, nature served to unify knowledge to a far greater degree than was the case for Blondel, who saw it as nothing more than a vague tutelary divinity.

Prony frequently used the theme of a nature which was unique, and which served to define the different sorts of knowledge as branches of one and the same tree. This reference to nature made it possible to transcend the distinction established by Diderot in his 'Système figuré des connaissances humaines',[31] between reason and imagination, mathematics being ranged under the former heading and design under the latter. Thus, the approach of the engineer entailed treating mathematics and design together, with their qualities deriving from the same source, and 'he will know that the art of discussing and analysing is not at all incompatible with that of depicting and of arousing emotion'.[32]

The quest for unity was now formulated quite clearly. The *Réflexions* were preoccupied with construction, defined as 'that long chain which begins with the most basic needs and which culminates in the most sublime and profound researches . . . from the machine which grinds the corn to the complicated structure which transports man across the seas'.[33] Construction thus considered the working of machines in the most general fashion possible, since buildings and boats were both subsumed within this category. The engineer was also supposed to respond to the needs of man in both peace and war. Like Blondel, Prony inherited such contemporary distinctions, but refused to accept any limitations in the scope of his subject. Construction was concerned as much with the military arts as with architecture. The adoption of so wide a brief obliged Prony to affirm both that the domain of construction was a unity and that more or less independent parts existed. Nature, the unification of knowledge and its divisions participated in one and the same approach.

Although Blondel still employed the venerable distinction between the mechanical and the liberal arts,[34] Prony had jettisoned it. However, both men referred to the same corpus of knowledges; mathematics and hydraulic architecture were taught at the Ecole des Arts, whereas the *Réflexions* entertained the possibility of a professor of 'decorative architecture'. Since the question of the origins of knowledge had ceased to be of any relevance, what now differentiated the architect from the engineer was the structure of this knowledge.

For his teaching, Blondel employed the same divisions as were operative in the *Cours*, and also adopted the model of a pyramid. At the top were the principles of architecture, which were applied in the first place to decoration 'which provides the most visible declaration of an architect's skill, and which requires the most knowledge of the theory of his art'.[35] There followed the art of planning, which had already been undermined by utilitarian considerations no longer amenable to the rules of *convenance*, the knowledge of the various aspects of construction, beginning with masonry, then a knowledge of a more instrumental kind, such as drawing or the art of modelling. In this system, mathematics was indispensable, or 'at least to a certain degree'.[36] 'A good architect', Blondel observed, 'is not at all an ordinary man, since, aside from the basic rules of his art, it is important that he be cognisant of the theory of those which are related to it'.[37] An architect should therefore know the rudiments of the sciences, understand the basic notions of hydraulics, but also those of astronomy and of medicine, whereas, for men of the world, the principles of architecture would suffice. Architectural knowledge was thus presented in a hierarchical form, with each person drawing from it the kinds of expertise which he required. The diversity of knowledge should not mislead one, however, into forgetting that it was the principles of architecture that were crucial, and it was these that Blondel reserved the right to demonstrate himself.[38] The other kinds of knowledge were arranged on the basis of these principles.

In Prony's *Réflexions*, on the other hand, hierarchical structure was replaced by a juxtaposition of courses. The professors, of whom there would be seven, would teach architecture, geometry, mechanics, hydraulic architecture, stone-cutting, naval construction and, finally, the art of fortification. This programme closely resembles that of the first Ecole Polytechnique, and one wonders whether there was some link between the *Réflexions* and the proposal for the founding of the Ecole.[39] Be this as it may, we shall note the emergence of a number of new preoccupations, which a few terms reflect quite accurately:

Representation

This would be tackled through the teaching of three-dimensional geometry. Shortly before Monge gave his first courses, we find Prony deploring the lack of a well-composed treatise on this subject, and the programme of study which he drew up closely resembles that of a descriptive geometry.

Calculation

Mechanics, the science of the balance and of the movement of bodies, was supposed to provide the engineer with the instruments for making effective forecasts.

Implementation

The teaching of stone-cutting, of hydraulic and naval architecture, and of military art belong to this category.

Then there was 'decorative architecture', which Prony placed at the head of the other six courses; he restricted it, however, to the study of 'Greek proportions', so that its object seemed futile when contrasted with the rest of his teaching.[40] More significant even than this relative marginalisation of architecture was the radical change which occurred in the whole structure of knowledge, as we observed above. As far as Prony was concerned, a science 'was strictly speaking nothing but a *tableau raisonné* or a philosophical history of such and such a part of the human mind'.[41] The whole range of information unified by nature was also a *tableau raisonné*. Instead of a pyramidal system, therefore, we find a 'horizontal' system of the sciences, in which each is connected to each by countless threads. Yet efforts were now being made to reorganise this system, so that, by representing, calculating and building, the organisation of knowledge would prefigure the anticipated rationality of production. This organisation gives one a sense of being a kind of repetition of the process which was to be set in motion at ground level. Whereas Blondel had arranged the various kinds of knowledge in a hierarchy, the absence of any visible hierarchy in Prony's work masked a new kind of order, one which was no longer founded upon a description by means of fragments, but upon the intuition that there were a series of stages, at once distinct from each other and integrated within one and the same approach. Nature was ultimately just the provisional name for this approach.[42]

APPRENTICESHIP

Forms of apprenticeship were directly affected by this difference in organisation and in knowledge. Thus, at the Ecole des Arts, Blondel dictated his course: 'on Tuesdays and Fridays, from three o'clock in the afternoon until six, lectures will be given on the theory of *convenance*, harmony, proportion, situation, etc., supported by quotations and by the authority of famous authors'.[43] But principles were only one aspect of an architect's studies. A student had also to learn, for example, how to frequent the Academies, the libraries, and the cabinets of men such as Réaumur or Daubenton, to visit the most beautiful buildings in Paris and its environs and, finally, to read wisely. In his *Discours sur la nécessité de l'étude de l'architecture*, Blondel compiled a list of these means of instruction, which ranged from the Academy of Architecture to the château of Clagny or of Saint-Cloud.[44] Books featured on this roll too, and the *Discours* mentioned those which were useful or even indispensable to the architects.[45] Repetition, familiarising oneself with the verities of art, the incessant to-and-fro between theory and practice were some of the key features of this trajectory, which was designed to multiply forms of knowledge and to cause the spirit of comparison and speculation to flourish.

A different set of features marked the trajectory of the engineer, in Prony's view at any rate. Construction was a 'chain', science a 'philosophical history', the adventure of knowledge was intended to be a continually unfolding trajectory, and Prony criticised those authors who, 'by an extraordinary contradiction, suppose the minds into which they wish to cast the first seeds of instruction to be already fully formed'.[46] One could thus see two kinds of trajectory emerging. The architect would be trained as a humanist, whereas the engineer would be poured into a mould at his own behest, in order that he might avoid 'the long and painful sequence of operations of the soul' which the elaboration of knowledge required.[47] Blondel and Prony were two generations apart. In the interval, the pedagogic conceptions of the engineers had been transformed, whereas the training of architects had barely transcended the model current at the Ecole des Arts.[48]

If architectural training developed so little, it was partly because of the imprecise character of taste and talent, which continued to be the chief qualities of the architect up until the end of the eighteenth century. Prony argued for the substitution of the notion of genius for that of taste, but what he had in mind was no longer the genius of someone like Jules Hardouin-Mansart, which had been based upon the felicitous taking of liberties: 'Imagination and genius are invariably to be found where new assemblages of ideas are formed, and where previously unknown relations are perceived'.[49] It is clear what divides Blondel from the engineers of the turn of the eighteenth century, who sought to replace the notions of taste and talent with a concept of genius which had a decidedly hybrid aspect. Ledoux, in his *Architecture*, also referred to genius, but without going so far as Prony. Genius, in an architect's understanding of the term, was still subordinate to taste, which Ledoux sometimes called 'the feeling one had for one's art', whereas genius for an engineer came to reside in 'assemblages' and 'relations', with the sciences and arts never representing anything other than 'combinations of ideas and of signs'.[50]

THEORY AND PRACTICE

Blondel, in his *Discours*, specified that 'by theory of architecture, I mean a knowledge of the history of this art, the study of the proportions of the Ancients compared with those of the Moderns; the most universally approved principles, as applied to the decoration, planning and construction of buildings'.[51] Theory therefore concerned general principles, whereas practice 'presupposed a familiarity with the different materials peculiar to the art of building, and the economy one should observe in using them, consideration of their weights and thrust, their bonding and their assemblage, the detail of uses and customs'.[52] Theory and practice thus complemented each other, facing each other without ever truly being opposed to each other, since apprenticeship in architecture entailed passing from one to the other.

As far as the engineer was concerned, however, theory now assumed a different meaning. It became synonymous with mathematical and physical calculation, and this very quickly posed an apparently insurmountable problem. As experiments proliferated, theory and practice seemed, after a particular stage had been passed, to contradict each other. Although Bélidor was still enthusiastic about theory in the 1730s,[53] Prony was already more cautious when writing his dissertations as a pupil, pointing out that one could not 'embark upon an important piece of work with nothing more solid to go on than one's faith in a formula'.[54] Aubry, also an engineer, observed that 'speculative knowledges, even those which are founded upon observations made in isolation, can give rise to many mistakes when they are not connected to experiments performed on the operations of nature itself'.[55]

If the engineers' experiments, which were 'performed on the operations of nature itself', differed from those of the architects, it was because they shed light upon the validity of a theory by drawing directly upon nature, instead of proceeding, as Blondel had done, by way of speculation and critique. Prony's insistence upon the unity of nature now appeared in a new light. Although they were often ineffectual, the sciences nevertheless represented, in the engineers' eyes, the quintessential domain of reason. The design, in its concrete dimension, was on the other hand an affair of the imagination. To compare reason and imagination, on the grounds that they were both a part of the 'operations of nature' of which Aubry had spoken, amounted to positing an intimate link between theory and practice. The engineers' experiments were bound to reinforce a direct connection of this sort, whereas those undertaken by the architects presupposed a certain distance between theory and practice. Blondel maintained that critique and design were situated in the gap between them. By seeking to eliminate the distinction between theory and practice, Prony was actually raising the problem of an engineer's science capable of mastering design, from conception through to execution.

This fusion of theory and practice naturally represented a desire rather than a reality. It is symptomatic of this situation that this wish first assumed the form of the paradoxical expression of a divorce between theory and practice, experiment serving precisely to accentuate the discrepancies. Yet this was undoubtedly one of the 'ruses of reason' customary with the engineers, who would combine like and unlike, continuities and discontinuities. Just as the knowledges juxtaposed in construction were immediately rearranged through a process having a practical purpose, so too did the divorce between physico-mathematical calculation and the results of experiments pave the way for a comparison between these two terms. Gauthey was to adopt a stratagem of this sort, as we shall see, in his report on Sainte-Geneviève.[56] Furthermore, Prony's *Réflexions* were written, a little before the founding of the Ecole Polytechnique, at a time when Lagrange had just finished formalising analytical mechanics,[57] which was to become one of the favoured tools of the nineteenth-century engineer. In the 1790s, calculation was still 'a torch whose light renders effects perceptible';[58] this torch was, however, soon to be transformed into an independent science.

'It has been our wish', wrote Blondel in his *Cours*, 'that the young architect should acquire those forms of knowledge which form the basis of an engineer's studies; in his turn, the engineer, since he may be entrusted with the majority of buildings in the sphere of civil architecture, such as a government building, hospitals, a basilica or a commissariat, should know the rules of the art'.[59] As far as Blondel was concerned, the engineer was still a kind of architect, one whose special expertise lay in utilitarian constructions and machines. Entrusted in the 1790s with the task of writing the volume of Panckoucke's *Encyclopédie méthodique* which was devoted to bridges and roads,[60] Prony wrote a *Discours préliminaire*, whose inspiration was quite different and which was related to the *Réflexions*:

> Experiment, which was the fruit of these great endeavours, gives us the impression that the engineer, charged with the task of supervising them was not simply an architect, a geometer or a physicist; this engineer had to have a particular kind of education, a system of theoretical and practical knowledge wholly specific to the walk of life he was choosing. This system embraces virtually all of the arts and sciences, and the branches which compose it, although belonging to familiar stems, form, either through the way in which they are assembled or through the way in which they are applied, a particular and new science.[61]

OPPOSED PRINCIPLES

Comparison between Blondel's *Discours* and Prony's *Réflexions* reveals two philosophies which differ as much in the forms of knowledge produced as in the modes of apprenticeship stipulated. For the director of the Ecole des Arts, knowledge presupposed the operation of universal principles, whereas the objects which it treated concerned the social order to varying degrees. The man of the world would thus be satisfied with those principles which *convenance* taught, whereas the architect was supposed to have mastered all the details of construction. For Prony, on the other hand, the principles from which knowledge was derived were only accessible to a tiny majority, and the universality of knowledge was revealed through its effects. This initial opposition gave rise to two kinds of organisation of knowledge, hierarchical in the case of Blondel, but based upon a juxtaposition of sciences in the case of Prony, a juxtaposition being ordered subsequently by the need to rationalise production in specific ways. Theory and practice played different roles here; their separation from each other made the project possible, while, conversely, the comparison of one with the other rendered the 'particular and new' science of the engineer a necessity.

In either case, it was a matter of controlling the disposition of space. Blondel laid down for this purpose some rules concerning the object that was to be built, defining at the same time their margins of tolerance; Prony, on the other hand, was more attached to the process than to the object.

Given these different approaches, the training of engineers gradually diverged from that of architects, which remained fundamentally empirical. Being defined in terms of representation, calculation and implementation, an organisation derived directly from that of production was to undergo still more transformations. Through outlining or advancing hypotheses, calculating or demonstrating, realising or concluding, one gradually passed from the project to that 'well-posed problem', for which Le Corbusier would subsequently envy the engineers.[62] But, by the same token, one had to abandon the old system of training, which closely resembled the architectural one. It was to be Prony who set this process in motion, reforming the Ecole des Ponts in 1798 so as to adapt it to the new pedagogic context occasioned by the Ecole Polytechnique.[63] Since Perronet's Ecole was no longer receiving, as it had done in the past, young persons instructed solely in the rudiments of mathematics, but pupils who had already received a general scientific training, it was now shown to be ill-adapted to current needs. Upon his appointment as the Ecole's new director,[64] Prony outlined a course of study which drew its inspiration from the subdivisions of his own *Nouvelle architecture hydraulique*, a work whose second volume, wholly devoted to heat-driven machines, had appeared just two years before, breaking both in structure and in content with the traditional engineers' treatises, which were basically architectural.[65]

Three chairs were created at the Ecole des Ponts et Chaussées, one in

TROISIÈME PARTIE.

TABLEAU SYNOPTIQUE D'UN COURS DE CONSTRUCTION.

MANDAR, PROFESSEUR.

ÉCOLE NATIONALE DES PONTS ET CHAUSSÉES.

AN VII.

LA SCIENCE ET L'ART de la construction peuvent être considérés.

- **EN GÉNÉRAL** ……
  - pour le bien-être des citoyens ……
    - réunis et ……
      - rassemblés dans les cités ……
        - où ils élèvent ……
          - pour l'usage particulier …… des maisons …… des magasins ……
          - pour l'usage général …… des établissemens …… des monumens ……
        - qu'ils rendent salubres ……
          - par des rues, des places, des jardins ……
          - par des conduits ……
            - pour apporter les eaux …… Aqueducs …… Fontaines ……
            - pour écouler les eaux …… Égouts …… Rigoles ……
        - Architecture civile.
      - occupés à des travaux relatifs ……
        - aux arts en général, et qui comprend ……
          - la recherche des matières premières au moyen de ……
            - la fouille des terres ……
            - l'extraction des minéraux …… Puits …… Galeries ……
          - la mise en œuvre ……
            - usines et fonderies ……
            - ateliers de tout genre ……
          - Travaux des mines.
        - à l'agriculture ……
          - en défrichant, en améliorant par des ……
            - canaux d'arrosement ……
            - canaux de dessèchement ……
    - voyageant et communiquant entre eux ……
      - par terre ……
        - Routes ……
        - Ponts ……
      - par eau ……
        - Navigation intérieure …… Canaux …… Rivières ……
        - Navigation extérieure …… Rades …… Ports ……
      - Ponts et chaussées.
  - pour la sûreté de la société ……
    - par les moyens de défense, qui donnent ……
      - sur les frontières ……
        - Les places fortes ……
        - Les lignes fortifiées ……
        - Les camps retranchés ……
        - Architecture militaire, ou fortification.
      - sur les côtes maritimes ……
        - Les forts et les batteries sur la côte ……
        - Les batteries flottantes ……
        - Les vaisseaux ……
        - Construction des vaisseaux.
- **EN PARTICULIER** pour l'architecture civile et celle des ponts et chaussées, qui comprend ……
  - la connaissance des matériaux et l'art de les employer pour l'exécution des ouvrages de ……
    - Terrasse …… Fouille et transport des terres.
    - Maçonnerie ……
      - L'exploitation et la fabrication des matériaux ……
        - Pierres …… naturelles …… artificielles ……
        - Chaux ……
        - Sable et ciment ……
        - Mortier ……
        - Plâtre ……
        - Lattes, clous, &c.
        - Qualité. Résistance. Durée, &c.
      - Moyens d'exécution …… Équipages ……
        - Échafauds.
        - Ponts de service.
        - Batardeaux.
        - Cintres.
        - Machines.
        - Ustensiles.
        - Outils.
      - Démolition des anciens édifices.
      - Plantation des bâtimens.
      - Arrangement ou disposition des matériaux dans la formation des murs ……
        - Système des anciens ……
        - Système des modernes ……
        - Appareil, taille et pose des pierres.
      - Manière de fonder les édifices selon les diverses circonstances ……
        - Terrains compressibles ……
        - Terrains incompressibles ……
          - hors de l'eau ……
          - dans l'eau ……
          - Radiers des écluses. Piles et culées des ponts. Digues, môles, formes, &c. Monumens, &c.
      - Manière de bâtir ……
        - les fosses d'aisance.
        - les égouts et les puisarts.
        - les aqueducs, les réservoirs, les bassins.
        - les puits, les citernes.
        - les caves et autres souterrains.
      - Manière d'élever les édifices, dont les principales parties sont ……
        - les murs ……
          - de soutènement.
          - de face, de refend.
          - mitoyens, de clôture.
        - les voûtes ……
          - de pierres.
          - de moellon ou meulière.
          - de briques, &c.
        - les planchers.
        - les colonnades.
        - les frontons.
        - les terrasses et les balustrades.
        - les cheminées, les fours.
        - les légers ouvrages ……
          - Pans de bois, cloisons.
          - Plafonds, corniches, carrelage.
          - Ravalement, &c.
      - Manière de bâtir ……
        - les serres chaudes.
        - les glacières.
    - Charpente ……
      - Bois …… qualité, résistance, durée.
      - Assemblages.
      - Pans de bois, cloisons.
      - Planchers.
      - Combles.
      - Dômes.
      - Voûtes.
      - Escaliers.
      - Pilotis et grillages.
      - Étaiemens, chevalemens.
      - Bois d'écurie, &c.
    - Couverture ……
      - Moyens des anciens.
      - Moyens des modernes.
    - Menuiserie ……
      - Bois.
      - Assemblages.
      - Mobile …… Portes. Croisées, &c.
      - Dormante …… Planchers, parquets, lambris, &c.
      - Treillage.
    - Serrurerie ……
      - Fer, charbon.
      - Pour la solidité …… gros fers. armatures.
      - Pour la sûreté …… à l'extérieur.
      - La commodité et l'embellissement …… dans l'intérieur.
    - Fondeur pompier.
    - Marbrerie ……
      - Marbres et brèches …… antiques. modernes.
      - Sciage, taille et polissage des ouvrages de tout genre.
    - Vitrerie, miroiterie.
    - Peinture, dorure, vernis.
    - Pavé ……
      - des chaussées.
      - des cours.
    - Plomberie ……
      - Pour la couverture.
      - Pour la conduite des eaux.
    - Poêlerie et caminologie.
    - Chaudronnerie et ferblanterie.
    - Sculpture, stuc.
    - Jardinage.
    - Divers moyens ……
      - Préservatifs contre ……
        - les torrens, ravines, lavanges, inondations.
        - les éboulemens des montagnes.
        - le tremblemens de terre.
        - les incendies.
      - Conservateurs des monumens. — Restauration des édifices.
      - Économiques en diverses circonstances.
      - Contre le grand froid, ou les grandes chaleurs de certains climats.
  - La direction ……
    - Rédaction des projets …… Arts des dessins ……
      - Levée des cartes et plans.
      - Lavis de l'architecture et de la carte.
    - Description et évaluation des ouvrages ……
      - Devis.
      - Détail.
    - Exécution des projets ……
      - Régime des travaux. — Organisation des ateliers, approvisionnement des matériaux.
      - Mode de comptabilité. — Régie, entreprise, fournisseurs.
      - Établissement, conduite et surveillance des travaux.
  - L'administration …… Exécution des lois et ordres du Gouvernement ……
    - Adjudications et actes y relatifs.
    - Affaires contentieuses ……
    - Surveillance comptable ……
    - Correspondance.
      - Attachemens, toisés, mémoires.
      - Vérifications, contrôle, réglemens.
      - Réception des ouvrages. — Liquidation.

applied mathematics, one in stereotomy as applied to the arts of drawing, stone-cutting and carpentry, and one, finally, in construction.[66] Masters' courses thus followed on from forms of self-education which were in principle very empirical. The content of these courses also assumed the form of branched synoptic tables (see Fig. 40) which linked together kinds of knowledge in terms of their various lines of descent.[67] Although these tables still bore a passing resemblance to d'Aviler's 'Système figuré de l'architecture', the analytic and genealogical spirit of the nineteenth century was, however, already discernible in them. Their subdivisions no longer corresponded solely to the different objects to which the knowledge was applied, but they also addressed themselves to the overall economy of the process of conception-realisation, since the classification of knowledge was often effected in terms of the types of problem encountered and the methods employed to resolve them. In mechanics, for example, Prony distinguished between precise and merely approximate solutions.[68] In the domain of construction, his approach became more specific still, with a survey being done of the general needs of society, the well-being of the citizens or the security of states, from which one might deduce the different types of edifice and the procedures involved in realising them.[69] The analytic method was bound to triumph over the confusion which for so long had surrounded the principles involved in the erection of buildings.[70] The old system of training, orientated exclusively around the practice of building, then gave way to a more structured kind of teaching,[71] which favoured the concatenation of knowledge and the resulting instruments of decision-making. Admittedly, the issue was far from being wholly resolved, and we will consider below the more traditional form of teaching, developed in reaction to the conceptions developed by Prony.[72] Institutions seem always to evolve at a slower rate than the ideas which will eventually transform them. Nevertheless, in the early years of the nineteenth century, a new figure emerged, that of the modern engineer, a man committed to the application of science, in whose eyes the order of knowledge and that of production ought to be compounded.

FIG. 40 G. Riche de Prony, *Plan d'instruction*, Year VII, 'Synoptic table of a course of construction'. This sort of presentation was common during the Revolution. Prony probably took his inspiration from the 'Table representing the organisation of the Central School of Public Works', which was conceived in the same analytic spirit.

7

STABILITY AND CONSTRUCTION

SOME GENERAL OBSERVATIONS ON THE ENLIGHTENMENT DEBATE OVER CONSTRUCTION

Relations between formal intuition and physico-mathematical calculation are nowadays regarded as essentially uncertain. This uncertainty may first have arisen in the course of the Enlightenment, when an opposition emerged between geometrical and visual solidity, founded on traditional principles, and new preoccupations regarding the mastery of construction through knowledge of the laws of physics. While architectural treatises were arguing for a still highly qualitative analysis of buildings, essays on the resistance of materials were proliferating.

These developments were themselves linked to a growing preoccupation with the rationalisation of production, especially, as we have seen, on the part of engineers involved in the development of the country. With their different vocations, architects and engineers increasingly diverged, and in this respect also the Enlightenment seemed to anticipate some contemporary debates. One should beware, however, of positing an excessively simple opposition, for innovation was not the monopoly of either profession. After a period of intense curiosity and of profound – although sometimes contradictory – reassessment of the problems involved in construction, a choice plainly had to be made between a visual notion of stability and the emergent science of construction. A structural approach, in particular, emerged, along with a more dynamic vision of the art of building. In order to complete the analysis undertaken above, I have sought to identify a number of the characteristic features of the period, and to study the lines of force, the convergences or divergences between Soufflot, Patte, Perronet or Prony.

Nevertheless, in many respects the eighteenth century inherited some of the unwieldiness of the classical age. It was not merely the possibility of a science of construction which was being decided upon; the often paradoxical character of the debate over construction in the Age of Enlightenment very probably arose out of this tension between tradition and innovation. This being the case, there is some point in beginning with a description of the constraints which weighed upon the construction of buildings.

THE PRODUCTION OF BUILDINGS

In the letter to Marigny, which serves as an introduction to Patte's *Mémoires sur les objets les plus importants de l'architecture* (1769), Patte described construction as 'the most useful, the most necessary and the most essential part of architecture'.[1] Construction was in fact central to architectural thinking, as is evident from the Abbé Laugier's ideas on the primitive hut and from his desire to rationalise the ordinance of buildings in terms of constructive data.[2] There was, nevertheless, a great distance between the theory and the reality of a mode of building whose complexity and fragmented character would seem to have hindered any attempt at reorganisation.

This complexity and fragmentation derived first of all from the existence of customary law. The country contained several hundred different systems of customary law,[3] all of which had some sort of impact upon the production of buildings. Since the end of the sixteenth century, the custom of Paris nevertheless enjoyed greater respect than did the others,[4] and was the object of numerous commentaries. The part of Desgodets' course which was devoted to it was published in 1748, with notes by Goupy, an 'architecte expert bourgeois'.[5] Listing the various easements, prescriptions and detainers specified by custom, the *Loix des bâtimens* considers, for example, the rules to which the construction of foundations, or of partition and party walls, were supposed to conform, rules whose minutiae would seem to reflect a degree of stasis in the technologies involved.

The production of buildings required a huge cast of actors. Thus, in his *Guide de ceux qui veulent bâtir* (1781), Le Camus de Mézières gave a shortened list, in decreasing order of importance, comprising masons, carpenters, locksmiths, tilers, plumbers, joiners, painters, marble-masons, glaziers and pavers.[6] Each of these trades had their own customs and, being organised in guilds, they had gradually codified even the most trifling jobs.[7]

Finally, there was the problem of the relations between architects, clients and entrepreneurs. Here too, a number of different combinations were possible, with the stone-masons sometimes arrogating the architect's privileges to themselves, workmen being paid either in a lump sum or at a daily rate. As the leading economic sector in the realm, the building industry was a labyrinth of regulations and prescribed practices.

In such a context, an architect had no choice but to respect whatever customs were in force. A mason could build a wall without any outside help, and the same was true, in their own domain, of carpenters and locksmiths. One therefore simply had to define the required operation with a degree of precision and to anticipate the eventual outcome, provided that such operations accorded with the stipulated rules.

Thus, at the end of the eighteenth century, building constituted a relatively settled sector of the economy. The mastery of traditional techniques seemed to rule out any sort of radical innovation (see Fig. 41). Architects and contractors had succeeded in finding some common ground, in spite of mutual recriminations. There had been confrontation between guilds and the administration since the end of Louis XIV's reign,[8] but architects were only marginally involved in this debate, being content simply to record the development of the situation. If seventeenth-century treatises still betrayed a desire for authoritarian supervision, those of the following period tended rather to ratify a *fait accompli*.[9]

LARGE EXPERIMENTAL SITES

This state of affairs was not, however, destined to last. A number of large sites – the unfinished colonnade of the Louvre, the western façade of Saint-Sulpice, the construction of the Madeleine and of the church of Sainte-Geneviève – raised issues whose scale transcended the traditional framework of production. Furthermore, these were the projects which Patte had touched upon in his *Mémoire sur les objets les plus importants de l'architecture*.

For instance, the work on the colonnade of the Louvre, which had been started up again by Soufflot in 1755, illustrated the problems involved in the construction of lintels with voussoirs, a problem of horizontal thrust to some

FIG. 41 P. Patte, *Cours d'architecture* (1777), 'Façade of a house built of stone'. An example of ordinary building construction.

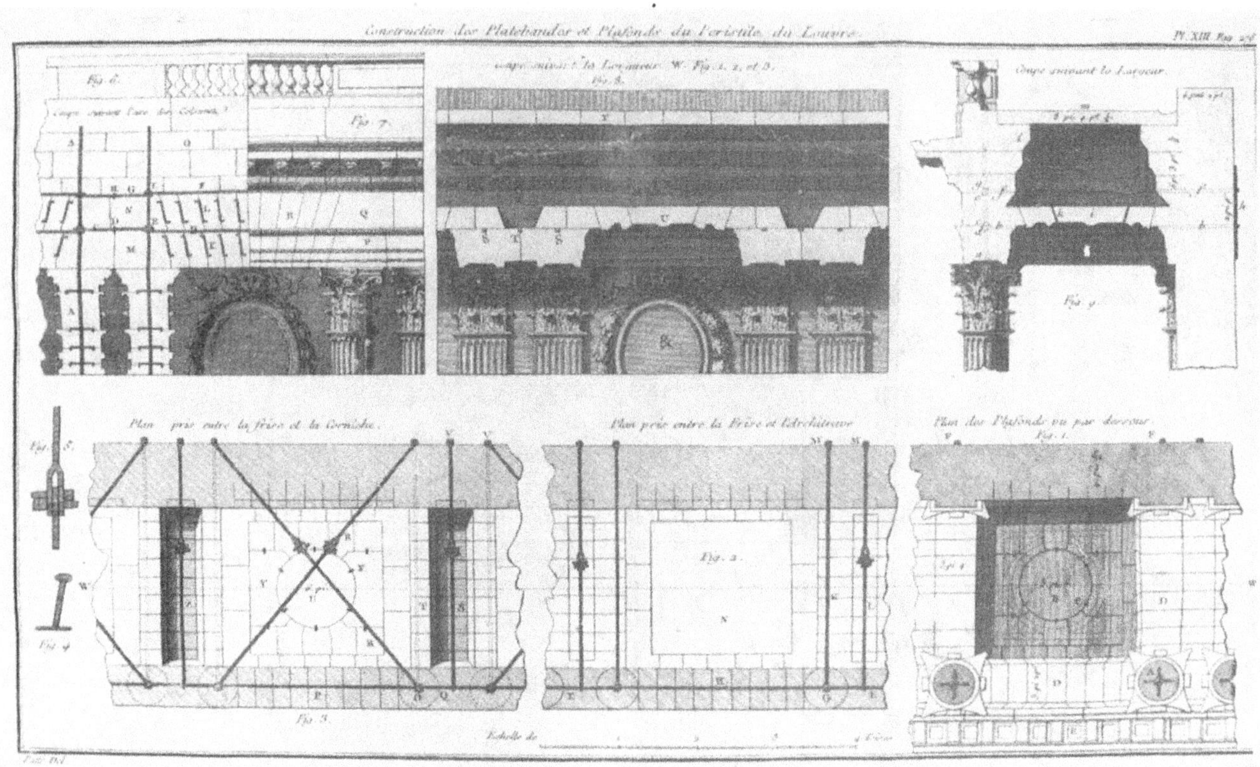

FIG. 42 P. Patte, *Mémoires* (1769), 'Construction of the flat vaults and ceilings of the Louvre colonnade. The iron frame of the building, with its bars and cross-ties, is plainly visible.

extent analogous to that encountered in bridge-building.[10] The rusting of iron reinforcements also served to show how difficult it was to use metal in stone structures in the Age of Enlightenment (see Fig. 42).[11]

In the case of Saint-Sulpice, which was still unfinished at Servandoni's death, the debate was concerned instead with the compressive resistance of stone, for it was feared that the weight of the great triangular pediment might overload the façade (Fig. 43).[12]

The foundations of Contant d'Ivry's church of the Madeleine gave Patte one of his first opportunities to exercise his talents as a polemicist. He criticised their depths, and their execution in his *Mémoires sur les objets les plus importants de l'architecture.*[13]

FIG.43 P. Patte, two designs for the façade of the great portico of Sevandoni's Saint-Sulpice. Patte was concerned with means of construction and the lightening of it.

Each of these problems, from lintels, the use of metals, the resistance of stone, to foundations, were epitomised by Sainte-Geneviève. We shall return below to the polemic between Patte and Soufflot, for it illuminates the issues raised by construction. The exceptional buildings mentioned above, to which one should also add Gabriel's buildings on the Place Louis XV,[14] together with Le Camus de Mézières' *halle au blé*,[15] allow us to consider preoccupations more general in nature than the questions which we have just tackled, although they are indissociable from each other. Engineers such as Regemortes, Perronet or Gauthey shared these preoccupations.

The crucial development was the emergence of a structural approach to architecture. This approach was primarily related to ecclesiastical building, from Perrault's design of 1675[16] to the ideas of Cordemoy, Laugier and Soufflot.[17] These theoreticians and experts displayed a concern with unity and legibility, that led them, for example, to recommend an isolated column supporting an entablature, instead of massive piers obstructing one's view. The economy of the building was thought to derive from a system based on the column and its entablature. Although Blondel, in his *Cours d'architecture*, mistrusted lintels, preferring arches, he was associated with this same tendency, which culminated in the work of Soufflot. The flat elevation of the seventeenth century yielded to a spatial structure which was unified by the repetition of the same constructive device. From this point of view, Cordemoy's *Nouveau traité de toute l'architecture*, Laugier's *Essai* and his *Observations sur l'architecture* allow one to define a homogeneous set of preoccupations, a rationalist current which informed Blondel's lessons too.

Architects had traditionally been concerned with the bonding of parts, especially as regards the foundations and the linking of the internal partition-walls to the façade of the building. Research into the construction of churches, however, altered the picture somewhat. The replacement of rectangular masses by isolated columnar supports served to concentrate the stresses transmitted to the foundations, thus entailing the risk of subsidences through an excess of point loading. Their bonding therefore became an absolute priority, the problem being one that closely resembled that of bridge piles.[18] More generally, the idea of structure was inseparable from that of articulations which one had both to express and to reinforce.

Architects, like engineers, were involved in the quest for homogeneity. Classical ashlar construction involved additional materials, which were usually concealed within the thickness of the walls or of the floorboards. Vaults were strengthened with beams.[19] Eighteenth-century architects resorted to such devices as never before, but were also repelled at the idea of using them. The question of metal, used in tie-beams, but also in the roofing of theatres,[20] was especially contentious, for iron rusts and therefore shatters stone. Homogeneity therefore became a central preoccupation, which bore not on the use of metal alone, since Perronet and Soufflot, for example, are known to have reduced the visible dimension of their joints at the pont Louis XVI, and at Sainte-Geneviève.[21] At the same time, the engineers were interested in

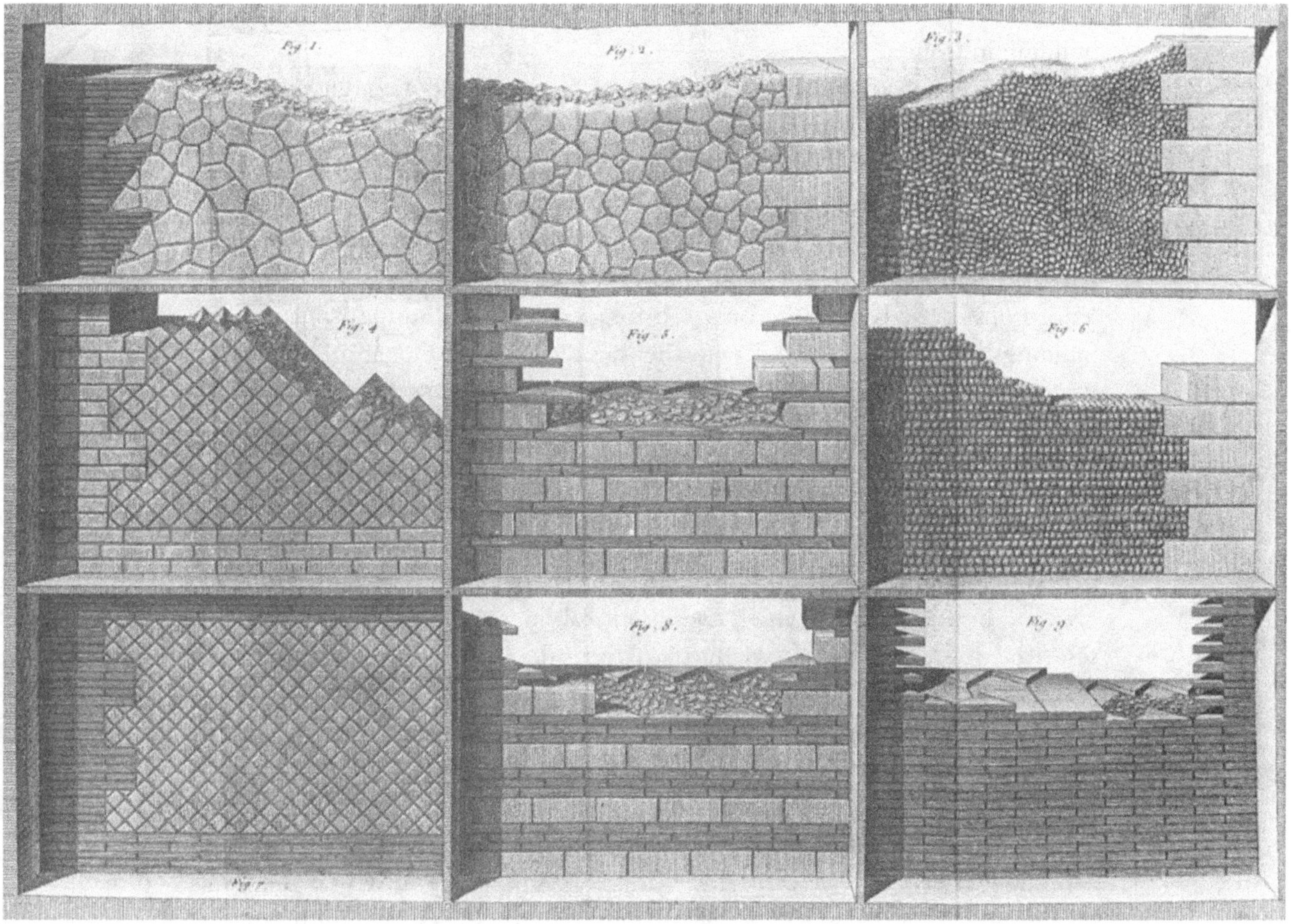

FIG.44 A.-J.-B. Rondelet, *Traité théorique et pratique de l'art de bâtir* (1802–17), different kinds of masonry, ancient and modern. There was unanimous admiration in the Age of Enlightenment for the strength of the mortar used by the Ancients in their masonry.

hydraulic binding materials (see Fig. 44),[22] a deputation was in fact sent to Italy to make a detailed study of the properties of pozzolana. Architects and engineers dreamed of a smooth, homogeneous material, which would lend itself to all uses, and which would express architectural structure in its purity. The cylinders and spheres which featured in Ledoux's *Architecture* may perhaps be related to this desire for a perfect material or, failing that, an ideal mortar which would enable one to free oneself from the constraints involved in building in stone.

From Laugier to Soufflot, the idea of structure now extended to the economy of material. The problem was not a novel one, but the interest shown in Gothic architecture, like the bold flights of Perronet, made it more pressing. In becoming a structure, the building was to be stripped of any unnecessary mass, save where this might undermine the impression of solidity, which an architect was still obliged to produce. Laugier was to criticise these 'enormous masses which are only too frequently to be found in our modern buildings',[23] contrasting them with the 'delicacy' of Gothic, and Patte was to write that 'The great art in architecture consists in only allowing as much thickness as solidity requires.'[24] This art of economy seemed to culminate in Perronet and Soufflot, before neo-classicism reverted to more compact forms.

Along with preoccupations such as structure, bonding of parts, homogeneity of the materials, and the quest for economy, there was a need for experimentation. Sensation and calculation represent, as we have seen, the extreme limits between which the empirical approach oscillated.[25] On one side, the source of all knowledge, on the other, the sophisticated play of reason; so it was that Condillac, after his *Traité des sensations* (1754), prepared a *Langue des calculs*, which was to be published after his death by Laromiguière.[26] Some years later, Destutt de Tracy was to stake out a path leading from 'the simplest perception' 'up to the most abstract ideas'.[27] The dyad of sensation and calculation also featured in relation to the structure and economy of matter. The economy was, for example, tested within the limits set by the eye of the spectator, which one should always treat with circumspection, and those assigned to it by calculation. I shall return below to the relations between sensation and calculation; at this stage I want merely to emphasise the evident desire to quantify and to calculate.

This desire was manifested in experiments of all kinds on the resistance of materials.[28] Architects, engineers and men of science all participated in such experiments. Perronet and Soufflot, for example, undertook a series of measurements of the expansion of iron,[29] while Gauthey developed a machine which enabled one to assess the resistance of stones in compression (Fig. 45), all in the context of the debate concerning Sainte-Geneviève.[30]

These ventures were marred, however, by the lack of a common conceptual framework, and by the continuing hold of certain hypotheses of classical science.[31] Although carried through in a piecemeal manner, the measures of the Age of Enlightenment served nevertheless to pave the way, as they

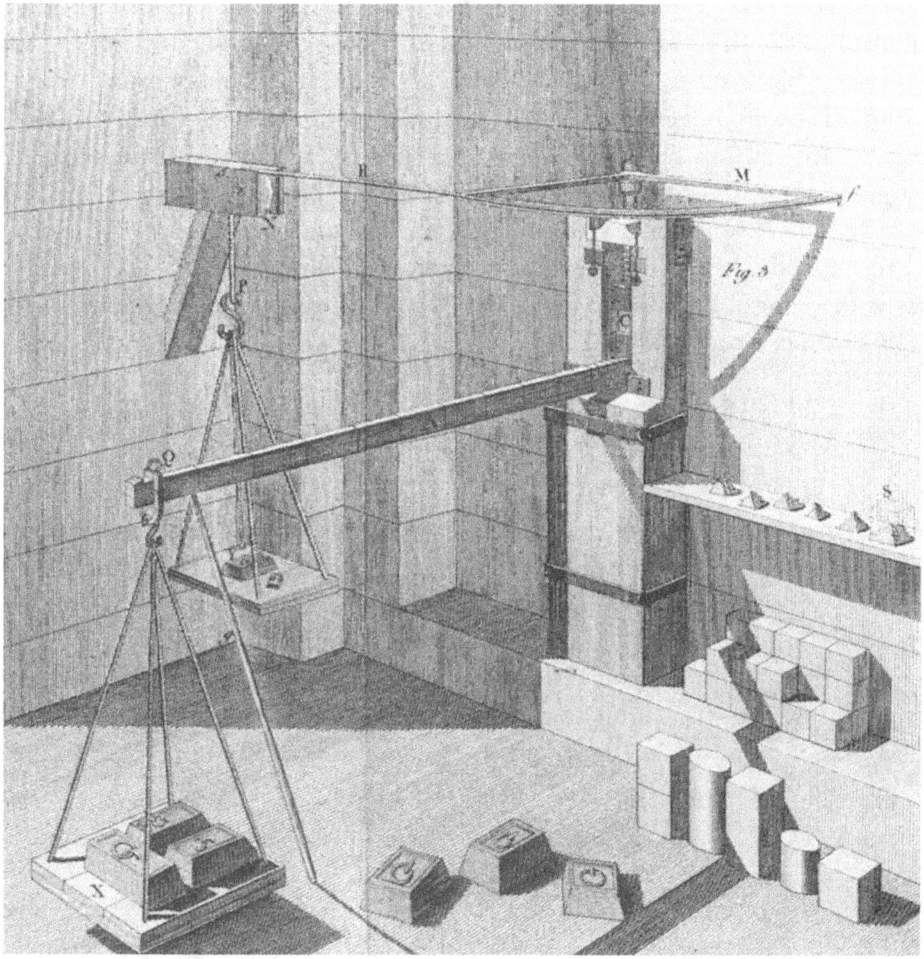

FIG.45 A.-J.-B. Rondelet, machine for testing the resistance of stone, used by Gauthey, Soufflot and Perronet when they were building the church of Sainte-Geneviève.

proliferated, for the reorganisation of knowledge at the beginning of the following century. In this context, the engineers proved more active than the architects, since they thought of experimentation as an essential aspect of their profession, and since it was not restricted to the resistance of materials. Coulomb, for example, studied the thrusts of earth,[32] while Chézy and Dubuat investigated the mechanics of fluids[33] and Prony the pressure of steam.[34]

Yet the crucial distinction between the architects and the engineers in the domain of construction was the scale of the problems tackled. If the great Parisian projects made it possible to push traditional technology to breaking-point, they were nevertheless exceptional, contrasting strongly with the day-to-day construction of buildings and incapable of causing any change to the latter. In the case of the engineers, the notion of breaking-point assumed a wholly different meaning, since their work, through its range, its often risky character and, finally, its cost, called for perpetual endeavour.[35] Development lay at the limit of the possibilities of the classical world, a limit which we shall now try to explore.

DEVELOPMENT PROJECTS

The road and the bridge represented the two poles of development in the Age of Enlightenment, with the former being characterised by its linearity, the latter by its singularity. Whereas the construction of highways was measured in leagues, all the technical difficulties imaginable might be concentrated in a few dozen *toises* of a bridge. The force of the engineers' position derived in large part from the powerful link uniting the banality of paving with the boldness of arch construction, whereas the Louvre colonnade or the church of Sainte-Geneviève had very little relation to domestic architecture.

ROADS

In the Age of Enlightenment, roads became a wholly independent means of transport. Although waterways still represented the main means of communication and exchange at the end of the seventeenth century, the situation began to change during the Regency.[36] The creation of the corps of the Ponts et Chaussées may partly be accounted for in terms of this change. Engineers believed firmly in progress, and they were identified with problems concerned with roads and the slow constitution of a national network.[37]

A number of technical innovations facilitated the development of roads. Trésaguet's research into the paving of highways did much to make the roads more viable.[38] There were, however, limits to this viability, and the upkeep of the roads remained precarious. Likewise, the network of routes set up by the engineers of the corps of the Ponts et Chaussées fell far short of meeting the country's needs. This network tended to follow the main axes across the country, and it was therefore not until the beginning of the nineteenth century that the internal regions of the provinces became less isolated. Around this same period, Navier, Schwilgué and Coriolis began to study the problems of haulage,[39] and thereby provided a scientific basis for the laying-out of roads.

In spite of their shortcomings, eighteenth-century roads nevertheless introduced a new dimension to the debate on the nature of building, beginning with that of the territory.

The techniques employed in the laying-out of roads in the Age of Enlightenment were almost exclusively geometric in nature. A road would generally be rectilinear, and would be directed at a noteworthy feature of the landscape, such as a bell-tower or château, which served also as a marker in the plotting of a map.

Road-building and cartography were intimately connected. When Orry, the Controller-General of Finances, stipulated in his *Mémoire instructif* of 1738[40] that the plotting of maps should follow a uniform model, he was very explicit about this connection. The engineer's first task was to explore the territory, to survey its riches, prior to considering how the proposed road might help to develop them. Likewise, the employees of the Drawing Office were, as we have seen, to delineate and rectify such maps, and take their inspiration from Perronet's directives, which instructed them to extend the embankments from 400 to 500 *toises* on either side of the highway. The building of roads and the description of the country were thus part of one and the same operation.

It is worth emphasising the importance of cartography in this respect, a point I shall return to below.[41] As far as the engineers were concerned, a map was a faithful image of the territory, considered in terms of all its wealth, whereas a road represented the development of this territory, or the possibility of exploiting it rationally. This rendered it necessary in a way that knew no parallel in the domain of architecture.

Labour was equally important here. For the *Mémoire instructif* of 1738, as well as being concerned with cartography, was designed above all else to establish the *corvée* in the *pays d'élections*.[42] Roads were generally built by the local peasantry, a work-force which, in contrast to the workers in the towns, was relatively unskilled. Since the roads engineer issued his own orders, and had a wholly free hand in enforcing them, the authoritarian supervision of production was from then on gradually extended, in a way that would have been unthinkable in the building trade.

A further novelty was the notion of working *en masse*. The volume of earth

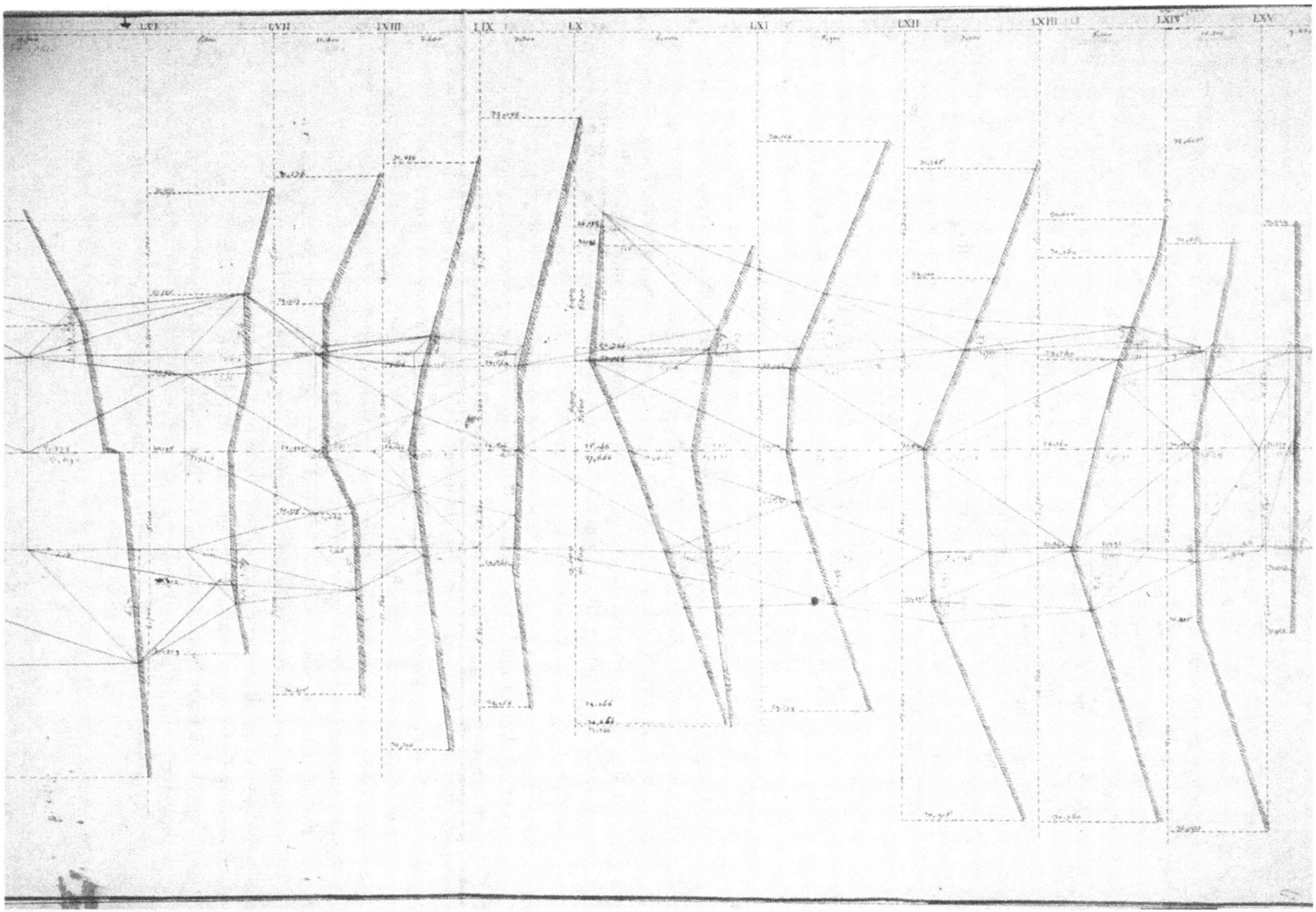

FIG. 46 C. Espagnol, projected road near Montmartre for the competition of Year VI, first prize, sections of the road and embankments, detail, ENPC, pen and ink. The road was altogether subject to calculation.

moved was enormous, and whereas architects had tended to address their orders to a handful of masons or carpenters workers might now be numbered in their thousands. In a context in which the slightest error might have unsuspected consequences, it became crucial to anticipate all eventualities.

Calculation therefore took on a new meaning. The construction of roads perhaps represented the only area in which the engineer did not run up against the shortcomings of scientific instruments. Levelling of the terrain only required the rudiments of arithmetic and of geometry; road projects were expressed in practice by addition and subtraction, and by the fairly simple measurement of whatever earth was to be moved. Calculation thus allowed one to plan the building of a road (Figure 46).

Furthermore, the road also enabled one to set about reducing the amount of human labour per given volume of earth, both of which could be expressed in figures and ratios. Through the *corvée*, the road represented, as we have just seen, a mass work effected by a relatively unskilled work-force. Through the application of a simple rule of three, the number of workers corresponded to the volume of earthworks. A mathematics competition on the eve of the Revolution proposed the following exercise: given two roads A and B, and given X as the number of workers at the engineers' disposal, how should one best distribute these latter across each of the sites, in order that work should proceed at the same pace in both?[43] Far more than was the case with other, more architectural applications, the eighteenth-century road heralded the homogenisation of the data of the project, and their gradual subordination to calculation.

THE PLACE OF DRAWING AND OF WRITTEN DESCRIPTIONS IN THE CONSTRUCTION PROCESS

It was to be a long time before the construction of lock-gates or bridges was to be handled in this manner. Nevertheless, some homogenisation of the data of the project was achieved through drawing and through written descriptions, which were ordered in such a way as to allow the levelling out of the kinds of information at the engineer's disposal.

Since the engineers were interested in rigorous definition, they tended to favour geometrical drawings and therefore neglected isometric projection, which deforms distances. The chief priority was to describe the work and the different parts of which it was composed. Much like an anatomical model, the plan displayed the different strata of the construction; the cross-section

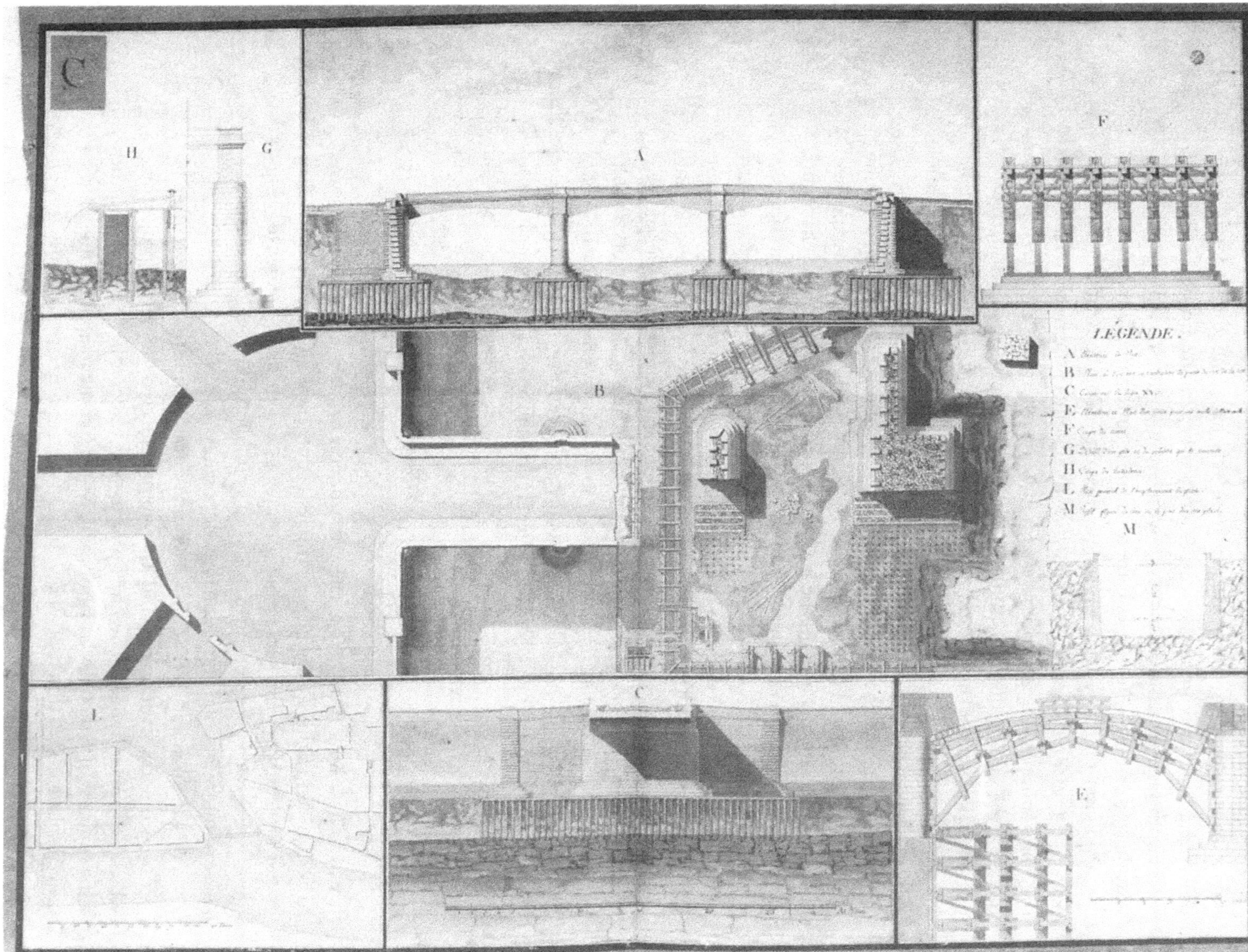

FIG.47 Design for a stone bridge between the Ile Saint-Louis and the Ile de la Cité, for the competition of 1792, ENPC, pen and ink, wash and watercolour, 110 × 315 cm. The drawing is arranged as stages, so as to illustrate the various phases in the construction of the bridge, from the laying of the abutment-pier and the piers to the laying of the roadway.

invariably featured the foundations, whereas the elevation enabled one to appreciate the proportions of the whole of the project.[44]

Although the drawing was of particular importance both at the conceptual level and for the purposes of communication, it was far from being the crucial aspect of the engineer's work. The bill of quantities proved to be equally significant, since it made it possible to reorganise and to render more precise the information furnished by the plans, cross-sections and elevations of the project.

The engineer's bill of quantities, which was described by Bélidor as 'a memoir providing information about all the parts of the work which one wishes to construct', serving to explain 'the order and the conduct of the work, the qualities and composition of the materials, and generally everything having a bearing upon the perfection of the works',[45] made its first appearance at the end of the seventeenth century.[46] Since it was accepted as a contrast, it made it possible for the construction industry to draw up agreements with enterprises, by supplying the latter with the details of the work and of its construction.

Finally, a detailed estimate would give some idea of the cost of the work, on the basis of standard prices for the materials and for the work-force. The estimate, like the bill of quantities, was a form of forecast, which therefore served as a means of controlling the site.[47]

Both drawings and written descriptions thus betoken an increasing mastery of the project. The engineer was charged with giving a faithful description of the built work, and nothing was of greater significance in this respect than the drawings representing, from one and the same viewpoint, a bridge at the different stages of its construction, from the laying of the piers to the installation of the roadway (Fig. 47). One can discern a new tendency in operation here, as the completed project was gradually replaced by the process leading from its conception to its effective realisation,[48] a process upon which Prony's *Réflexions* were based. The engineers' drawings, like their meticulous bills of quantities, in fact paved the way for a correlation of conception and realisation, a correlation which would indeed be one of the chief objectives of the nineteenth-century science of engineering. There is also good cause to link the appearance, in the practice of Enlightenment engineers, of the process of conception-realisation, to the much more general, technical imaginary, which sought to organise and to quantify flows and movements of every kind, from road traffic to the problems of water flow in rivers or canals.[49]

A dynamic vision of the problems encountered in the domain of construction gradually emerged, in place of a more static analysis of the parts and proportions of works.

Architects drew up bills of quantities and estimates also.[50] Yet they almost certainly lacked any awareness of a dynamic and continuous process of conception-realisation, based upon homogeneous data. The engineers' strategy, on the other hand, required explicit representations of this same process; they had yet to reduce the project to its underlying determinations. Engineers' projects were never so 'architectural' as they were at the end of the eighteenth century; the balance between form and technical reasoning was, however, fragile, as the example of the bridge proves.

THE BRIDGE AS MODEL

The bridge was the engineering project most closely associated with architecture. Although bridges were mainly designed to be a means of crossing, they were also monuments, and their ordinance and decoration had their importance too. 'We have thought that the architecture of bridges might be subject to variety', Prony observed in his *Description des projets et de la construction des ponts de Neuilly, de Mantes, d'Orléans, etc.*, 'so as not to opt for too slavish an imitation of the forms of ancient bridges.'[51] Thus, at a time when architecture was gradually acquiring a sense of its own history, bridge-building was making some advances in this direction also. This progress was nevertheless not restricted to the architectural aspects of the project. The debate concerning construction was also of crucial importance, as the numerous engineers' treatises appearing in the course of the eighteenth century show. Gautier's *Traité des ponts* (Fig. 48), Bélidor's *Architecture hydraulique* and the works of Regemortes and Perronet echo the experiments attempted in this domain. Even if the architectural model was still a real presence, the engineers were showing clear signs of wishing to emancipate themselves. This desire found expression in innovations both in the conception and in the building of bridges. Decoration of bridges, on the other hand, posed far fewer problems. Solidity, boldness and lightness were the defining features of an aesthetics which few cared to formalise very much. Bridge construction was the main issue, whereas decoration was merely a matter of taste.

In order to illustrate the development of bridges in the Age of Enlightenment, I shall refer to a number of different built works, namely, Gabriel's bridge at Blois, the bridge at Moulins built by Regemortes, the bridge at Saumur realised by de Voglie and, finally, Perronet's great bridges.[52] These

FIG. 48 Gautier, *Traité des ponts* (1716, second edition of 1765), frontispiece.

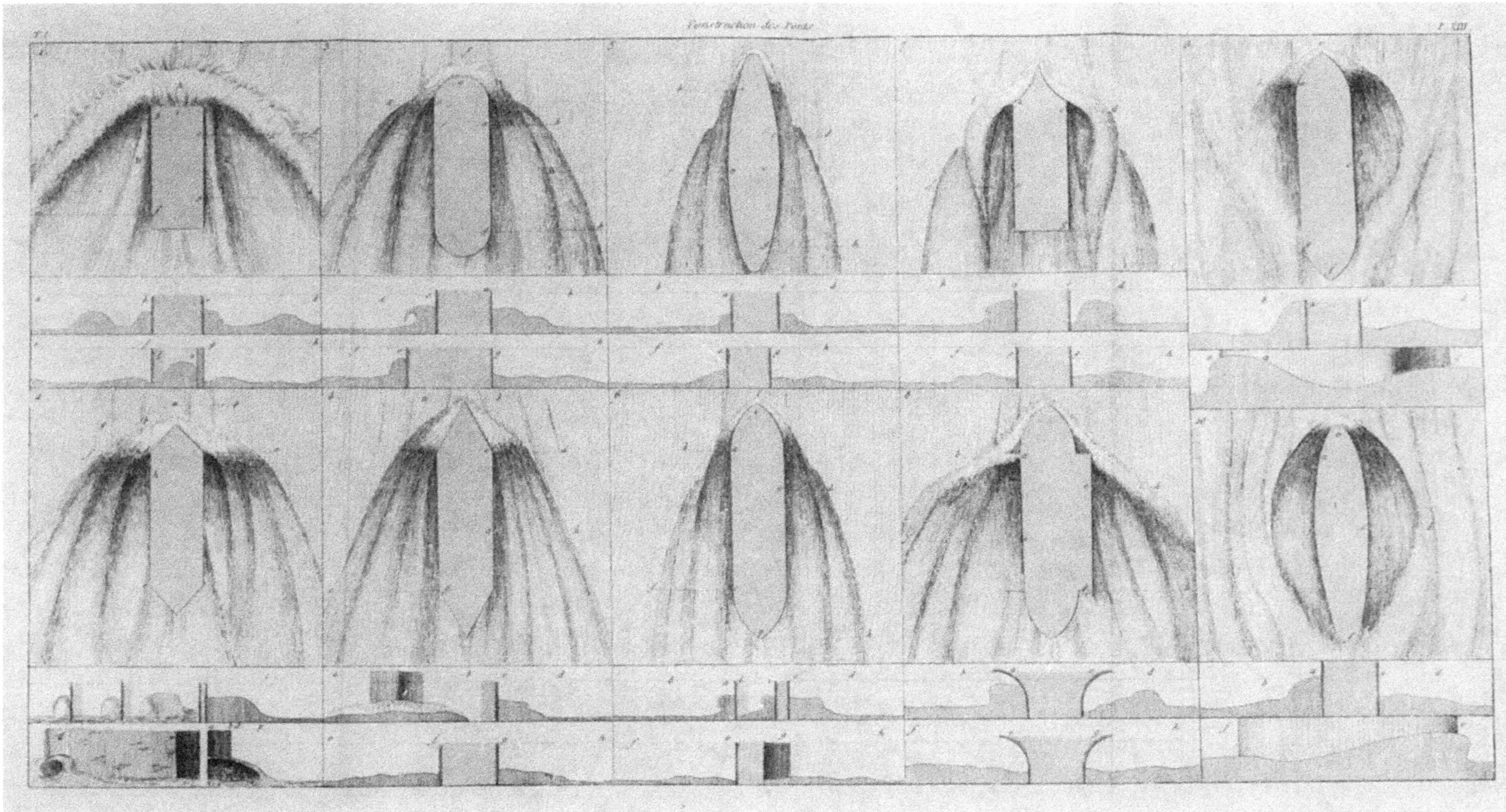

FIG. 49 E.-M. Gauthey, *Œuvres* (1806–9), various profiles of piers, together with the flow of water produced in each case. The development of bridges in the eighteenth century had much in common with the hydrodynamic experiments of the period. The question of the best form to give to the piers, in particular, calls to mind the problems of resistance of boats to sailing which were studied by d'Alembert, Bossut and Condorcet in 1775.

works played a similar role in their own domain to that of the great Parisian projects discussed at the beginning of this chapter.

The new bridge at Moulins, built by Regemortes as a replacement for a bridge by Mansart which had been swept away in a flood in 1710, clearly demonstrates the main problem in the building of such works of art. In the eighteenth century, rivers were still subject to violent floods,[53] which bridges might not always survive, especially if debris carried down from the river banks blocked up the arches. More generally, bridges were intended to resist the turbulence occasioned by severe variations in the flow of water, as well as the effect of suction, which might at times wear away the foundations.[54]

In the past, this problem had been resolved by giving the bridge a massive character. The width of the piers and the reduced span of the arches were designed to enable the bridge to resist floods. The Pont-Neuf represents a fairly good example of the results thus obtained.

Towards the end of the seventeenth century, this pattern was reversed. The pont de Blois, which was built by Jacques Gabriel in 1716, followed principles wholly different from those outlined above. Whereas medieval and Renaissance bridges resisted through their mass, through the width of their piers and the narrowness of their arches, bridges in the Age of Enlightenment were designed to offer the least possible resistance to the flow of water (Fig. 49),[55] and were therefore characterised by piers with smaller cross-sections and by an increased span for each of the arches.

The engineers therefore argued for a change in the form of arches, which became increasingly flattened, so that 'bucket-handle arches', obtained by means of arcs of circles with different radii at a tangent to each other, now made their appearance. The pont de Blois used just three radii, but it was not long before the pont de Neuilly was built using eleven radii, with the arches assuming approximately the form of elongated ellipses. Meanwhile Perronet had invented another system, that of 'stretched arches'.

With the decrease in width of the piers, the need was apparent for stronger foundations than had existed in the past. Between 1756 and 1770, the pont de Saumur saw the introduction of the technique of floating caissons, and that of a machine for sawing off piles under water (Fig. 50).[56] A general revetment was employed at the pont de Moulins.[57] Perhaps more important still, the classical techniques were perfected, conferring a greater stability upon bridges.

The main consequence of these transformations was the appearance of a new model of what a bridge was. The flattened arch and the reduction in the piers increased the horizontal thrusts, which were transmitted from arch to arch. The traditional bridge, which had been framed by girders all in equilibrium, as if juxtaposed, was replaced by a dynamic model, with arches counterbalancing each other by shifting the stresses on to the abutment-piers.[58]

These new principles in structural work were plainly related to developments involving a system in which a column supported an entablature constructed with voussoirs[59] and those which, more generally, involved architectural structure.[60] The bridges of the Age of Enlightenment derived from preoccupations fairly close to architecture. With Perronet, nevertheless, the model of a bridge assumed a new scope which transcended this shared framework.

THE SYSTEM OF 'STRETCHED ARCHES': FROM THE SPATIAL MODEL TO THE CONSTRUCTIVE MODEL

Perronet's approach was at first no different from that of other engineers, except in so far as it reflected a desire to unify problems and to treat them as a whole.[61] This coherence, rather than any specific technical detail, was the defining feature of Perronet's work, which was itself a synthesis of various forms of disconnected research, through a process bearing some similarity to the phenomenon of 'concretisation' described by Gilbert Simondon.[62]

'Stretched arches' only featured in Perronet's last built works, for example, the pont Saint-Maxence (Fig. 51), built between 1771 and 1786, and the pont Louis XVI, his masterpiece, whose building proceeded in stages between 1787 and 1791.[63] The pont de Neuilly (Fig. 52), finished in 1774, still had 'basket-handle' arches, whose variously curved outlines are reminiscent of an ellipse. The 'stretched arches', on the other hand, were achieved by means of a single arc of a circle with a large radius, which was butted against the pier at an acute angle. This arrangement, which was designed to increase still more the span of the arch, by placing it above the level of the highest waters, was sometimes criticised by contemporaries. By introducing a sudden change of direction in the spring of the arch, where the pressures were strongest, Perronet was in fact defying one of the traditional principles of stonemasonry, a principle of homogeneity which the 'basket-handle' arches still respected.

The diminution of the width of the piers was equally spectacular. Whereas Bélidor had recommended that this width be about one fifth of the span of the arches, Perronet stipulated that, in the case of the pont de Neuilly (Fig. 53), it be only one tenth, and he also gave some thought to the best contour to give

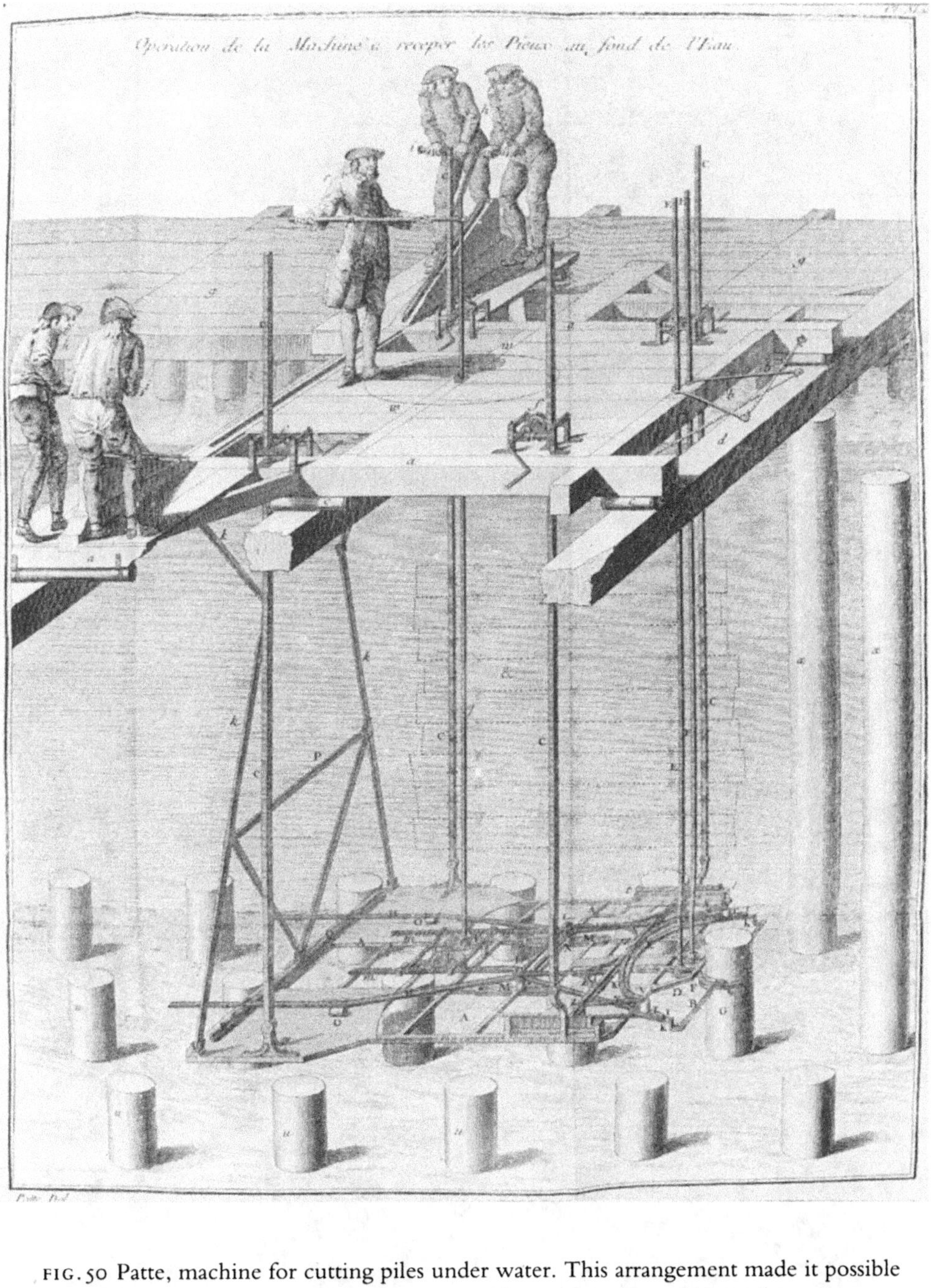

FIG. 50 Patte, machine for cutting piles under water. This arrangement made it possible to cut off the heads of the piles, which were blunted by repeated blows from the ram, without having recourse to costly drainage. After the cutting down of the piles, the next step was the laying of the platform designed to bear the first stone course of the piers.

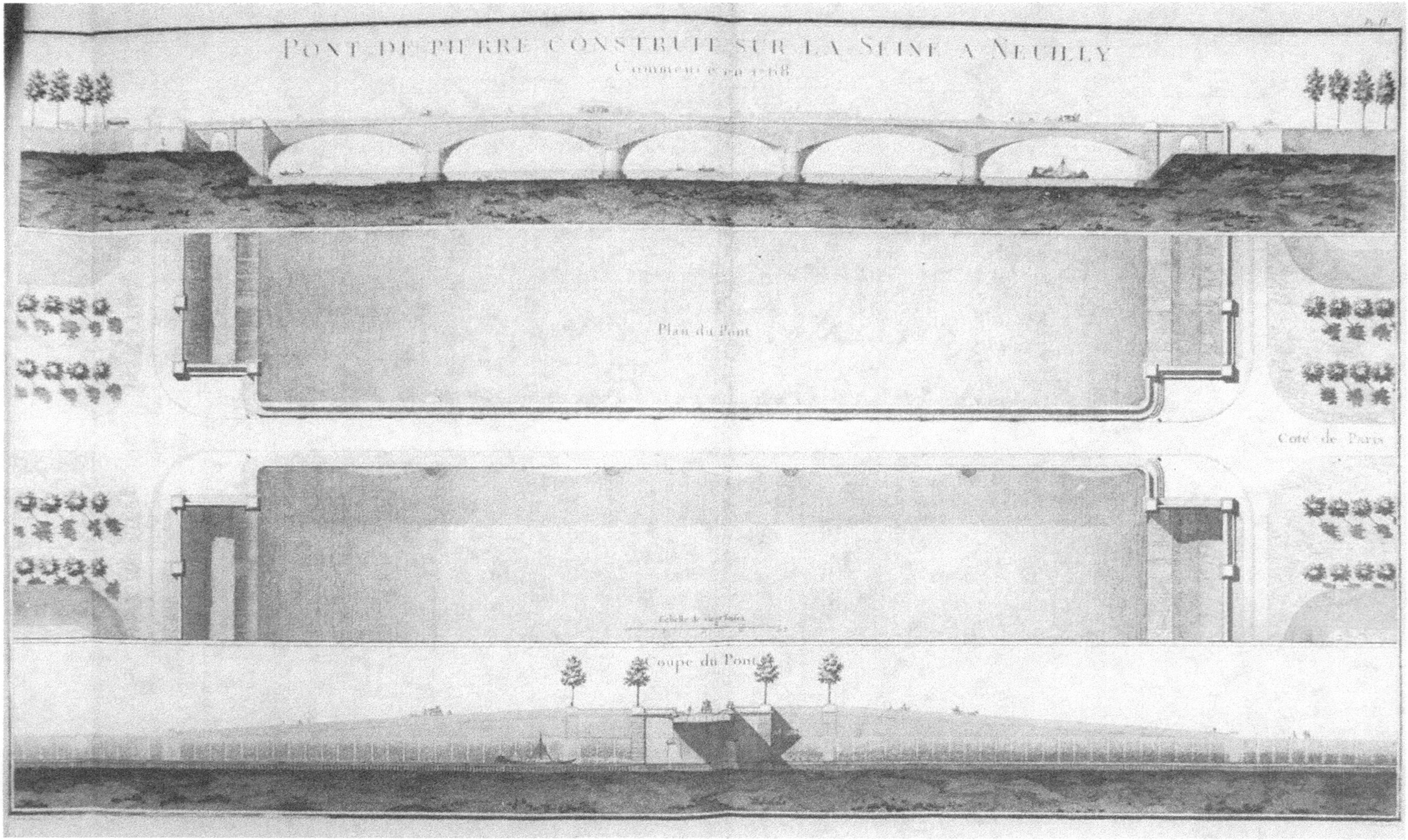

FIGS. 51, 52 J.-R. Perronet, *Description* (1782–3), pont de Neuilly and pont Sainte-Maxence, 'basket-handle arches' and 'stretched arches'.

Dessin du Pont projetté sur la Riviere d'Oise,
a Pont Sainte Maxence
Coupe sur la Ligne A B.

FIGS. 53, 54 J.-R. Perronet, pont de Neuilly and pont Sainte-Maxence. One should note the substitution of rows of columns for continuous piers at the pont Sainte-Maxence.

to the piers.[64] All of these features – 'stretched arches', reduced piers, shaped contours – were aimed to ensure the least possible resistance, and thereby to facilitate the free flow of water. With a similar aim in mind, Perronet designed the piers of Saint-Maxence (Fig. 54) as rows of columns, and he suggested that the same solution be adopted in the case of the pont Louis XVI. However, the Assembly of the Ponts et Chaussées expressed such doubts about this notion that he had no choice but to abandon it.[65]

The system of 'stretched arches' represented a considerable improvement

VUE PERSPECTIVE DU NOUVEAU PONT
DE PONT SAINTE MAXENCE,

in bridges, primarily as regards the formal qualities of the construction. The presence of sharp angles at the spring of the arch, and the desire to replace the piers with isolated columns represented an approach to problems of structure which permitted the expression of a certain number of contradictions. The first contradiction was between the movement of the river and that of the bridge. Perronet abandoned the old kinds of pier, which resembled ships, and which enabled one to reconcile on a symbolic plane the flowing of the water and the transverse passage of men. The construction no longer echoed the pontoon bridge; it was now free of any origin which declared its kinship with the river.[66] In the work of Perronet, the tension of the bridge between its two abutments, the series of flattened arches which conveyed this tension, contrasted with the succession of thin, vertical piers. The articulation of the continuous line of the bridge and of its supports was effected by means of a sharp angle, which served to emphasise the clash of two conflicting logics. For Perronet, therefore, the structure arose out of the overcoming of problems.

Yet Perronet did not simply set out to define a spatial structure which, although based upon a series of contradictions, was coherent. His perspective, which owed much to the rationality of the manufactures,[67] led him to treat form as indissociable from the process which had produced it. The construction of his major bridges saw innovations both in the organisation of labour and in the various practical arrangements which were adopted (see Fig. 55). The strictest discipline was enforced on his sites; a whole set of previously separate stages were now coordinated. The use of cocked centres, in particular, brought about the decentering of all the shuttering (see Fig. 56).[68] The transition had thus been effected from a predominantly spatial structure to a new model of the bridge, which was 'constructive', for it now integrated the different phases of construction, together with the formal apparatus required to resolve any difficulties which might arise (see Fig. 57).

In such a context, calculation assumed a new importance as an instrument of control. Prony's role was especially significant here. His collaboration with Perronet dated from the 1780s, a period when the pont de Neuilly was the subject of harsh criticism, the narrowness of its piers seeming to be contrary to all recognised principles of construction.[69] In 1783, Prony wrote his *Application de la mécanique des voûtes à la construction du pont de Neuilly*, which was designed to prove the soundness of the bridge, and he followed it with his *Calcul sur la stabilité des voûtes du pont Louis XVI*, which was meant to serve the same purpose.[70] Prony's contribution to the art of bridge-building was not truly original.[71] The theory of arches had not developed far in the eighteenth century,[72] aside from the new attention paid to the problem of friction.[73] Adopting much of La Hire's hypothesis regarding the break-up of the arch into three parts,[74] Prony's contributions nevertheless impressed his contemporaries with their brilliant formalisations, which were based upon finite difference equations.[75] Appointed to oversee the site of the pont Louis XVI (Figs. 58 and 59), Prony epitomised the new, scientifically based values of order and method. His role had been limited, admittedly, to putting Perronet's

FIG. 55 J.-R. Perronet, works at the pont de Neuilly, draining the waters in order to lay the foundations for a pier.

conceptions into practice; formal intuition still tended to prevail over calculation, which functioned as a posteriori verification. Nevertheless, the engineers' innovatory spirit and classificatory impulse were soon to take precedence over tradition.

Perronet's work was not wholly orientated towards the future. His approach, like that of the other engineers of the period, represented a mixture of the old and the new. For example, a highly monumental conception of the bridge survived in Perronet's work, and this conception found its apogée in his curious *Mémoire sur la recherche des moyens que l'on pourroit employer pour construire de grandes arches de pierre de deux cents, trois cents, quatre cents jusqu'à cinq cents pieds d'ouverture*, in which technical invention came close to megalomania (see Fig. 60).[76] An extremely complex form of stereotomy, which was akin to the structures of Delorme and the Mansarts, also survived.[77] The pont de Neuilly and the pont de la place Louis XV, although very much admired at the end of the eighteenth century, nevertheless represented a kind of impasse.[78]

As we have already emphasised, if Perronet now appeared to be an innovator, it was more through the coherence of his approach than through its details. What mattered was the emergence of a model of construction which transcended the framework of architectural thinking. As far as Perronet was concerned, the model of the bridge was no longer limited to a spatial structure, but integrated conception and execution, objectivising and summarising their various phases. The model gradually elaborated by Perronet should be seen as a response to Laugier's ideal church, whose structural shortcomings Frézier, as a specialist, had taken some delight in denouncing.[79]

Being dynamic by nature, Perronet's model arose out of the contradictions which it had overcome; it effected the homogenisation of a whole series of data. Form and process of construction were fused, just as work was put on the same footing as materials. Like the workman, stone was seen as 'working' in the arch of the bridge; it bore both the stamp of its geological origin and of the efforts involved in dressing it. The constructive model also represented a compromise between geometrical intuition and that of the forces traversing matter or, to be more precise, it concerned the arrangement or the combination of these forces, which were channelled by both form and calculation. On the threshold of the nineteenth century, the constructive model thus assumed a metaphorical value. One would soon have to 'work' society in its very depths in the same way that the engineer worked, or set others to work, upon material.

CONCERNING THE DISPUTE OVER THE PANTHÉON

If we are to arrive at a clearer understanding of what it was that separated the thinking of the majority of architects from the structural intuitions of figures such as Perronet, it may be of interest to revert to the Church of Sainte-Geneviève, which became the Panthéon during the Revolution, and to study the famous dispute between Patte and Soufflot over a monument which

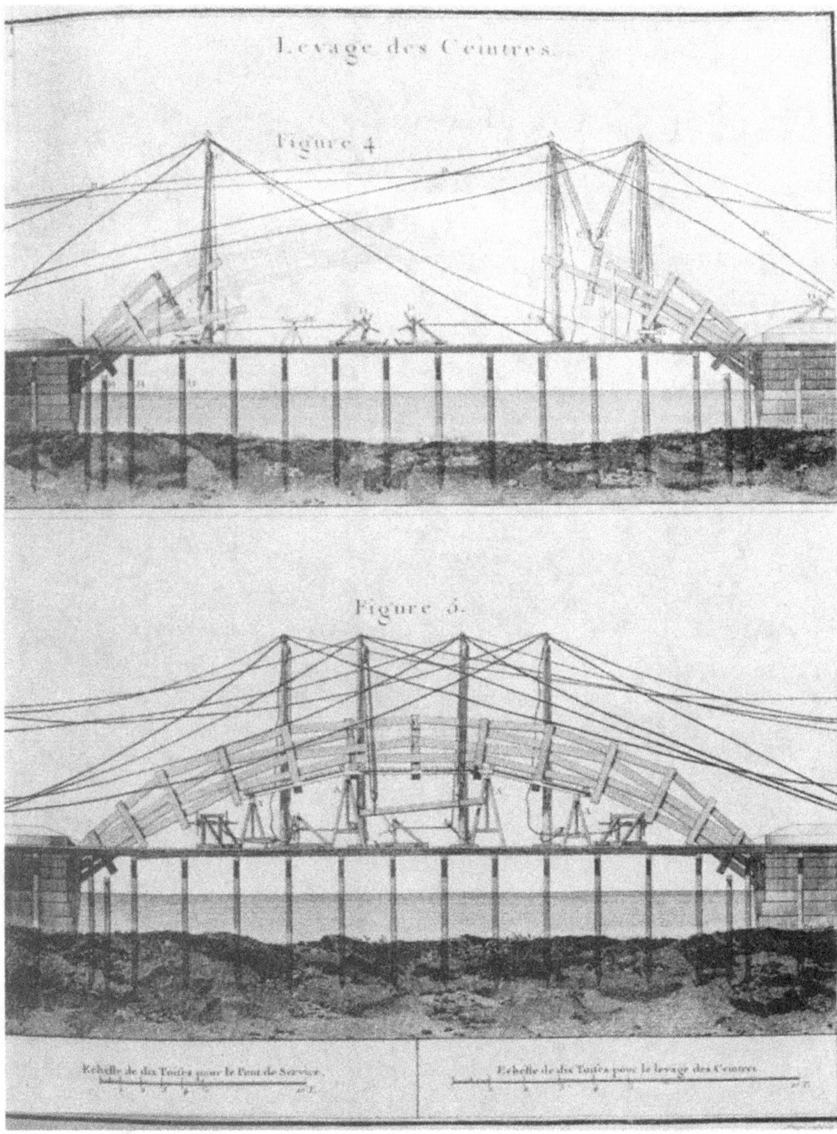

FIG. 56 J.-R. Perronet, installation of cocked centres. The use of cocked centres was an especially delicate matter. Because of their elasticity, they had to be compressed while the arches were being constructed, and this was done by placing mobile weights at the top. Furthermore, they all had to be decentred at once.

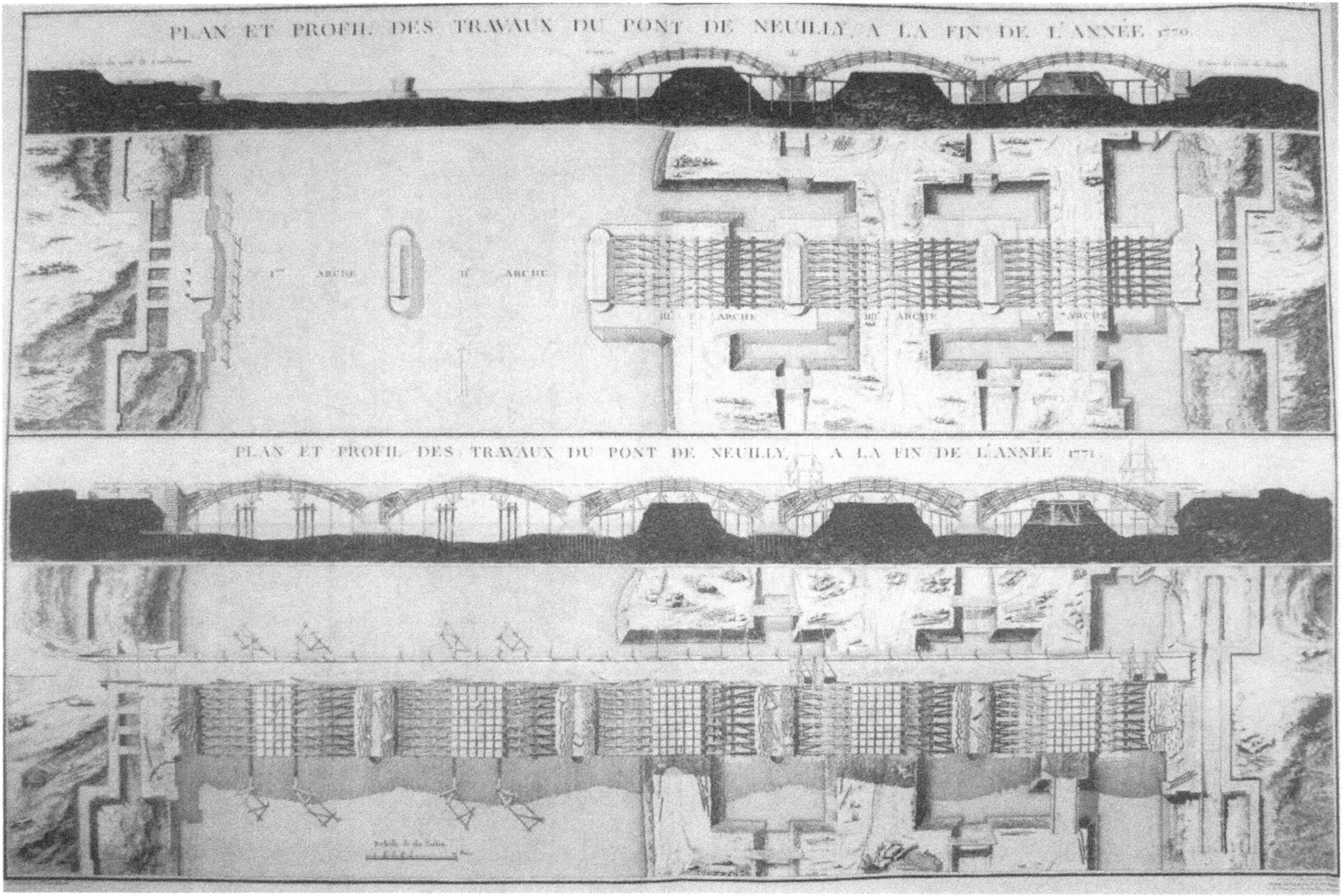

FIG. 57 J.-R. Perronet, work at the pont de Neuilly in 1770 and 1771. The manner in which the bridge is represented emphasises the idea of a continuous process of construction, taking its inspiration from the rationality of manufacturing.

served as a focus for most of the issues of the period. Perronet, and Gauthey and Frézier also, participated in the polemic which Patte launched around 1770.

As is well known, Marigny had commissioned Soufflot in 1755 to build the church of Sainte-Geneviève. The plans were published in 1757, and the laying of the first stone took place in 1764. A year later, in his *Monumens érigés en France à la gloire de Louis XV*, Patte was to write that: 'The precautions which have been taken in order to ensure the perfect execution of this monument should render it eternal.'[80] In his *Mémoire sur la construction du dôme projeté pour couronner l'église de Sainte-Geneviève*, however, which was written in 1769, his viewpoint had altered. Following a journey to England, in the course of which he had the opportunity to meet Chambers and to study the dome of St Pauls in London,[81] he considered if the piers Soufflot had designed to support his dome were too light. A comparison between the domes of St Peters, St Pauls, of the chapel of the Sorbonne, of the Val-de-Grâce and of Les Invalides proved, according to him, that the size of the piers supporting the drum of a dome should be reckoned between one quarter and one seventh of the inside diameter, whereas Soufflot had opted for one tenth of the diameter.[82] The reader will immediately grasp the resemblance between this dispute and the debate over Perronet's bridges, which were also criticised because they failed to respect the traditional relation between the size of vertical supports and span. Soufflot, who numbered Perronet among his friends, asked him to draft a reply to Patte's arguments. The *Mercure* of April 1770 published two letters from Perronet to Soufflot. In the meantime, Patte had been granted permission by Marigny to publish his *mémoire* (Fig. 61), and the arguments which it contained were thus aired in public. In order to counter Perronet, Patte then appealed to Frézier, who began by judging in his favour but later retracted.[83] Gauthey brought out his *Mémoire sur l'application des principes de la méchanique à la construction des voûtes et des dômes* in 1771, which took Soufflot's side; Cochin, the Abbé Bossut and d'Angivilliers each intervened as well. Without entering too deeply into the finer points of a debate whose various phases have been reconstructed very precisely by Mae Mathieu,[84] it may be worth considering the attitudes which it reflects, beginning with those of Patte, a man whose exact intentions are not always easy to grasp.[85]

The importance of calculation in this dispute is at any rate clear enough, as we can see from Patte's attempt to base the demonstration contained in his *Mémoire sur la construction de la coupole projetée pour couronner la nouvelle église de Sainte-Geneviève* on a mathematical formula taken from Bélidor.[86] Supposing

Pl. XLVII

PROJET D'UN PONT SUR LA RIVIERE DE SEINE VIS A VIS LA PLACE LOUIS XV.

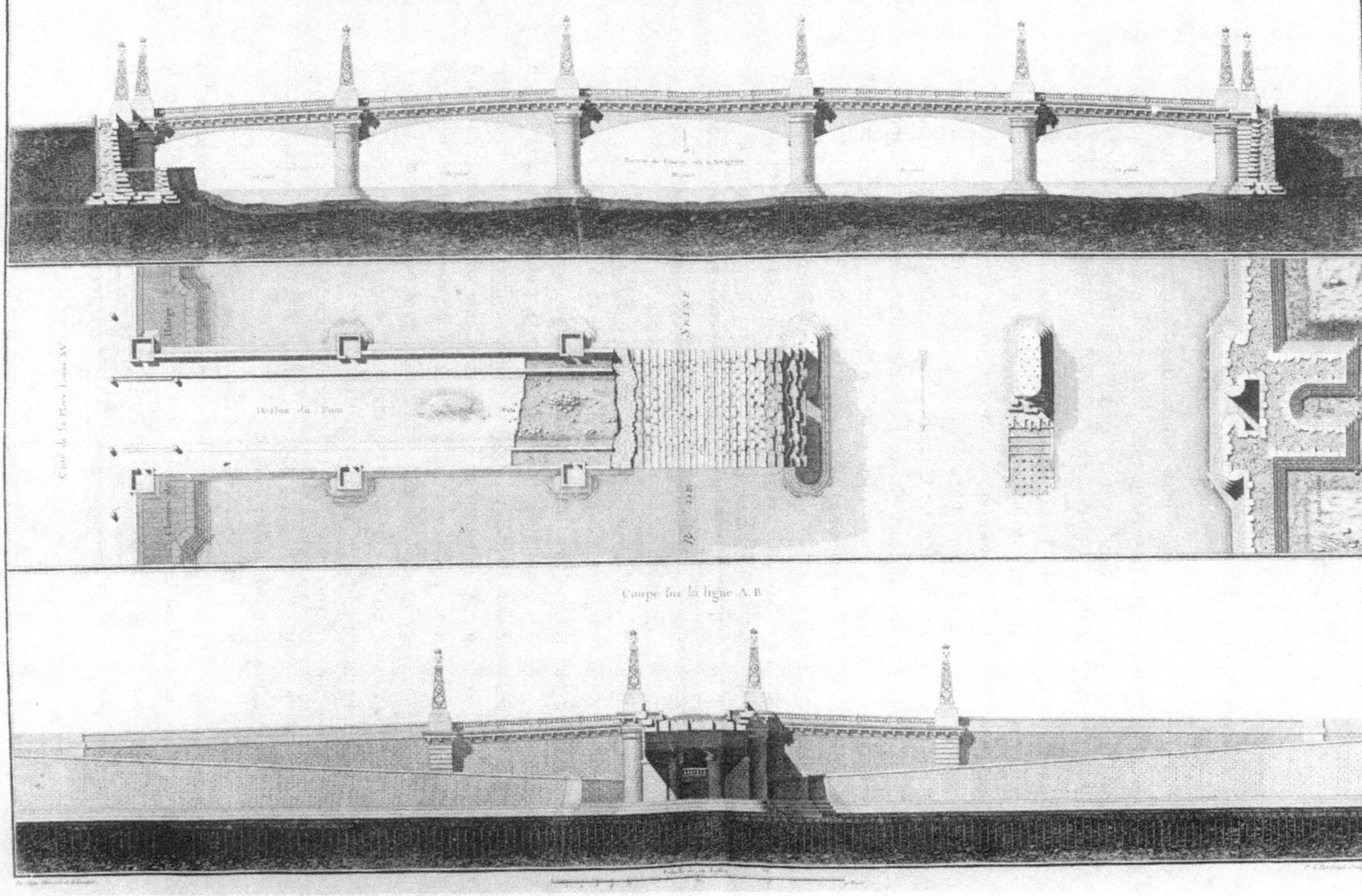

it to be a 'problem addressed to all learned societies, to engineers, to architects and to those who are known for their work in construction',[87] this *mémoire* begins with a profession of faith which a man of science would certainly not have disowned:

> It is one of the obligations we owe to the progress of the sciences that it has enabled us to appreciate in advance an infinite number of operations in which one used only to be able to proceed as if groping to feel one's way, or in allowing chance to dictate success or failure. Of all the arts, there are few so suited to being illuminated as construction. Since it is invariably a matter of raising bodies one above the other, of the bearing of loads, of bracing thrusts from arches, as well as supporting either inclined planes, horizontal or perpendicular surfaces, it follows that, because all these objects are subject to relations or considerations relative to the laws of equilibrium and of gravity, they belong of necessity to mathematics, and chiefly to mechanics, or, in other words, they can be assessed in terms of these rules.[88]

In the opening pages of his *Mémoire*, Patte thus painted a bold picture of a science of construction which would be able, through mathematics and mechanics, to assess the actual details of execution of a design. If one looks more closely, however, the arguments which he then proceeded to develop were of a wholly different order, and referred to the principle that the strong should support the weak;[89] the piers of Soufflot's dome seemed to him too small. In fact, Patte's position, like that of the majority of engineers of the day, was the result of a compromise, whose subsequent development might already be foreseen, between tradition and calculation. Tradition corresponded of course to the kind of solidity which a decorous size and a fine execution conferred upon a building. With a degree of subtlety, Patte disguised tradition beneath 'considerations respecting the laws of equilibrium and of gravity'. From an architect's point of view, these laws were simple and determinist, and construction was assumed to be a matter of loads and thrusts.

At Soufflot's request, Perronet took part in the dispute over Sainte-

FIG. 58 J.-R. Perronet, pont Louis XVI. The monumental aspirations of the project are evident enough here. While the piers look like squat columns, and a cornice serves to emphasise the roadway, the ironwork lamps, being more slender, confer a 'pyramidal' quality on the bridge. The tension thus expressed between the solidity of the foundations and the transparent nature of the balustrade is reminiscent of the procedure employed by Soufflot at Sainte-Geneviève, where the massive Order of the crypt supports the columns of the nave, wholly 'Gothic' in their lightness.

Geneviève, although at first with reluctance, the project still being fairly unfamiliar to him.[90] In his letter of 26 January 1770, however, he offered some words of encouragement which, although vague, nevertheless reveal a great deal as to his attitude to problems of construction. After praising the plans and sections of the future church of Sainte-Geneviève, he endorsed the scheme to reconcile Gothic lightness with Greek elegance – a notion which is known to have inspired the architect[91] – and then he went on to compare Gothic construction to the skeleton of an animal: 'The magic of these latter buildings consists chiefly in their having been built in some sense in imitation of the structure of animals, with the columns tall and slight, the crossvaults and ribs with beams, the pointed arches and secondary ribs like their bones, and the small stone and webs which are only four to five inches deep in section being like the flesh of these same animals. Such buildings subsist, like a skeleton, or the carcasses of ships which seem to be constructed after similar models.'[92]

This comparison must have occurred quite frequently in Perronet's conversation, since it cropped up again a little later in the writings of Prony, his faithful disciple.[93] It is also of interest to observe the differences that there were between the vocabulary of Patte and that of Perronet. The former spoke of 'raising bodies one above the other', and even employed the term 'load'. Perronet, an engineer, used a wholly different idiom. Thrusts and forces, whether active, destructive or resistant, constituted a repertoire in which force and work, although as yet distinct from each other, nevertheless imparted a strong sense of dynamism. Patte's approach to structure was therefore phrased in terms of weight, whereas that of Perronet involved a system of forces and resistances which counterbalanced each other. Yet it would be wrong to lapse into caricature. Patte had a perfectly good understanding of the problems of structure. He provided a precise analysis of the pont de Saumur in his *Mémoires sur les objets les plus importants de l'architecture*,[94] and he later described the pont de Neuilly in the treatise on construction which was appended to Blondel's *Cours d'architecture*.[95]

How should one interpret Perronet's curious comparison between architectural structure and the animal skeleton? What he may perhaps have had in mind was the idea of the economy which, in contrast to the basic assumptions of the tradition, enables one to realise one's conception, an idea which was missing in Patte (see Fig. 62). The architect's approach rested upon the intuition of continuity of structural solutions. The piling up of two blocks of stone or the building of Sainte-Geneviève basically derived from the same principles and, above all, from the same basic assumptions. Perronet's

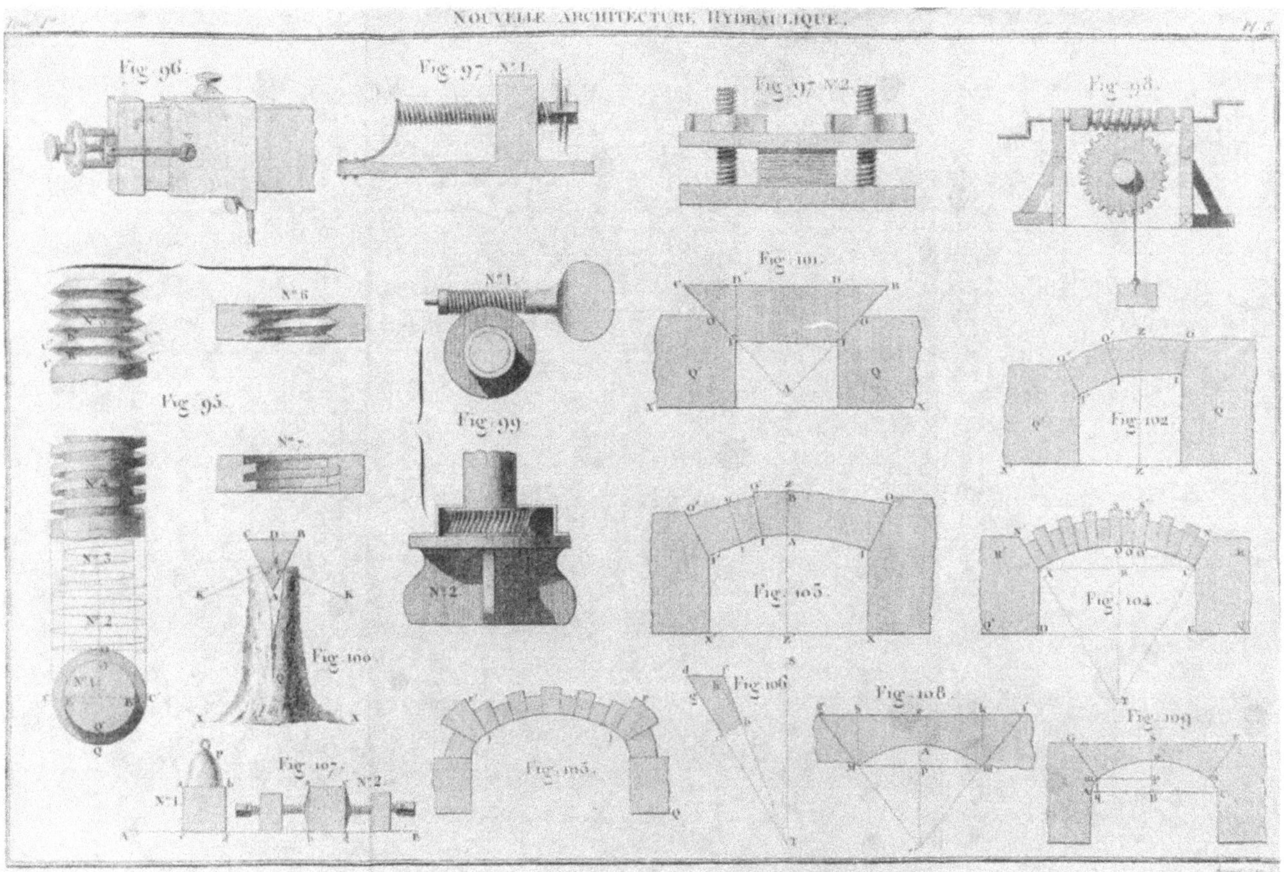

FIG. 59 G. Riche de Prony, *Nouvelle architecture hydraulique* (1790–6), simple machines, the vice, the wedge, and their use in the theory of arches. Perronet's concepts are realised here through calculation.

thinking, on the other hand, was characterised by a radical discontinuity in the problems of conception. 'In imitating nature in our structures', Perronet wrote, 'one can, with much less material, make some very lasting works',[96] for such imitation results in arrangements which are irreducible to the springing-stone, just as the skeleton of an animal cannot be seen as a mere accumulation of bone. Factors of scale and of complexity are therefore linked to the continuity of the process of conception and realisation, dividing it up into zones of distinct pertinence.

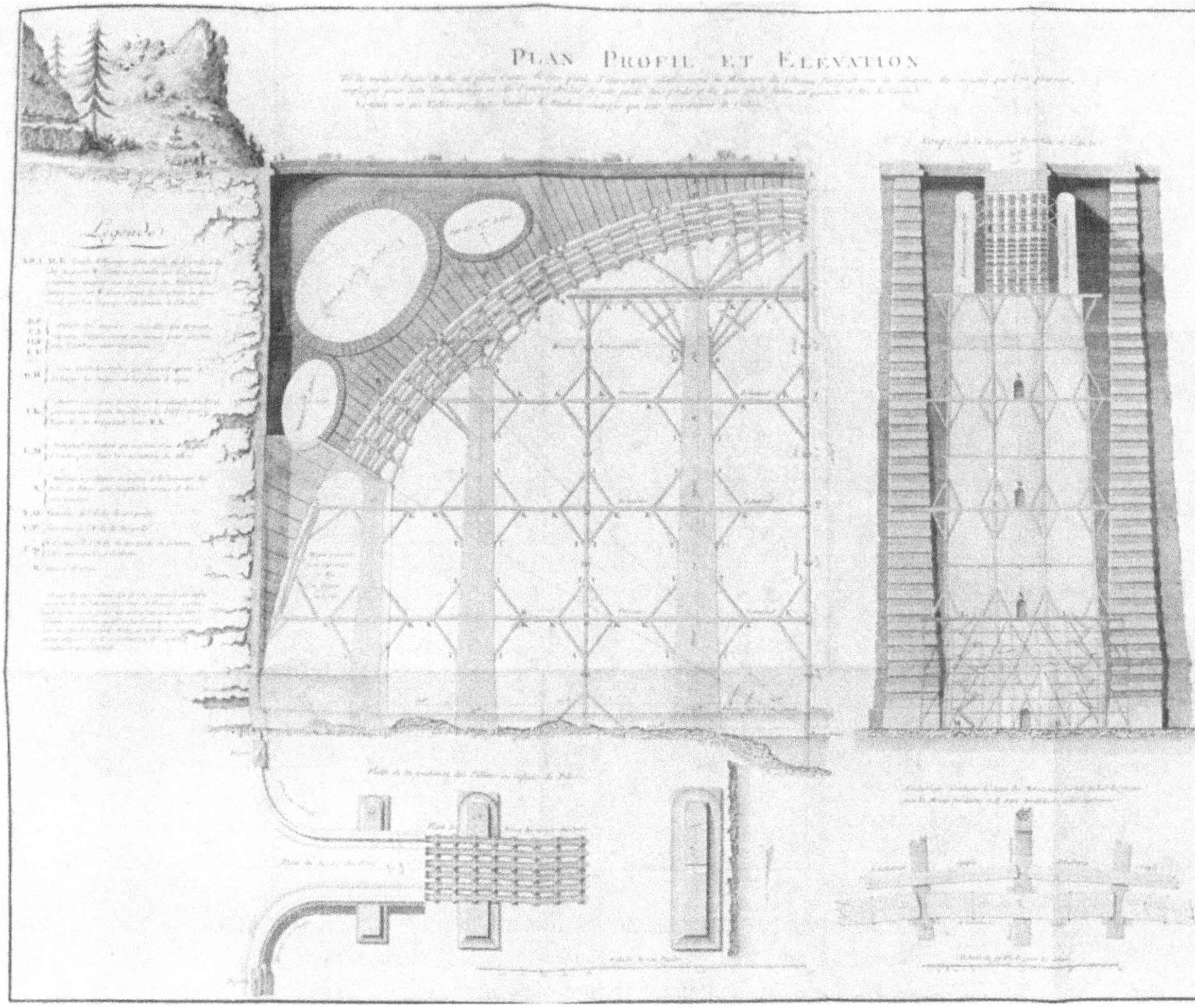

FIG.60 J.-R. Perronet, *Mémoire sur la recherche des moyens . . . pour construire de grandes arches de pierre* (1793). The oval openings were intended to lighten the huge arch which the engineer had conceived.

FIG. 61 P. Patte, *Mémoire sur la construction de la coupole projetée pour couronner la nouvelle église de Sainte-Geneviève à Paris* (1770), plans and sections of the dome of Sainte-Geneviève.

Another characteristic of the structural model thus emerged. Architectural structure, as Patte conceived it, presupposed that, in contrast to the fragmented nature of the mode of production, causes and effects were continuous. The structural model, on the other hand, depended upon a continuous process, leading from conception to actual execution, a process within which breaks of a new kind then arose.

These breaks probably corresponded to a change of attitude towards physical laws.[97] Whereas Patte believed that such laws might still be condensed into formulae which were universally applicable, whatever scale of problem was involved, Perronet regarded natural laws as expressing any number of different dynamics, which calculation might one day account for. Yet in the meantime one had to transcend the rules and the formulae of the past. This was Gauthey's purpose in particular, when he criticised the results achieved by theory in his *Mémoire sur l'application des principes de la méchanique à la construction des voûtes et des dômes*,[98] in order to undermine the arguments developed by Patte. Nevertheless, Gauthey launched himself into calculations which enabled him to conclude that the dome of Sainte-Geneviève[99] was in fact stable – the 'torch of mathematics' was not averse to a contradiction.[100] What Soufflot, Perronet and Gauthey nevertheless sought was a new type of dialogue between mathematical formalism and experiment,[101] a dialogue involving, somewhat paradoxically, both a fresh challenge to the traditional assumptions of the science of construction, and a whole series of gropings and compromises: to calculate but to distrust calculation, to innovate but to rely upon tradition.

Such paradoxes had been the rule among engineers, as we have already seen. Yet Patte was to prove much more rigorous in his approach. Indeed, it was this rigour which in the end served to restrict him, at a time when a fresh approach to constructive problems, based upon a combination of the continuous and the discontinuous, was being developed, one which presupposed a continuity of laws and of processes, and a discontinuity of problems of conception, of rules and of formulae. The bridge served in some sense to symbolise this situation, since it expressed both the continuity of the crossing and the discontinuity of the ravine which it had crossed. The contradictions which the bridge's structure had overcome referred to the overlapping of road and river, to a confrontation between horizontal and vertical logic, control of which was becoming more and more necessary.

By combining comparisons and contrasts, the engineers' conceptions seemed finally to open the way to the analytic approach whose economy they

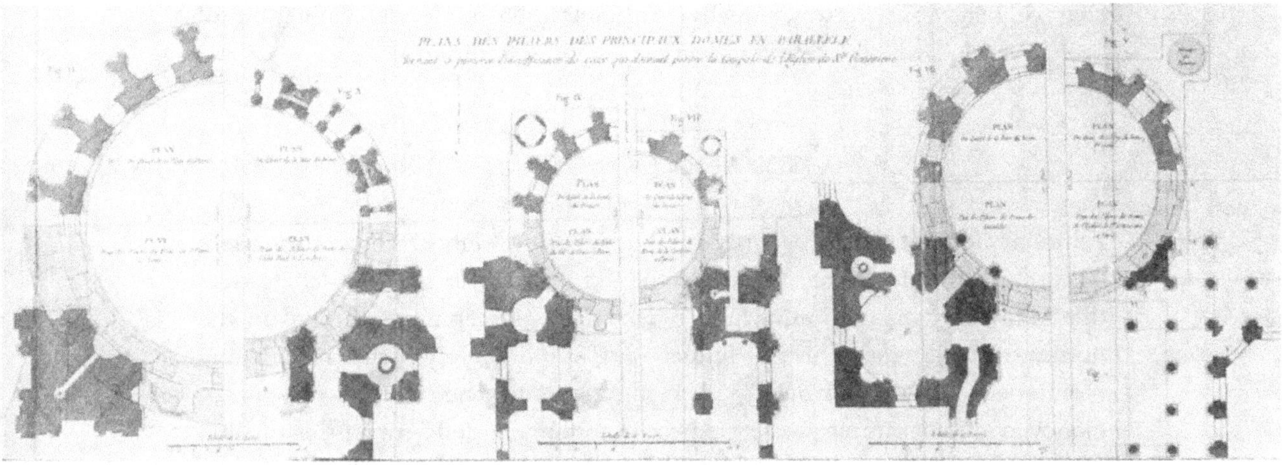

FIG. 62 P. Patte, 'Plans of the piers of the principal domes set alongside, so as to demonstrate the inadequacy of those designed to bear the dome of Sainte-Geneviève'. As far as Patte was concerned, the same dimensioning formulae could be applied to buildings of a wholly different scale.

had anticipated. With the works of Euler, d'Alembert or Lagrange, mathematical analysis made significant advances in the Age of Enlightenment,[102] which enabled it to go beyond the 'Vitruvian paradigm',[103] evident in Bélidor or Patte. Traditional technical knowledge was based upon a listing, and a consideration of similar relationships between the whole and its parts. Where Euclidean geometry, which had been so prominent in the classical treatises, argued in a generally synthetic manner, analysis was soon to manipulate thresholds and limits, to handle convergences or divergences and to address the infinite by means of the finite. As a method which characterised the continuous by means of the discontinuous, and which studied general properties locally, analysis was to prove well suited to the engineers' control plans. In this respect, the Panthéon dispute took place at a time when an approach to construction based upon geometrical reasoning and upon formulae of empirical dimensioning was all but exhausted. Although they were not always clearly conscious of it, Perronet and Soufflot's desire for innovation led them to argue for a study of the laws of matter. In areas such as hydraulics or the resistance of materials, the search was now for the boundary parameters for uniform flows or for the elasticity thresholds of solids. With their validity hypotheses, laws of behaviour were gradually replacing the simple determination of average values, which led to a stability attested by tradition.

STABILITY AND CONSTRUCTION

Although horizons had broadened by the end of the eighteenth century, there was still much uncertainty regarding the use of calculation. The debate over the nature of construction, which reflected this uncertainty, turned on the notion of stability, a notion of fundamental importance which I have barely mentioned so far. Perronet's great bridges and the church of Sainte-Geneviève, for example, were held to be somewhat deficient in this quality. The designer's approach should be inspired by a concern for stability.

In classical architectural theory, stability had two main aspects. First, it involved a correct dimensioning and a good execution of the work, although it was also concerned with the impression produced on the spectator.[104] In the Age of Enlightenment, the notion of character[105] added to the confusion, which dated back to the Renaissance and to the rediscovery of the Orders. Since the width of a column bore some relation to its resistance, stability might seem to be an extension of this principle to all architectural members. Stability and ordinance were thus indissociable.

Yet stability was also an expression of the traditional relations between architects, workmen and clients, and it therefore involved everything from the work of dimensioning undertaken by the architect, to the quality of the artisans' execution and the appreciative glance of the client. Stability therefore stood for the smooth running of a system of production based upon negotiation between the various partners involved.

A number of developments in the course of the eighteenth century served to upset this balance. First, there were, as we have seen, the engineers' attempts at rationalisation. In this respect, it is worth emphasising how ambiguous their approach to stability was. No matter how much Perronet declared his allegiance to it, his was very probably a highly personal definition, or so reactions to the pont de Neuilly would suggest. More generally, the engineers saw stability as an a priori instrument of verification rather than an unshakeable dogma. To diverge from prescribed rules constituted, as far as they were concerned, a degree of risk but was in no sense heretical.

The majority of architects disagreed. Their science of construction rested upon the identification and labelling of the objects, practices and customs of the building trade, together with the principles of dimensioning legitimised by stability. This notion thus concealed a desire to control production through a formal definition regulated by a nomenclature.

Yet the architects gradually became aware that the scientific foundations of stability were shaky, and this caused them to turn their attention to sensation, the only alternative to mathematics capable of reinvesting architectural form with some degree of technical efficacy. Two options thereby became available.

Rationalists such as Blondel, Patte and, later on, Viel, put the emphasis upon the verisimilitude of structure, which they saw as preferable to those *tours de force* which aroused the idea of danger in the soul of the spectator. It was very probably this attitude which, in the course of the second half of the eighteenth century, robbed stereotomy of some of its attraction (see Fig. 63),[106] with many architects therefore being prepared to endorse Blondel's judgement regarding its 'more singular than remarkable'[107] character, which aroused the same anxiety as did Gothic structures. This quest for verisimilitude, with its desire to combine appearances with a necessary rationalisation of the construction, was the counterpart of a wish to socialise stability, so as to provide it with a new basis.

However, other architects rejected this patching-up of classical theory and chose instead to explore the whole gamut of architectural effects, from the most surprising to the most prosaic, their aim being to arrive at a kind of combinatory of sensation as rigorous as the calculation of the engineers. The reader will probably think of Boullée and Ledoux's experiments here.[108] Revolutionary architecture thereby presented itself as an attempt to recover stability by means of sensation, which constituted an expression of the ultimate regularities of the world. Boullée's simple bodies or Ledoux's cubes and spheres were connected to natural laws determining both the spectators' emotions and the behaviour of the material. Their foundational violence was supposed to blend with genuine stability. Boullée may have distinguished the conception of architecture from its execution,[109] but the latter term was wholly contained within the former.[110] For the revolutionary architects, a design which was ordered in terms of the development of sensation could and should be built as an instrument of technical control.[111]

As we have seen, the question of calculation and/or sensation was still a long way from being settled. The Ecole Normale's programme for the Year III featured a course on the 'analysis of the understanding', which was to be taught by Garat, together with lectures by Monge, Lagrange and Laplace.[112] The *idéologues*, the heirs of Condillac, whose thought swept all before it in the early years of the Revolution, regarded sensation as the source or foundation of all knowledge. There could therefore be no question of opposing sensation and calculation. Numerous engineers were of the same opinion, and foremost among them was Prony, for whom 'the art of discussing and analysing is in no way incompatible with that of exciting the emotions'. The philosophers were not alone in advancing a programme which posited the union of the arts and the sciences. Alongside the teaching of mechanics, Prony's *Réflexions* also recommended, as we have seen, the study of 'Greek proportions'.

The balance between calculation and sensation was nevertheless a fragile one. One is also bound to wonder whether the engineers did not put these two terms on the same footing in an attempt to confer legitimacy upon their activities, as if to respond to the objections which had previously been levelled at the universalist pretensions of physico-mathematical theory. In the 1750s, Diderot in particular had criticised the abuses of Newtonian physics, which had reduced nature to the status of a machine devoid of all sensibility, in spite of the fact that the mode of existence of living beings seemed to contradict any excessively general and excessively simple mechanistic hypotheses.[113] Likewise, in the *Encyclopédie*, Venel drew an implied contrast between the intuitive art of the chemist and the dryly deductive procedures of the Newtonians.[114] The engineers' response to this set of objections may perhaps have consisted in seeking to make calculation the index of the secret life of nature which the play of forces traversing matter revealed, or that of chemical affinities. Perronet's reference to the 'structure of animals' may also be interpreted in this way. At any rate, at the end of the eighteenth century, philosophers, architects and engineers converged around the idea of transcending the opposition between mechanics and the organic world, reason and nature, through an active 'engineering', the very one in whose eyes 'the art of discussing and of analysing is in no way incompatible with that of describing or of exciting the emotions'. It remained to be seen which of the two arts would ultimately prevail. Architects and engineers held differing opinions on this latter point.

At the same time, the engineers were in fact redefining strength in terms of the current reappraisal of knowledge. Lagrange's *Mécanique analytique* was published in 1788; the geometrical stability to which Perronet in part referred was to be overlaid with another dependent mathematical formula establishing a reality in a more and more rigorous fashion. If one is to be altogether accurate, this change was spread out over several decades, and it was only from the 1820s and 1830s that the engineers were really in a position to base their work upon a knowledge of the strength of materials, owing to the researches

FIG. 63 C.-A. d'Aviler, different examples of stonecutting. Stereotomy was prominent in d'Aviler's *Cours d'architecture*, but it gradually disappeared from the architectural treatises during the second half of the eighteenth century.

of Navier, Cauchy, Lamé and Clapeyron.[115] Nevertheless, certain architects had a presentiment of this change at a fairly early stage, and the opening years of the nineteenth century saw a polemic between architects and engineers over the true nature of structural stability.

In the years around 1800, Charles-François Viel thus proved himself an indefatigable polemicist. His best-known work is undoubtedly *Décadence de l'architecture à la fin du XVIIIe siècle*. In 1805 he published a report of a similar nature entitled *De l'impuissance des mathématiques pour assurer la solidité des batimens*, to which was appended some 'Recherches sur la construction des ponts'. The argument was simple enough, and could have been borrowed from Patte, were it not for the fact that the latter had declared himself to be a fervent supporter of mathematics. Viel countered the engineers' convoluted calculations with the simple and convincing reasons of the craft of building, and he pitted the 'direct forces' of the tradition against the 'indirect and powerful forces' of the structures built by the corps of the Ponts et Chaussées.[116] The examples adduced by Viel – the portal of the Sorbonne, the Porte Saint-Denis – were just as telling, and were in striking contrast to the sometimes meaningless inventions to which the exclusive application of calculation gave rise. The same line of argument led him to prefer the Pont-Neuf to contemporary bridges, which were almost Gothic in their lightness.

Poyet was later to complain bitterly of the rejection he had suffered by the General Council of the Ponts et Chaussées of his project for a 'bridge in wood and cast iron', with a braced roadway[117] which Prony and Navier had thought unstable. In his response to the Council, written in 1821–2, Poyet wrote that 'the engineers claim the right to practice architecture and yet they are reluctant to let the architects build bridges; as if bridges did not lie within the domain of architecture'.[118]

These texts might lead one to suppose that the dialogue between architects and engineers was finally disrupted. It was, however, to be reopened, albeit with a different set of premises from those of the Age of Enlightenment.[119] If, meanwhile, the polemics of Viel and of Poyet were so vehement in tone, it was because their context was a transitional one. The debate over the nature of stability had posed a challenge to the actual status of architecture. Subsequently, the nineteenth century was to be characterised by a balance between geometrical strength and the strength of mathematics. Despite the dispute between architects and engineers, collaboration based upon a sharing-out of tasks involved in a project continued. But engineers increasingly subordinated the practice of building to calculation, form giving way to the process by which it was established and an aesthetic of form yielding, in Prony's case, to an embryonic aesthetic of method.

On the eve of the Revolution, this aesthetic borrowed, as we shall see below, from the categories of the sublime and the picturesque.[120] But, above all, it coincided with the 'specific and new' science of the engineer which Prony had foreseen. It was no accident that the *Réflexions* were actually concerned with construction, and anticipated a process of conception and realisation which was both continuous and divided into separate sequences;

road projects or the bridge model made this process possible. In matters of structural knowledge engineers of the Age of Enlightenment had reached almost exactly the same point as architects, as the liveliness of the debate between Patte, on the one hand, and Soufflot and Perronet, on the other, testifies. But the planning and execution of built works involved so great an investment that one had to conceive of new methods of rationalisation which were designed to minimise risks and costs[121] and which ruled out all leeway. The engineers' attitude, their submission to the despotic dictates of necessity, conditioned this requirement. If it is sometimes hard to identify radical shifts during the period concerned, it is because what was involved, in the last analysis, was a series of slippages. The building process was superimposed upon built form; fruitful acts of daring ousted geometrical stability and problems requiring a solution took the place of a designation of the principles and parts of architecture. Within the continuous world of classical construction there arose systems of oppositions which one would soon be able, through mathematics, to account for. As far as construction was concerned, the situation in the eighteenth century was akin to a period of waiting in which one might discern, in a sometimes paradoxical form, an intimation of transformations yet to come. This mitigated balance-sheet reflects the ambiguities of the classical world in its decline.

8

PIERRE PATTE AND THE CONCEPT OF THE RATIONAL TOWN

ARCHITECTURAL THEORY AND URBAN PROBLEMS

The town as such rarely featured in the treatises of classical architects and engineers. However, one should not conclude from this that professionals were wholly indifferent to urban problems, for in fact, alongside the vocabulary of Orders and the analysis of techniques of building, the town was invariably present. To be more precise, the town constituted a kind of horizon in architectural theory, even if this theory pretended to ignore it and to devote all of its attention to an exposition of the principles of decoration, distribution and construction.

In more concrete terms, urban reality was introduced into theory at several different levels. First of all, in the domain of decoration, the discourse on the Orders was also a discourse on the town and on its hierarchies. In classical architecture, considerable thought was given to the monumental aspect of cities, and the Age of Enlightenment persisted with this line of enquiry, albeit in a minor mode, as it developed an interest in such unprecedented phenomena as merchants' residences and tenement buildings.[1]

The town was still more evident in matters involving planning, since the irregular form of a plot of hand had an especially restricting effect upon the work of the architect. Progress in the domain of planning was inseparably linked to the desire for a rational use of urban space. Such progress presupposed the constitution of a form of knowledge regarding the town which aimed to come to terms, as far as was possible, with its accidents.[2]

Finally, there was construction, in which the town might seem to play no part, since the designing of the building and the accumulation of architectural details were on a wholly different scale. But, as in the case of decoration, so too classical construction was based upon the existence of a dimensional coordination and of technical logic which served to constitute, at the deepest level, urban space. This aspect was never so clear as when one touched upon the ways and customs of the art of building. If structural knowledge operated with the agreement of the different partners involved, from the owner to the architect and the mason, by the same token it expressed that of the building with the context in which it was inscribed. Custom therefore decreed that specifications regarding joint-ownership, for example, reflected a constant interaction between urban morphology and the design of the building.[3]

Urban issues thus permeated architectural thought. Nevertheless, classical treatises tended to make no mention of the town in its universal aspect. Whereas details of the architectural Orders or questions concerning construction were lavishly illustrated, it is symptomatic of the situation that representations of the town were scarce. In the second half of the eighteenth century, Laugier and Blondel's writings provide evidence of a theoretical study of towns that was far more specific than had been the case in the past. But the town only really became a preoccupation of architectural theorists with the publication of Patte's *Monuments érigés en France à la gloire de Louis XV* and his *Mémoires sur les objets les plus importants de l'architecture*.

THE SHORTCOMINGS OF THE CLASSICAL TOWN

The lines of argument advanced by Laugier, Blondel and Patte in fact belong to a far more general questioning of the classical town, a town which was still medieval, and the disorder and incoherence of which were universally condemned. 'Our towns', wrote the author of the *Essai sur l'architecture*, 'are still what they were, a pile of houses heaped up pell-mell, without any system, economy or plan.'[4]

A direct consequence of such disorder in towns was the lack of hygiene and of public facilities. Sébastien Mercier's *Tableau de Paris*, which was published between 1781 and 1788, depicted the capital in an apocalyptic light, dwelling upon the filthiness of the streets, the unhealthiness of the houses and the obstructions of traffic. Patte was even more forceful in his descriptions of the city and its shortcomings. If one were to take him at his word, Paris would be no more than refuse, smells from hospitals and cemeteries, infected gutters and wells, the mud and trampled ground of the helpless citizens.[5]

These apocalyptic visions were undoubtedly exaggerated.[6] Yet the tone was commonplace, shared by Laugier and Blondel who, in more moderate form, said much the same thing. At the same time, other discussions on the town were being offered. Thus, Voltaire, in his *Temple de goût* (1733), sung the praises of urban life, and of the luxury and the conveniences which it procured.[7] Guides to the capital, listing the pleasures to be found there, proliferated, among them Lesage's *Géographie parisien* and Thiery's *Voyageur à Paris*.[8] Whether one saw the town as a place of degradation and a stink of stews, or as a Babylon adorned with a thousand attractions, this opposition basically referred to the same increasing awareness of the potential of cities. The search for a harmonious development of the town was chiefly the concern of the municipal officers, the scholars and the technicians.[9] One had to rethink the overall structure of the city, modify the lay-out of the roads,[10] dig the sewers and build quays and bridges, in order that air, water, men and vehicles might circulate freely. These preoccupations also related to the collective unconscious of a period which believed in the benevolent or malevolent influence of fluids of every kind, in electricity or in Mesmer's tubs,[11] in the air and in the poisonous exhalations of cemeteries.[12] The movement of such forces had become synonymous with hygiene and with health; water itself only remained pure if it circulated freely,[13] as Parmentier was to emphasise in his *Dissertation sur la nature des eaux de la Seine* (1787).[14] The dream of unimpeded circulation was shared by all the enlightened elites of the period, as in the economic doctrine of *laissez-faire*.

LAUGIER AND THE CRITERIA OF IMPROVEMENT

In the domain of planning, the dream was far from reality. If the town was to be transformed, one would face the problem of pursuing coherent urban policies in a context which was still largely a classical one, characterised by the fragmentation and interlinking of the various authorities.[15] Innovatory experiments had certainly taken place in Rennes, Nantes, Bordeaux and Toulouse,[16] and Paris was full of new developments, from the royal mint (*hôtel des monnaies*) to the corn market (*halle au blé*).[17] There were buildings going up almost everywhere, land was being divided up into plots, avenues and quays were being created; but people hesitated before taking drastic measures and overturning the existing structure.

This tentative approach was known as improvement.[18] The improvement of a town generally involved the laying out of a royal square, the cutting of several streets or the planning of entries to the city. Improvement did not entail a thoroughgoing renewal, but affected the interstices or fringes of urban structures.[19] It might involve nothing more than the building of a monumental gate or the refurbishment of a public building.

Laugier, in his *Essai sur l'architecture* (1753), devoted a whole chapter to the improvement of towns, his main criticism being that it was too restricted: 'The taste for improvement has become general, but it is to be wished that, for the greater progress of the arts, this taste should endure and be perfected. But this taste should not be limited to the houses of private persons, it should be extended to whole towns.'[20] Patte later embraced this same idea, and argued, in his *Monumens érigés en France à la gloire de Louis XV*, for the 'wholesale improvement' of Paris.

But Laugier did not restrict himself to such general observations. Since he believed that 'the beauty and magnificence of a town chiefly depends upon three things, its points of entry, its streets and its buildings',[21] he proposed to limit the town by giving it a regular form, by planning its points of entrance, by straightening out and extending its streets, before going on in the light of this to consider the architecture of its buildings. But this would require a plan, which Laugier saw as being akin to that of a garden:

> A town may be likened to a forest. The streets of the former correspond to the roads of the latter, and should therefore be cut in very much the same way. What constitutes the chief beauty of a park is the sheer number of roads it contains, their width and their alignment, but that is not sufficient: one needs a man like Le Nôtre to draw up a plan, so that it is tastefully and thoughtfully executed, and so that it is characterised by both order and imagination, symmetry and variety.[22]

Laugier again resorted to this parallel in his *Observations sur l'architecture* (1765).[23] When the town became, with the publication of Laugier's *Essai*, an integral part of architectural theory, a number of crucial problems had then to be addressed. One had, for example, to redefine the relationship between buildings and urban spaces,[24] by means of a plan conceived on the model of a forest or a park. There has been much debate surrounding such references to garden plans which, in Manfredo Tafuri's view, heralded the advent of bourgeois naturalism.[25] Although there is much to recommend this interpretation, it may perhaps be a little hasty. For the comparison between town and garden depended above all else on the art of planning, the importance of which we have seen in the work of Blondel.

BLONDEL AND THE PLANNING OF TOWNS

It was in the chapter of the *Cours d'architecture* which dealt with planning that Blondel considered the question of the town, along with gardens and apartments,[26] and the Abbé Laugier's *Essai* likewise discussed the improvement of gardens immediately after that of towns.

Furthermore, Blondel's position was imbued with realism, so that, separating himself from Laugier's suggestions, he specified that 'one must not expect to find . . . these projects extending as far as the eye can see, as idle onlookers have from time to time proposed'.[27] Blondel merely outlined his projects in Metz and in Strasburg, explaining their essential features.[28]

Summoned to Metz in 1761 by the Duc de Choiseul in order to build the abbey of Saint-Louis, Blondel was entrusted with the task of redesigning the town centre (Fig. 64). The *Cours* therefore presents us with the 'Plan in terms of masses of the new buildings and new roads, built since 1764'.[29]

It is worth noting, first of all, that the title of this plan refers both to buildings and to roads. At Metz, Blondel combined all the means at his disposal in order to transform the town, ranging from the construction of public buildings (such as the Parlement, the Town Hall and the Episcopal Palace), the cutting of new thoroughfares and the creation of regular squares. In so doing, Blondel invoked the classical function of improvement, which was designed to express the sovereignty of the prince by taking charge of the town and moulding it in his own image.[30] In more concrete terms, it was a matter of defining a series of monumental sequences whose decoration served to add to the overall coherence of the scheme. From the portal which Blondel added to the cathedral to the Orders serving as decoration of the various buildings, the architectural vocabulary emphasised the link between improvement and spiritual and temporal power.

But Blondel's plan transcended, through its universal nature, such a framework. Indeed, the improvement of Metz was no longer dictated solely by the rules of *convenance*, for Blondel was at least as much concerned with the convenient planning of buildings within the town. Blondel made a series of observations of this sort in the *Cours*, reflecting, for example, upon the need

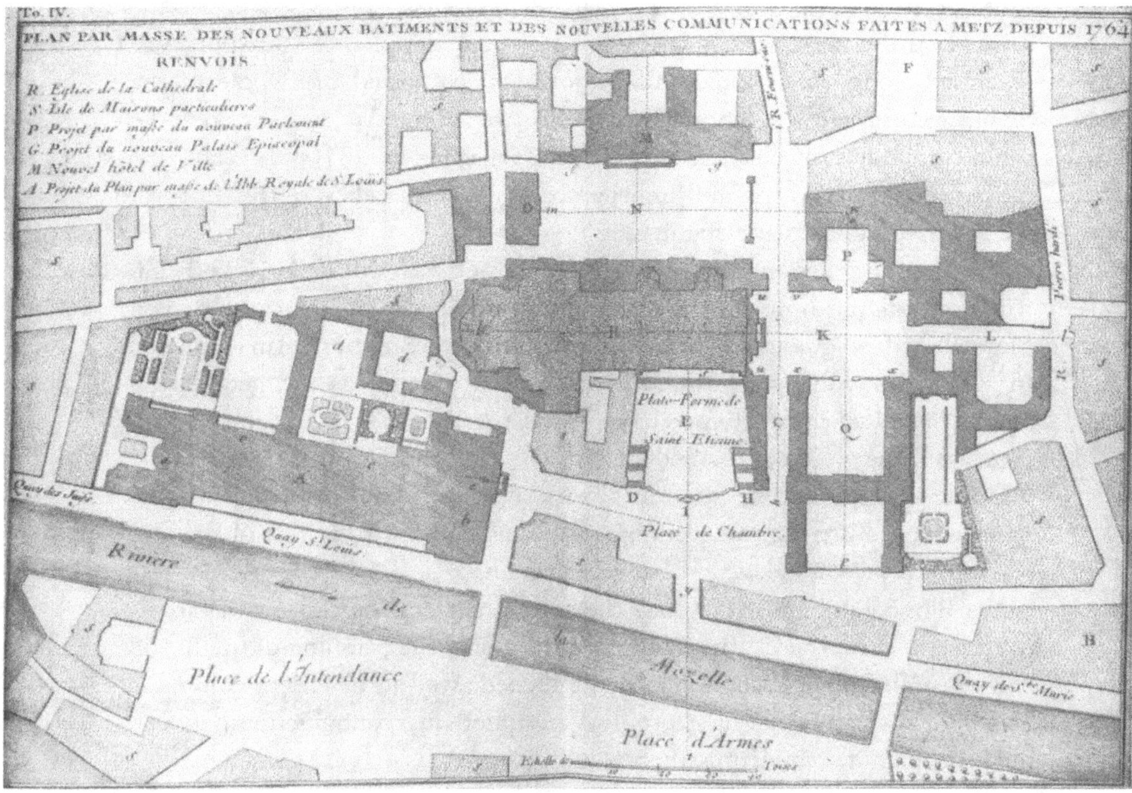

FIG. 64 J.-F. Blondel, 'Block outlines of the new buildings and new ways of building effected at Metz since 1764'.

to shift cemeteries and hospitals to the outskirts, or upon the appropriate connections between the chief public buildings.

Understood in this way, improvement raises a question of a somewhat different nature, namely, that of the town conceived as an entity to be planned as rationally as possible, that of the town as a design. For, if the town could be planned like gardens and apartments, it was precisely because it partook of the whole realm of planning. This rethinking raised a number of questions, foremost among them the continuing relevance in such a context of the traditional tools of the architect.

Although Blondel avoided this question in the *Cours*, the project he designed for Strasburg (Fig. 65), a few years after that of Metz, provides the basis for an answer. Whereas the plan for Metz was organised in a static manner, as if it were merely a sequence of sites, the plan for Strasburg was organised along the Route d'Allemagne, which crossed the whole of the town, and formed a continuous sequence including the corn market, the parade ground, a royal square and the new senate house. This plan served as the backbone around which Blondel embarked upon a radical restructuring of the town, using vistas and alignments.

The Strasburg design saw the introduction of a new vocabulary. To begin with, one should note the role played by the Route d'Allemagne, which Blondel used as a guiding thread. More generally, the Strasburg plan employed an approach involving solids and voids, curves, alignments and geometrical figures organised in a network based upon the Route d'Allemagne. It thus seems as if the architect had recourse to concepts close to those used in the planning of the country.[31] Blondel's work seems in fact to prefigure in some respects the future, for the planning of towns bore a strong resemblance to planning, as the engineers understood the term.

Yet Blondel still refused to draw the appropriate conclusions as far as architecture was concerned. When all is said and done, the planning of towns still had a marginal place in the *Cours*, being tucked away at the end of volume 4. This should clearly be seen as proof of an underlying contradiction between the principles of architectural theory and urban reality.

PATTE AND THE CONCEPT OF THE RATIONAL TOWN

Patte was prey to this contradiction too. Yet, of all architectural theorists, the author of the *Mémoires sur les objets les plus importants de l'architecture* was the one who devoted the most attention to urban problems, providing in this respect the most complete expression of the desire for rationalisation in the whole of the Age of Enlightenment.

THE IMPROVEMENT OF PARIS

The *Monuments érigés en France à la gloire de Louis XV*, which appeared in 1765, represents a first outline of the ideas developed a few years later in the *Mémoires*. The former book, published at the time of the competition for the siting and arrangement of the place Louis XV, which was to be built in Paris,[32] began with a survey of the various royal squares which had been constructed

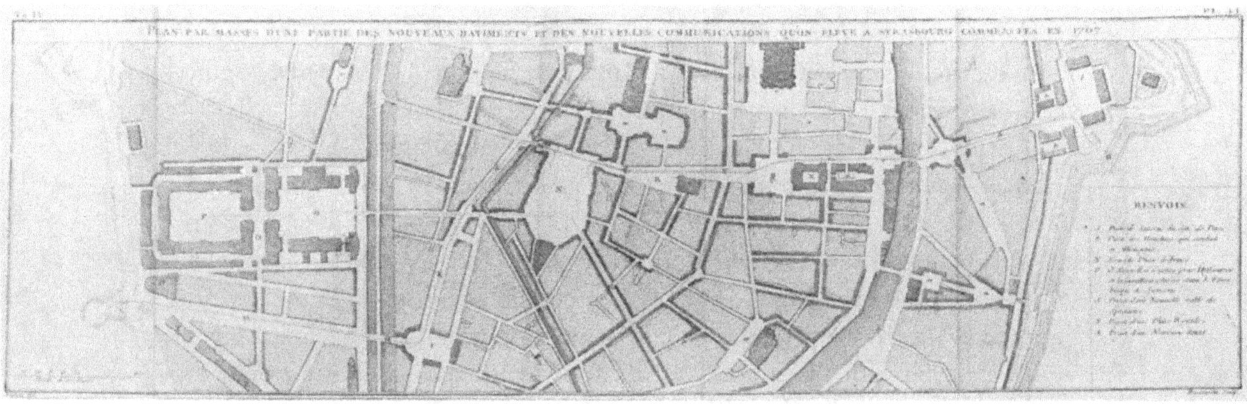

FIG. 65 J.-F. Blondel, 'Block outlines of some of the new buildings and new ways of building at Strasburg from 1767 onwards'.

in the provinces, and then went on to consider the projects of Boffrand, Servandoni, Gabriel and Soufflot. In order to coordinate the results of the competition, Patte had his famous plan of Paris engraved (Fig. 66),[33] which includes all the improvements all the architects had suggested. From the linking-up of the île de la Cité to the île Saint-Louis proposed by Soufflot to the proposals for rectangular or circular squares, this plan juxtaposed the full set of arrangements which had been proposed, with a polemical aim similar to that of the Abbé Laugier. At the same time, Patte was in fact developing the idea of an 'overall improvement' of Paris, which was designed to integrate the various features into a renewed urban tissue. Patte even envisaged a total reconstruction of Paris, which would only spare buildings of primary importance, such as the Louvre, the Tuileries, the Luxembourg, the Palais Royal, the chief public monuments and the most beautiful town-houses.[34]

With his unrelentingly practical spirit, which was never to desert him, even when he was defending the most outrageous opinions, Patte reverted to the financing of the operation. He foresaw the need for the setting-up of a 'fund of four or five millions in order to begin the improvements of Paris', a fund which would make it possible to knock down the most 'defective' quarters, such as the Cité, the quartier Saint-Martin or the faubourg Saint-Marceau, but which would then be recouped through the resale of lots marked out on the terrain freed. If one were to proceed in this fashion, with a series of expropriations and division into lots, as determined 'by the external decorations dictated by the total project', the architect reckoned that 'in forty or fifty years Paris might be in large part improved'.[35]

Without providing an exhaustive description of an approach whose basic principle is rehearsed in the *Mémoires*, it is nevertheless worth noting the new scope of this 'overall improvement' of Paris. To the universal design recommended by Laugier, Patte added a much more strongly asserted normative dimension. Naïve although his outlook in certain respects was, Patte saw the importance of the problems posed by landholding and by the financing of works of improvement. There was still, however, something fundamentally ambiguous about his plan. There was some conflict between the ideal character of the reconstruction of Paris which the architect envisaged and the actual presentation of the various designs for a royal square, which were superimposed upon the already existing urban structure. One was thus faced with something which was little better than a collage of disparate elements. Indeed, Patte's approach still closely resembled that of Blondel; the *Monuments* were moreover contemporary with the latter's projects for Metz and Strasburg. In either case, their thinking turned upon the question of the solid and the void. It was a matter of creating, within the dense weave of the old town, spaces described by means of a monumental architecture, spaces which one would enumerate and then link up, by means of fixed alignments and vistas.

Like Blondel, Patte was not yet able to imagine all the implications of this initial reflection upon the town. Like his juxtaposition of projects for a royal square, Patte's improvements of the quartier de la Cité, which is enlarged in the top right hand corner of his plan of Paris, betrays the still uncertain nature of his concept. 'Let us give free rein', wrote Patte, 'to our imagination. Let us represent to ourselves if we can, as if in a beautiful dream, the stupendous effect that improvement would have upon the quartier de la Cité.'[36] In emptying out the urban space, with its two symmetrical squares forming a casket around the new cathedral which was formed at the point of the île de la Cité, this arrangement expressed a wish to revitalise the centre. As Bruno Fortier has emphasised, doing this involved a reversal of the relation to space.[37] Whereas the classical squares represented a form of sumptuary expenditure, serving to interrupt the tissue of the town in such a way as to glorify the monarchy, the spaces opened up by Patte were designed to increase wealth by facilitating traffic and exchanges. The architect also dreamed of installing

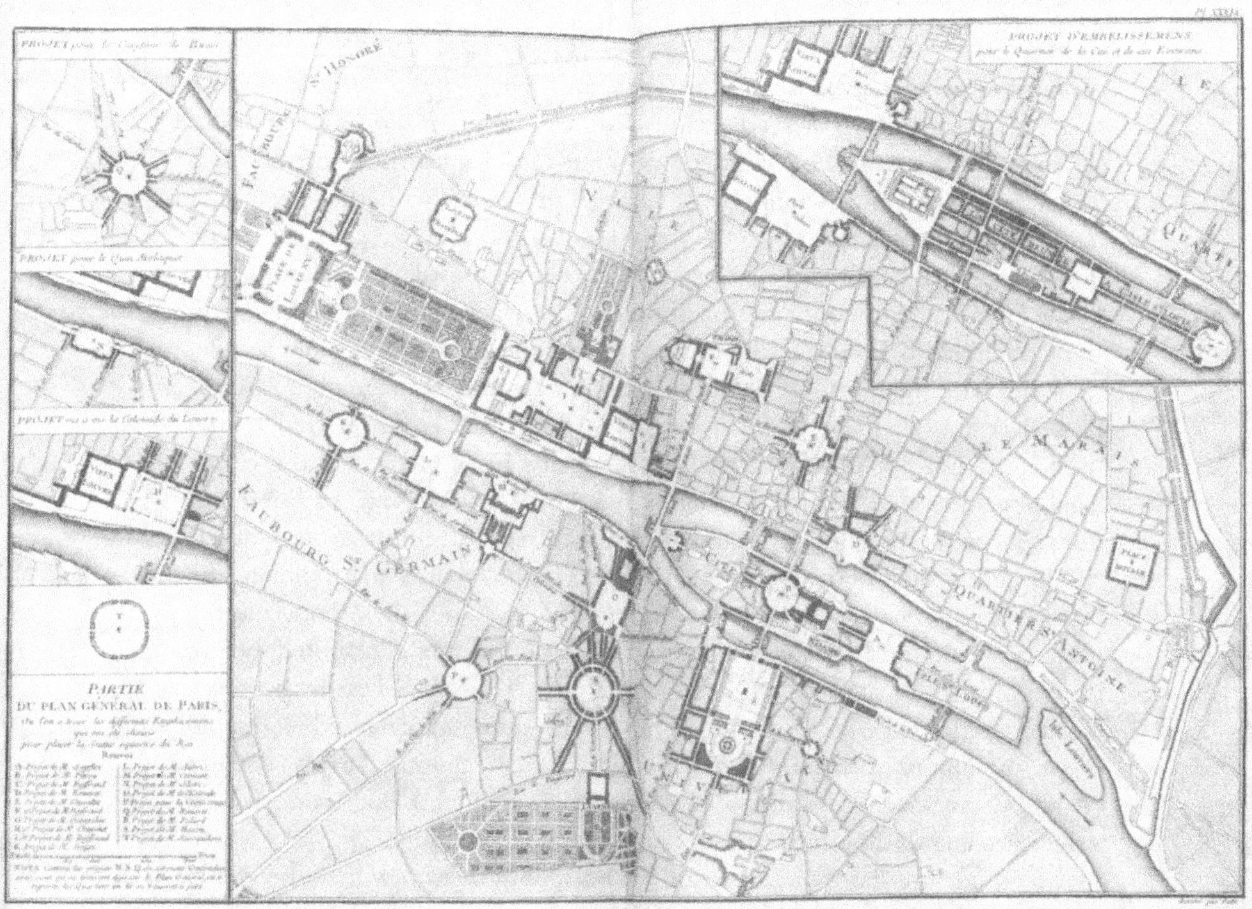

FIG.66 P. Patte, *Monuments* (1765), 'Part of a general plan for Paris, illustrating the different sites proposed for an equestrian statue of the King'. In the top right-hand corner is featured the architect's own 'project for the improvement of the quartier de la Cité and its environs'.

public granaries around the city. However, in spite of its innovatory character, this project remained an isolated one, with no effect on the existing urban structure. It thus made one dream of an immeasurably expanded royal square. This expansion plainly represented a basic change;[38] but this 'beautiful dream' of the improvement of the capital was still clearly linked to baroque town planning.

THE IDEAL TOWN AND ITS PLANNING

There was a noticeable change of tone, however, in the *Mémoires sur les objets les plus importants de l'architecture*. This time it was no longer a question of transforming Paris alone, but of defining the characteristics of the ideal town, a town which would be conceived in accord with the dictates of reason.

Reason dictated, first of all, that one choose the most favourable situation for a town. Careful consideration should be given to both site and climate. One should also bear in mind the wealth of the surrounding countryside, together with the means of communication likely to encourage trade.[39]

Where the planning of the town was concerned, Patte referred to the principles already outlined in the Abbé Laugier's *Essai*. He aimed to give the agglomeration a regular form, roughly hexagonal or octagonal, to minimise, for reasons of policing, the distance from one extremity to the other. Boulevards would be laid around the town, with activities that polluted the environment, such as tanneries, laundries or butchers' shops being cast outside. A canal would form a second ring around the suburbs thus constituted, before flowing into the river which crossed the city. The entrances to the town would be adorned with triumphal arches and monumental gates, which would serve to impress spectators and announce a great nation. Regular squares would be associated with these entrances, each of them being transected by streets leading to the centre and to the chief public buildings.

In the picture Patte presented of a wholly improved town, he revived a number of the characteristics of the baroque town. The perspectives, for example, were to culminate in 'interesting features such as fountains, obelisks, standing or equestrian statues'.[40] However, in contrast to baroque town planning, the geometrical lay-out of the plan was ultimately of relatively little importance: 'the crucial thing . . . is that all of its approaches are easy, and that there are sufficient connections from one quarter to another for the transportation of merchandise, and the free circulation of vehicles'.[41] Thus, the town lost that quality of organic wholeness which improvement aimed to confer. Patte's plan would simply be convenient; one merely had to ensure that traffic flowed freely and that 'monotony and an excessive uniformity in the total arrangement'[42] be avoided. With regard to this latter point, Patte was less adventurous even than Laugier, who had gone so far as to demand 'oddity'; in the *Mémoires*, however, the town could not really be compared to a garden, with its large regulating outlines and its lively details. Rather, to the network of its streets was appended a set of technological requirements and regulations which conditioned its functioning.

TECHNICAL REQUIREMENTS AND REGULATIONS

It is in the domain of technology that one may take the measure of Patte's ingenuity. The *Mémoires* are full of every kind of description, observation and suggestion. His conception of the town resembles nothing so much as a catalogue of urban technology.

This catalogue consisted, first of all, of building regulations, which were dictated by a concern for hygiene and for convenience. Patte advised against building houses too high, because it gave them too little sun, and he therefore favoured buildings of around three storeys. He also advised against building houses on bridges, since they prevented the free circulation of air and prevented passers-by from deriving their full enjoyment from the river.[43] In order to combat the risk of fire, he recommended the use of brick, and the replacement of wooden-framed roofs with flat roofs concealed behind balustrades.[44]

The *Mémoires* also listed a number of measures designed to improve streets and highways.[45] Patte thus proposed rounding off the corners of buildings at crossroads so as to improve 'the movement of traffic'. The highway would be divided into three parts, namely, a central raised strip for vehicles, with two sidewalks on either side, separated from the vehicular area by bollards at ten foot intervals (Fig. 67). Patte was concerned also with techniques of paving, and throwing up of mud during rainy weather, for which he suggested a system of retractable blinds.

Patte's deepest concern, however, was in the management of water, a current preoccupation. Paris and the other major towns lacked water. The Academician Deparcieux read, in 1762, his *Mémoire sur la possibilité d'amener à Paris . . . mille à douze cents pouces d'eau, belle et de bonne qualité*, which was designed to prove that it was possible to tap the waters of the Yvette for the benefit of the capital.[46] There was soon to be a fierce debate between the supporters of a system of this type and those who favoured a supply based upon the use of pumps.[47] Patte himself envisaged, without due thought, a series of reservoirs to collect rain water for public fountains.[48] But Patte also proposed something well in advance of its time, namely, a genuine sewage system. Being much concerned with the question of sewers, he advocated a systematic building policy, such as would only really be initiated in the nineteenth century.[49]

For the 'ordinary passage of water', Patte studied the qualities and defects of channels made of wood, lead or cast iron, before opting for varnished tile, which was less expensive and so more easily replaced. Upon this occasion, he considered the possibility of combining pipes for the distribution and for the disposal of water, illustrating his proposal with one of the very first cross-sections of a street (Fig. 68), a theme which would become popular subsequently. His cross-section includes the houses on either side of the street in the centre of which there is a fountain in the form of an obelisk, then the underground sewer also containing the water-mains set on wall-brackets.[50]

The chief interest of this engraving consists in the close connection it posits between the buildings and those systems which assure their efficient functioning. The houses, the highway, the sewer and the various channels are almost machinelike. The houses also offer arrangements designed to prevent fires, with a flat roof on the right and a vault on the other side of the street. Finally, suspended above the left bay of the road, there is a cover protecting passers-by from the rain. In his 'section of a street', Patte summarised his basic ideas; nothing was missing, not even bollards separating the vehicles from the pedestrians.

It is worth emphasising the relationship between this plate in the *Mémoires* and contemporary drawings by engineers. In either case, technological objects were represented entire, including the hidden parts. Patte's cross-section was also designed to illustrate a model for the distribution of water; it thereby evoked the principle of the bridge, an immobile form of architecture below which the river flowed. Just like a productive work, the street became a machine regulating a set of flows: the flowing of the water and the traffic of vehicles.

Patte's ideas and those of the engineers were linked in other ways too. After the *Mémoires sur les objets les plus importants de l'architecture* had expounded the principles of town planning, it treated the Orders, and then the erection of buildings. Under this latter heading, Patte ranged, as we have seen, the great Parisian sites such as the Louvre or Sainte-Geneviève. But he was also interested in the innovations introduced in the course of building the pont de Saumur, noting that he derived his information on this from Cessart and Perronet.[51]

Like the engineers, Patte accorded a dominant place to structure, even going so far as to subordinate decoration – whose rules were weakened as a consequence – to it. In the *Mémoires*, the 'Dissertation sur les proportions générales des ordres d'architecture' refers to the primitive hut with as much

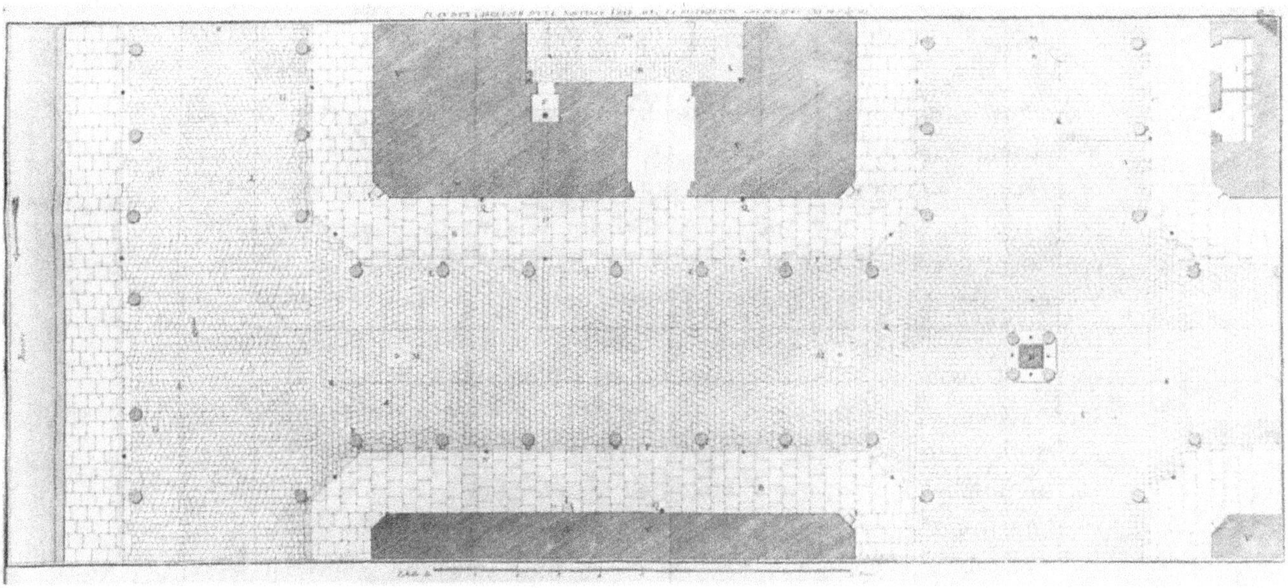

FIGS.67, 68 P. Patte, *Mémoires*, plan and section of a street.

insistence as the Abbé Laugier's *Essai* had done.[52] Since Patte was motivated by a strong desire for rationalisation, he proved later to be more of a supporter of Perrault, whether it was a question of recognising the diversity of the proportions given by the Ancients to their monuments or of emphasising the fundamental role of use in architecture. As a contemporary of Jacques-François Blondel, Patte could not, however, radically question tried and tested recipes.[53] Yet this prudence did not prevent him from recommending a degree of standardisation. There were therefore, according to Patte, only three Orders, Doric, Ionic and Corinthian, and their extremely simplified measurements were reminiscent of those given in Perrault's *Ordonnance*.

This simplification related to building and to the fine execution of details by workmen which constituted one of the major obsessions of the architects of the classical age. Through his evocation of the primitive hut, with which his essay on the Orders had begun, Patte also sought to base decoration on rules which were sure, and as 'positive' as the art of building itself. It was also a question of redefining the relation between building and the town, the rationality which was present in the planning of the streets now being reflected even in the façades of buildings. The attempt was now being made, by way of construction, to formulate a universal and coherent approach to urban problems.

A THEATRE OF MACHINES

One might not agree entirely with Georges Teyssot's claim that Patte was a theoretician who proposed 'a town plan which functioned just like the machines he had designed'.[54] In spite of the machinelike character of the street, or of this or that technical appliance, Patte was not ultimately able to apprehend urban reality except in terms of the double register of very general principles of organisation and of fragmentary applications, applications whose occasionally incongruous character calls to mind the scenes of a writer such as Raymond Roussel, with their sequence of absurd situations. Thus, the *Mémoires* frequently lapsed, somewhat surprisingly, into anecdote.

On the matter of cemeteries,[55] for example, Patte mixed his genres, invoking the practice of the Ancients, that of Oriental peoples, canon law and the history of Western Europe, in order to prove that burial within the town precinct was not a religious obligation, rather a bad practice developed through time.[56] Then, changing key, he tried to imagine the consequences of such a measure upon funerals, giving a forceful and detailed description of the itinerary of the dead person, who was first of all transferred from his place of residence to a chapel of his parish church, in which 'there would be an easily raised wooden tomb, under which one would cut a burial chamber whose vault would be pierced by a pipe, in order to carry the cadaverous exhalations above the roof'.[57] The funeral service would be held as usual in this chapel, then, 'after having satisfied the divinity, one would satisfy public health', by taking the corpse through a trap-door communicating with the burial chamber beneath the chapel. Finally, 'at a fixed hour, such as two o'clock in the morning, a waggon drawn by two horses covered in a funeral cloth, would arrive from the parish cemetery to take the bodies from the burial chamber; it would be accompanied by gravediggers, each of whom would carry a lamp, and by a trustworthy priest'.[58] Patte then gave an equally circumstantial description of the cemeteries which he proposed to build on the outskirts of towns. He planned, in imitation of the Campo Santo in Pisa, enclosures with high walls, which would serve to lead the smell from the corpses upward and prevent infection. In order to economise on space, the tombs would be placed one above the other; the centre of the cemetery would contain a chapel, in which Mass would be said every day.

Anecdotal though Patte's presentation was, it nevertheless marks the emergence of a new dimension to town planning. Disposing of dirty water or corpses was seen in much the same light by Patte, but he still felt it right to keep up appearances, hence the artifice of the wooden tomb. The corpse of the dead person would leave by a hidden door. Patte was already placing technology in an illusionist perspective designed to make it acceptable.

Similar observations are scattered throughout the *Mémoires*, so as to render the general economy of the town more precise. One should note, in particular, the emphasis placed upon security and surveillance. Patte thus envisaged the creation of 'little known public buildings', which would be designed for the safekeeping of 'the fortunes of citizens, together with the deeds which record their state and condition'.[59] At another level, he recommended that one avoid the building of too many porticoes and galleries, in whose shadow villains might lurk.[60] The rational town was to be extensively policed.

The theatrical and droll character of these observations is very probably the peculiar characteristic of inventors who were by turns theoreticians and *bricoleurs*. Patte was forever inventing lighting systems,[61] giving a staging to his production, as in the case of burial. Yet his anecdotal style also resulted in a degree of fragmentation; the process of invention did not produce a unified system.[62] Patte saw the town as a catalogue of juxtaposed technologies akin to

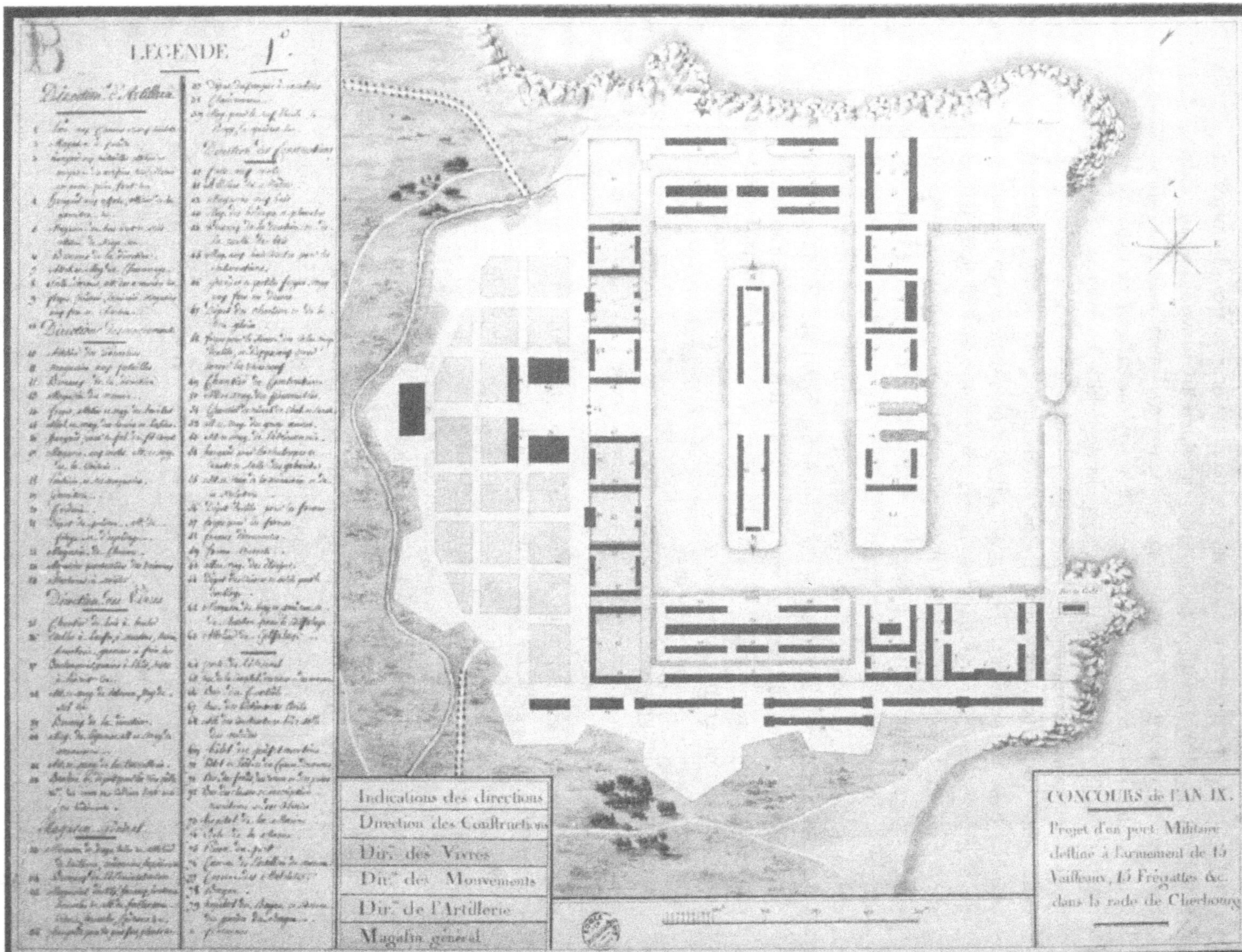

FIG.69 N. Le Payen, project for a military port in the bay of Cherbourg, for the competition of Year IX, first prize, ENPC, ink, wash and watercolour; 45 × 57 cm. The machinelike geometry of the design results from the analysis of the functions of the port listed on the left.

the parts of the building in Blondel's *Cours d'architecture*. Like the latter, Patte took his bearings from a classical knowledge, which proceeded by means of tableaux and listings. In his case, however, technical regulations and machines replaced the details of ordinance and of planning. The *Mémoires* also bore a resemblance to treatises written by contemporary engineers. Patte's writing reminds one of Bélidor, of his *Science des ingénieurs* and of his *Architecture hydraulique*. The work of this engineer kept fairly close to the 'Vitruvian paradigm', as the use of a module in the discussion of the dimensioning of the lock-gates proves;[63] technology, as Bélidor and Patte conceived it, was organised in terms of a systematic exposition of a diversity which was difficult to recombine. The town was not so much a machine as a collection or theatre of machines which made the stories possible.

The town's lack of unity might now be read at another level, in terms of the fact that no overall representation of it existed. In the *Mémoires*, the town continued to be as visually anonymous as it had been in the treatises of the classical age, even if the plan and cross-section of the streets featured there. We know merely that its form was regular, whereas its planning involved an alternation of grand perspectives and picturesque views. There is a long way to go before we get to the design of those arsenals (Fig. 69) which the engineers would later see as model manufactures, machines whose importance in the development of urban theory has been demonstrated by Bruno Fortier.[64] If, to quote his formulation, one moves very quickly in this domain 'from an urban policy of signs to a spatialised administration of things'[65] – one of the most striking manifestations of the latter being the various engineers' projects for Cherbourg – Patte had indeed restricted himself to a policy of signs, even if these signs already belonged in essence to the world of technology.

Nevertheless, Patte's conception of the town as a system of movement and flow is still premonitory. If I may look ahead, the author of the *Mémoires* sometimes reminds one of Hénard. Both in the *Etudes sur les transformations de Paris*, and in the work of the Enlightenment architect, the town is littered with a host of technical appliances, the street itself becoming one of these appliances,[66] perhaps the most complex. The immobile signs installed by technology in fact serve to circumscribe movement, to suggest possibilities of regulation, and to bear witness to an always possible future of the urban tissue.

Our comparison between Patte and Hénard makes it easier to appreciate what divides the 'serviced town'[67] of early nineteenth-century engineers from the one described by the author of the *Mémoires*. Patte's Utopia is not a mechanical town working with perfect precision; like Hénard, he wanted a town that was endlessly adapted and readapted to its conditions of use by means of construction. Everything was constructed, perfected and modified, from monuments to drainage systems. Between a mechanical perfection of an ensemble and the prevailing urban disorder, the *Mémoires* sought to find a middle way, one which was based upon reconstruction on the site.

As the last observation implies, Patte was always keen to indicate how such reconstruction might be effected. Complementing the list of elements which had been presented in the *Monumens érigés en France à la gloire de Louis XV*, the *Mémoires* lavished advice in this respect. Any activity that involved unwholesome or polluting elements, such as cemeteries or hospitals, should be confined to the suburbs, before one set about realising a plan for the whole of the town involving methods akin to those specified in the 'overall improvement' of Paris. Patte's attention to detail was so scrupulous that he reproduced *in toto* the regulation issued in 1763 by Catherine the Great of Russia for the improvement of St Petersburg, a regulation adapted according to him to the composition of ideas preliminary to the plan to be put into effect.[68] As in the case of the *Monumens*, the financing of these alterations was to be achieved by the resale of expropriated plots of land and by the agricultural surplus-value thus realised.

Patte's reformism caused him, however, to neglect a crucial aspect of the question. Like Blondel, the author of the *Mémoires* was committed to the principles of classicism, and to the ideal of *convenance*, and he therefore refused to contemplate a radical overturning of the relationship between architecture and urban space. Construction might help to simplify decoration, but, as far as Patte was concerned, it was not the instrument to effect revolution. A few years later, the revolution was to take a wholly different path. Assuming an embryonic form in Ledoux, the typological approach was to triumph in the work of Durand, who sought to turn architecture into a precise combinatory. The relations between the building and the town then came to involve the subjection of both to rules for the control and management of space. Large-scale lay-outs and typology of services would both serve to bring about a complete reorganisation of urban functions.[69]

THE UNPRODUCTIVE TOWN

One of the basic weaknesses of the doctrine developed by Patte can now be seen, namely, the indifference displayed in the *Mémoires* towards those effects of contrast so beloved of the Abbé Laugier. Where the author of the *Essai* advocated oddities, Patte was content to invoke the classical principle of variety. There was no real disorder in the town which the architect described, but rather a hierarchical order which might readily be compared with the one prevailing in those gardens which Blondel designed at the same period.

This was because Patte's concept of the town was not as yet natural in the sense in which the engineers, for example, understood the term when speaking of the country. Being composed, even the details of the town were seen as artificial, although the subsequent discovery of the need to distribute such universal benefits as ventilation and lighting brought it closer to nature.

The artificiality of the town also found expression in its unproductive nature. The physiocrats, who enjoyed real political influence, regarded agricultural land and mineral resources as the sole source of wealth, so that, as far as they were concerned, urban space was simply a place for exchange and for consumption. The town described in the *Mémoires* was not exempt from this rule; its planning was exclusively in terms of traffic and the requirements of production were wholly neglected.

This restriction placed on the role of the town goes a long way to explain the priority accorded by the administration to the planning of the country. The Age of Enlightenment was content to improve the existing urban structure in relation to the exchange which it might allow. In its dream of a town that was wholly renewed, in spite of calculations which tended to prove that this renovation could be effected by immobilising a fairly weak capital and by depending upon the mechanisms of agricultural surplus-value, Patte's thought may be said to have been Utopian in its underestimation of the importance of technological constraints and in its failure to have produced a wholly convincing economic analysis. At the end of the eighteenth century, urban development clashed with the unwieldiness of the classical town. In any case, it was only much later that agricultural surplus-value was to constitute an effective tool on a large scale. Nevertheless, Patte clearly had some presentiment of the importance of networks in the constitution of this surplus-value, although this intuition was always on a theoretical level.

MACHINE-LIKE TOWN, FOREST-LIKE TOWN AND LIMITLESS TOWN

If one considers the real problems of towns, the engineers' proposal to organise their arsenals as machines seems just as theoretical as Patte's schemes, since their plans for a rigorous geometry ruled out any notion of change over time. Yet their emphasis upon the importance of exchange in fact necessitated some reference to such change; being ultimately destined to develop, the circulation of men and of commodities occasioned a break with the limited town of the classical age.[70] A notion of progress emerged in the course of the eighteenth century which was applicable as much to human knowledge as to economic wealth. Being crossroads, turn-tables and sites for the accumulation of such wealth, the towns could not avoid this growth, even if it remained slow. What could be more telling in this respect than the endlessly reiterated motif of urban expansion?

The development of towns was subsequently to oscillate between geometrical rationality and topographical realism, between the wish to plan and an acceptance of the constraints of the site and of the 'natural' laws governing the growth of agglomerations. Ledoux, in his *Architecture* (1804), was already employing these two registers, with his ideal town of Chaux being based around the saltworks of Arc-et-Senans, 'a machine-for-producing' as Anthony Vidler has called it,[71] whereas the residential quarters were scattered over an area which was apparently less structured but which, in reality, was controlled by the two axes of traffic and services. The interlocking of these two kinds of organisation demonstrates how necessary it was to have new tools for structuring space. With a clarity which had been lacking in both the designs of Blondel and the work of Patte, Ledoux also raised the problem of production and of its control, although, it must be admitted, he enclosed it within a narrowly architectural definition which served to restrict the scope of his proposals.[72]

The choice was thus between a regular town or an irregular agglomeration; in a sense, it was still the Abbé Laugier who had shown himself to be the most realistic in insisting upon both the grand lay-outs of Versailles and the disorder of the forest. The use of the term 'oddity' might well seem to anticipate a kind of urban *laissez-faire* very different from the designs of someone like Le Nôtre; the criticisms which the author of the *Essai* made of the park of Versailles are evidence of this.[73] Admittedly, an overall plan was required, but should it be a traditional composition? The interaction between order and disorder, the emphasis upon the effects of contrast seem to indicate a town whose natural qualities are alien to classical rule.

Nevertheless, it is hard to see the *Essai* as a truly revolutionary document. Laugier had probably merely glimpsed the undefined potential of urban development. The town like a park or forest is not yet that of the engineers of the following century, but its ambiguities may well have paved the way for it.

With the advent, at the end of the eighteenth century, of axes of constraint and a domain of freedoms, the town had become an essentially contradictory object. Inasmuch as it was conceived as an overall design, it could be said to be a machine-town, and yet the existence of a sequence of apparently unconnected structures made it seem a forest-town also. The town had therefore split, appearing to be both a general design and a territory, in the sense that this territory demanded a specific approach, different from that of architecture.

These two conflicting models were nevertheless closely linked. To have grasped this contradiction, and the complementarity which it masked, would have meant acknowledging that architecture was not equipped to tackle the new object which it was now seeking to make its own. Being a disciple of Blondel, Patte was bound to reject the alternative which was offered to him, and to take refuge in technical minutiae which would preserve the fundamental principles of classicism for the time being. The *Mémoires* therefore present us with a town whose separate parts are listed, but this town is never reconstituted, since Patte was unable to name the product of this recomposition. The service-town of the engineers, on the other hand, would be defined in terms of the coexistence and confrontation of the previous models. Its possibilities would be explored through the rationalisation of the arsenals, but also through a systematic description of the prevailing disorder. It would refer both to the factory, the garden and the forest, each of these figures referring to specific rules of organisation.

THE ENGINEER AND THE TOWN

It would be meaningless to contrast the concepts of the architects and those of the engineers at the end of the eighteenth century, in so far as the two professions adopted virtually identical positions in relation to the town. We have already had occasion to emphasise the kinship between, for example, Patte and Bélidor. For the engineers, the town was also a site for traffic and for exchange, which had to be improved as much as was possible.

A number of fine distinctions should, nevertheless, be drawn. The first of these nuances concerns the minimal place which engineers' treatises allocated to urban considerations. Still loyal to the classical model, the engineer generally said nothing about improvement.

In practice, however, engineers were a real presence. Saget, who was engineer for the Estates of Languedoc, developed Toulouse,[74] creating

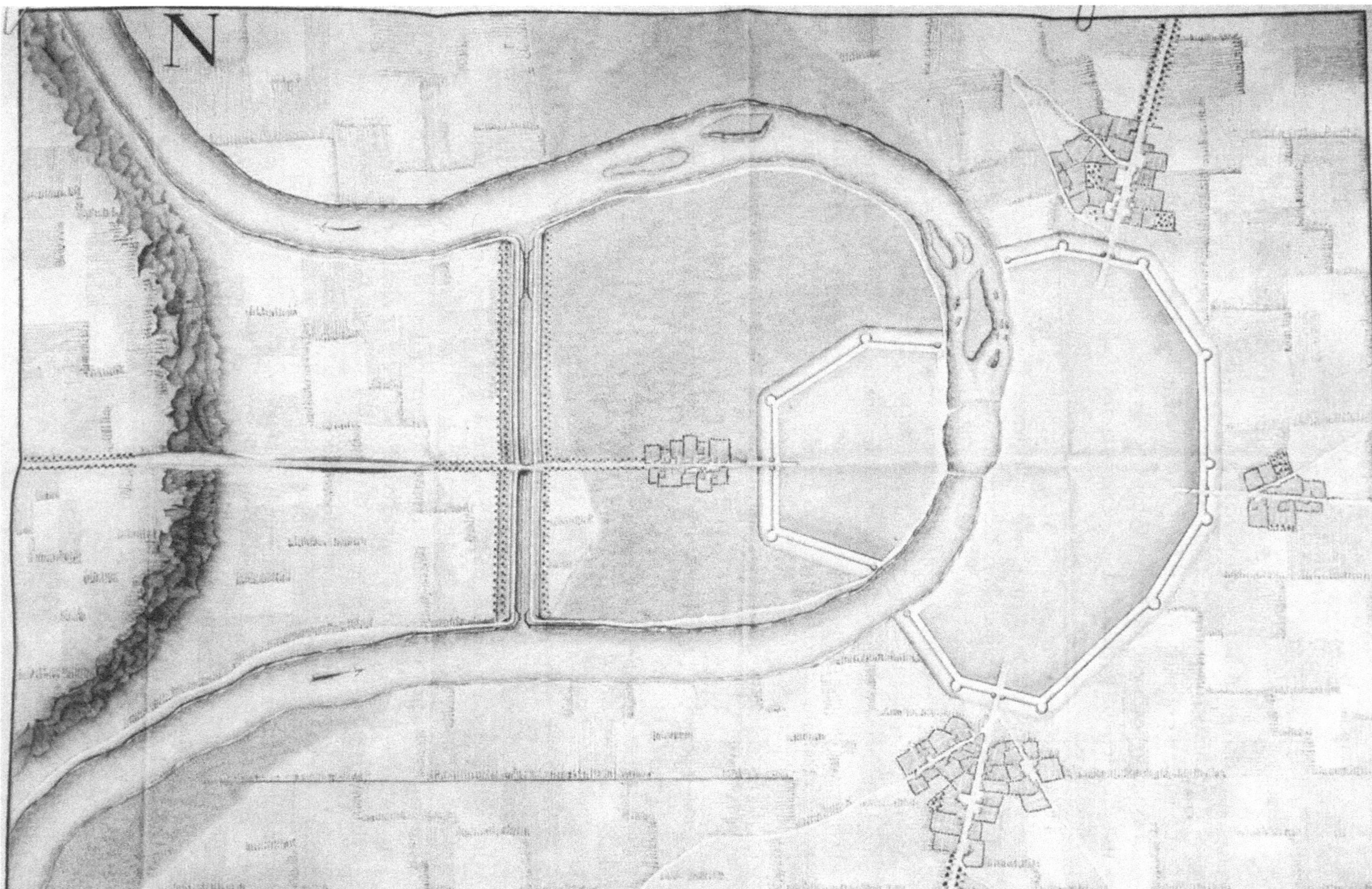

FIG. 70 Simon, project for a headrace for the 1756 competition, site plan, ENPC, ink, wash, watercolour; 64 × 103 cm.

regular squares and quays. Legendre, who was Inspector General of Bridges and Highways, drew up the project for the Royal square of Reims, of a type that was still very close to Blondel's teachings.[75] Perronet began his career as an employee of the town of Paris prior to entering the corps of the Ponts et Chaussées. In 1778 he drew up a project for the extension of the town of Nantes;[76] subsequently he was to be entrusted with the pont Louis XV, the project designed to complete the most prestigious monumental ensemble in the capital.

Nevertheless, engineers did not regard town planning as the crucial aspect of their work. This concerned especially the land, which was to be organised and run in terms of the dictates of reason.[77] Since agriculture and mines rendered the land productive, it was, as far as the engineers were concerned, in sharp contrast to the town. The pure geometry of the great road lay-outs was in strong contrast to the dense network of streets and alleys. Since the town had a complexity which resisted all attempts at far-reaching rationalisation, it gave the appearance of being to some extent the negative of the territory, and thus inspired the engineers with a degree of mistrust. A competition at the Ecole des Ponts et Chaussées in 1756 reflected this mistrust.[78] Pupils were asked to plan a road and a canal crossing each other at right angles, close to a town traversed by a river (Fig. 70). On the site plan, the town was represented as an area of clear colour with irregular contours, an indistinct mass beside which the engineer disposed the elements of territorial development, namely, road, canal and lock. The legibility of the structures situated in the open country was contrasted to the compactness of the urban space. Like a bridge design, the lock was represented at the different stages of its realisation, through a diagram combining visual order and systematic description of the tasks involved in its construction.

This competition also indicates the sort of approach to the town being adopted by the engineers. Their aim was initially to link up town and country. Thus, the crossing of the road and the canal in the competition for 1756 was related to the old town, so as to link it to the country. The town planning of the Enlightenment engineers anticipated a time when urban space and territorial reality would be tackled by means of the same range of technologies. Nothing was of more significance here than the numerous bridge-building programmes at town exits, whether at Saumur, Moulins, or even in the case of the pont Louis XV in Paris. Starting with bridges,[79] or with other boundaries, such as quays or the final stretch of the major road networks, it would soon become possible to begin to restructure the urban fabric.

This restructuring, which had been anticipated by Blondel in his project for Strasburg – with the transformation of the route d'Allemagne into a monumental sequence – assumed a somewhat paradoxical form in the early nineteenth century, since two conflicting models coexisted, that of the machine-like town and that of the forest-like town. The town as the engineers conceived it was, however, unified by a similar concern with control. Like the territory, the town had to be explored and described, so that no lacuna remained. An exhaustive sequence of surveys and programmes was intended to reduce town and country to a single quantifiable space, a conflation that had already been achieved in an embryonic form in the *Ancien Régime* attempt at a medical topography, which had related mortality rates to regional areas,[80] or again in the enquiries into the operation of various infrastructures, beginning with the hospitals.[81] Houses were soon to be numbered, and notions of hygiene and the economy of a habitat emerged, these too being based upon the application of surveys and of numbering.[82] This was the price to be paid for moving from the immobile, limited town to the limitless town. What finally undermined Patte's approach were statistics and his lack of a statistical apparatus.

9

A PRODUCTIVE COUNTRYSIDE

The countryside, which was a fundamental point of reference for the engineer, was a complex entity functioning at several different levels. Since what was real and what was imaginary frequently intersected in the country, I shall be concerned in the present chapter with such points of intersection, which may even be discerned in the most concrete of applications. To be more precise, I shall begin by studying the country in terms of cartography, before going on to analyse its natural form, which caused the engineer to treat it as a sort of garden. It is but a short step from the land to the design, for it was from the land that the design derived its necessity and its efficacy. Yet this necessity contained the first intimations of a reversal of classical aesthetics, through an expression which was by turns sublime and picturesque (and which I shall try to investigate more deeply below). One then has to tackle the productive aspect of the country, perhaps its most essential quality, and, finally, the divisions and recompositions effected under the aegis of the state, which was attempting to rationalise the management of space. From the cartography of the Age of Enlightenment, and from the productive 'garden', we move on to the quantification of experimental data, and to the statistical mastery of the country, which paved the way for the nineteenth century's 'conquest of space'.

THE MAPPING PROGRAMMES OF THE ECOLE DES PONTS ET CHAUSSÉES

The engineer's most important tool was the map. Since maps enabled one to 'reduce a considerable space to a very restricted space',[1] they proved wholly indispensable to the definition of coherent planning policies. Allowing an overall perspective, they also constituted the nearest thing there was to a theoretical approach to the problems raised by the land. The most explicit reflections of this sort were to be found in cartography manuals. Whereas the treatises of someone like Bélidor or the works of Patte regarded the land as a given, Buchotte's *Règles du dessein et du lavis* or Dupain de Montesson's *L'art de lever les plans* advanced a far more elaborate analysis of territorial reality.

This analysis seems to have been the outcome of a general shift in emphasis, for cartography underwent an unprecedented development during the Age of Enlightenment.[2] Quite independently of ventures linked to 'the' map of France, hydrographic maps and plans of towns and of fortifications proliferated. Engineers did not enjoy a monopoly, but their works were characterised by an unrelenting concern for accuracy. Learning to draw maps was in fact learning the rudiments of the engineer's profession, for no one could actually make projections without a plan as a basis or without reference to the topography. But practising cartography involved learning the landscape, its overall logic, along with the most fortuitous detail.

By means of an apprenticeship of this kind, cartography was related to theory. It enabled one to link theory and practice more precisely, fulfilling one of the most crucial conditions of training, as engineers understood the term. Given such circumstances, it was not surprising to find that the Ecole des Ponts et Chaussées had originally organised its instruction around the drawing of maps. We have seen how Perronet's pupils drew up plans of roads which were designed to be built subsequently in the regions. Up until the 1750s or 1760s, the drawing of maps continued to be indissolubly linked to problems of road construction, because of a concern to locate the immediate surroundings of the project.

Once the road map of the Ponts et Chaussées was completed, however, cartography began to break free of that context. Thus, Turgot's *Instruction* of 1775 set up an annual map-drawing competition, a competition which soon resulted in displays of virtuosity.[3] Plans of real situations were no longer represented, but ideal territories were imagined instead, and these allowed one to imagine the full range of possible combinations. There was a shift from territorial reality to a more symbolic register, in which *trompe l'œil* devices or puns were not overlooked. The subject of the exercise was still a map, but it was a map upon which other maps, or a set of documents mentioning the word 'map' or evoking this magically resonant term, were superimposed. Thus, playing cards,[4] instruments for surveying a plan,[5] and even those

FIG.71 'Ichal map with a blue knot', a map for the 1784 competition, ENPC, ink, wash, watercolour, gouache highlights; 48.5 × 65.5 cm. Geometrical instruments frame a medallion representing Perronet in profile.

butterflies known at the time as 'geographical maps' because of the markings of their wings,[6] might feature (Figs. 71 and 72).

Although often described as Utopian, the map competitions of the late eighteenth century provided an opportunity for listing the techniques of engineering. Nothing was missing, from roads and canals to fortifications and rational extensions of the old towns. Maps were used in such a way as to parade all possibilities, from *trompe l'œil*, to the superimposition of planes, the use of colour or that of simple black and white drawings. Such play with language and representation betrayed cartography's real ambition, that of evoking a world whose diversity might readily be described in its essential

FIG.72 'Trompe l'œil with butterflies', map for the 1793 competition, ENPC, ink, wash, watercolour; 56.5 × 43.5 cm.

kinship with the artifices of the drawing. Much like the *Encyclopédie*, a project which was then in its final stages, the engineers conceived of the map as an image of a whole. Even if an infinity of plates were required to provide an exhaustive account of the world, cartographic depiction served to guarantee that an indissoluble link remained between words, drawings and things. Through the fantasies of the Utopian maps, it was the word 'map' that was represented, and this word signified the encounter between reality and its double, its image serving as a support for the operations of the engineer (see Figs. 73 and 74).

The map was thus part of a descriptive strategy akin to that involved, as we have seen, in the project.[7] If it proceeded by way of the imaginary, this had been merely to allow the various planning techniques to be listed. It may be

FIG.73 'Trompe l'œil with the Knave of Spades and the King of Diamonds', map for the 1792 competition, ENPC, ink, wash, watercolour; 32 × 44.5 cm. A sheet from the Cassinis' map, some playing cards, an invitation card (*carton*), serve to show the many meanings of the word *carte*. Wholly in line with Rousseau's reflections on musical notation, the score in the upper right hand corner recalls the arbitrariness of systems of signs. The map assumes a position between nature and writing, between the bird's eye view and invariably artificial conventions.

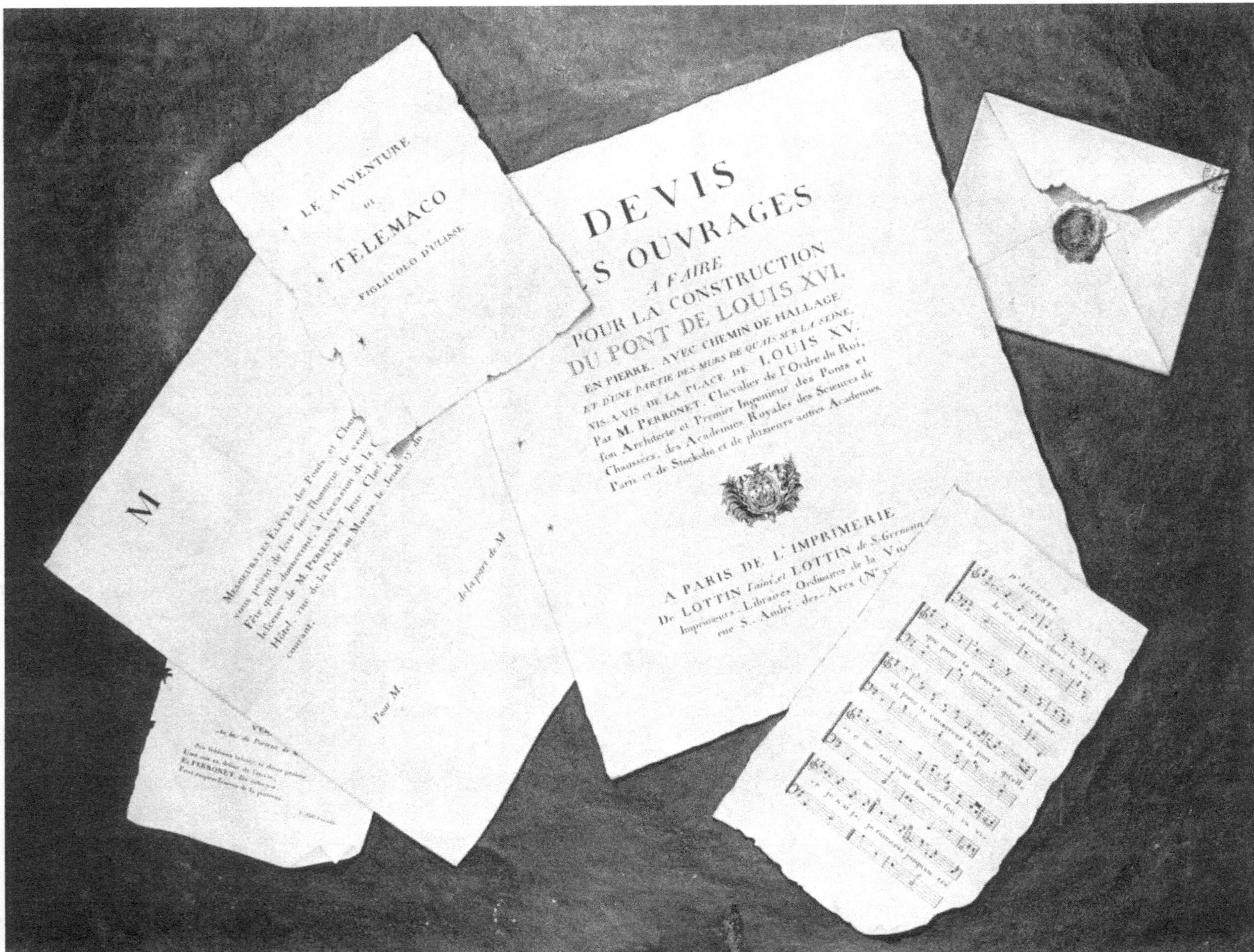
LE AVVENTURE
DI
TELEMACO
FIGLIUOLO D'ULISSE
DEVIS
DES OUVRAGES
A FAIRE
POUR LA CONSTRUCTION
DU PONT DE LOUIS XVI,
EN PIERRE, AVEC CHEMIN DE HALLAGE
ET D'UNE PARTIE DES MURS DE QUAIS SUR LA SEINE,
VIS-A-VIS DE LA PLACE DE LOUIS XV.
Par M. PERRONET, Chevalier de l'Ordre du Roi, son Architecte et Premier Ingénieur des Ponts et Chaussées, des Académies Royales des Sciences de Paris et de Stockolm et de plusieurs autres Académies.
A PARIS DE L'IMPRIMERIE
De LOTTIN l'aîné, et LOTTIN de S.-Germain
Imprimeurs-Libraires Ordinaires de la Ville
rue S.-André-des-Arcs
M
Messieurs les Elèves des Ponts et Chaussées
Fête qu'ils donneront à l'occasion de la
Hôtel rue de la Perle au Marais le Jeudi 15 du courant.
de la part de M.
Pour M.
Pour M. Perronet

worth pointing out in this respect the kinship between the works of pupils at the Ponts et Chaussées and the illustrations contained in cartography manuals[8] or in those 'study maps' in which, for the purposes of study, the widest possible range of representations of the territory were assembled.[9] On the eve of the Revolution, it was to the territory that engineers mainly referred. While they still lacked the necessary instruments for calculation and statistics, cartography continued to be the sole mode of reasoned exposition of territorial diversity.

Maps drawn up by pupils from the Ponts et Chaussées had a disturbing resemblance to the art of gardens, albeit an art of gardens which no longer had anything to do with the designs of someone like Blondel, belonging rather to the heterogeneous world of follies. Partly because of the scale which was used, and partly because of the large lay-outs and the meandering qualities of the relief, towns and villages came to resemble fantastical parks, modelled upon those pleasure spots favoured by the grandees of rank or finance (see Fig. 75). The entries of the student engineers to the various competitions remind one of the lay-out of Bagatelle or of Chanteloup, of the conceptions of Carmontelle or of the Marquis de Girardin.[10] Map after map betrays the same obsessions, namely, Versailles and its avenues, road, woods and rocks, the contrast between the regular and the irregular, or between the natural and the artificial. These recurring motifs enabled one to attempt a description of the territory. If the map sometimes assumed the aspect of a picturesque park, it was undoubtedly because the land, having been crossed, conquered and tamed, could be compared to a garden, with the engineer as its foreman.

THE COUNTRY AS THE ENGINEER'S 'GARDEN'

What is most striking about eighteenth-century cartographic manuals is their punctilious nature. Learning to draw a map involved, first and foremost, learning to read nature, and to decipher the signs which she lavished upon an attentive observer.

FIG. 74 Map for a competition at the end of the eighteenth century, ENPC, ink, gouache; 42.5 × 56 cm. To the multiple meanings of the word 'map' there is here added the theme of the journey, through an allusion to the adventures of Télémaque. Maps and structural projects were intimately linked, as Perronet's estimate for pont Louis XVI had emphasised.

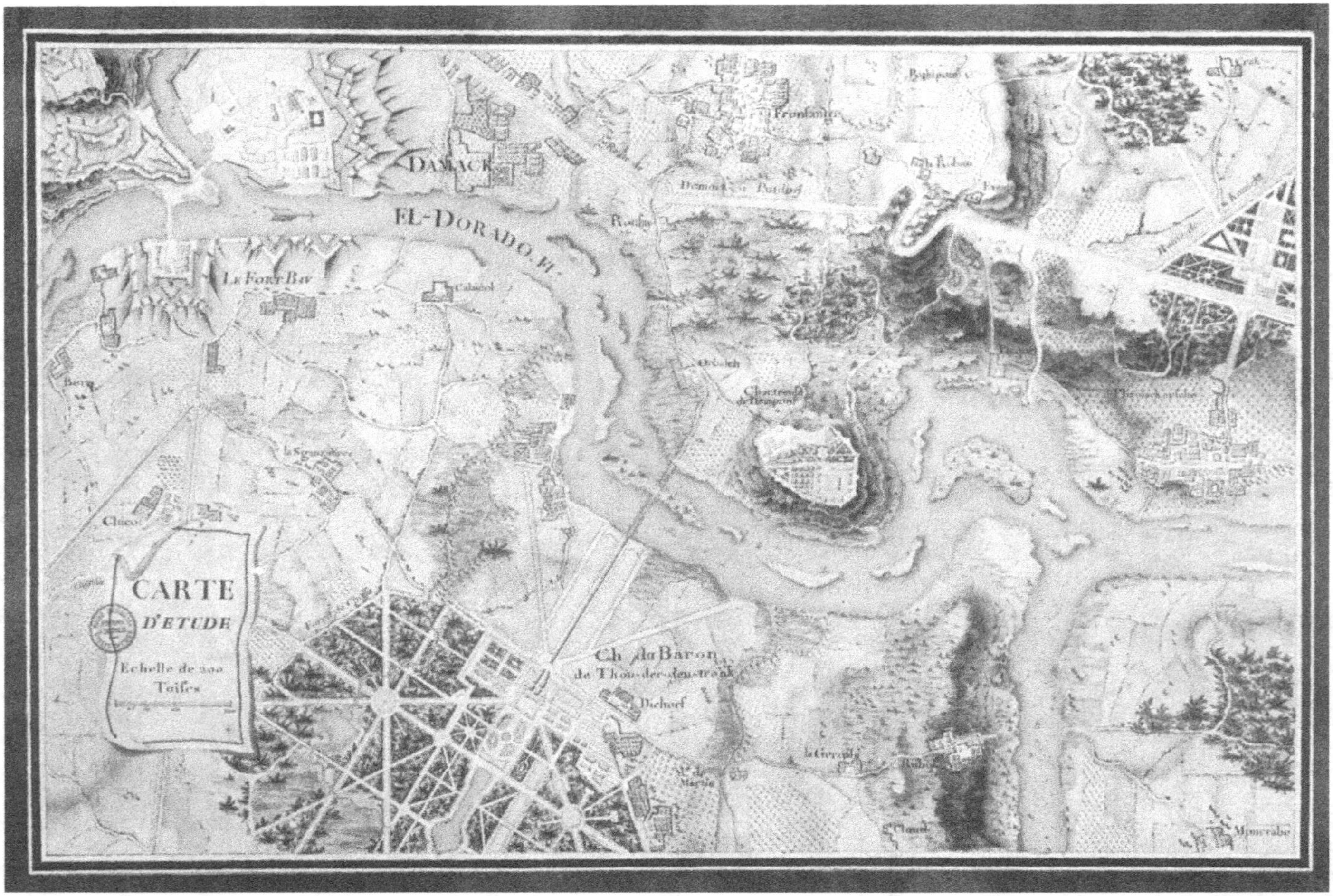

FIG.75 Brulé, map study intended to serve as a model for the pupils at the Ponts et Chaussées, 1793, ENPC, ink, wash, watercolour; 39.5 × 56 cm. Through the proliferation of detail, the map resembles a picturesque garden; it also represents an 'invitation to a voyage'. Here the river Eldorado flows in front of the château of Thunder-ten-tronckh, inspired by Voltaire's *Candide*.

The graphic codes used by the engineers were generally designed to reproduce these signs. To represent grass, one used a pale green; to represent cultivated fields, one used washes of green or red finely streaked with brown; for rocks, one used Chinese ink mixed with carmine, although one then had to imitate the effects of erosion by adding veins and fissures. The representation of forests and vines, of sand and water, was equally realistic.[11] In fact, the cartography of the Age of Enlightenment endlessly oscillated between the need for conventions and the desire to imitate nature (see Figs. 76–8).[12] As a manual in use at the Ecole des Ponts et Chaussées stipulated: 'A plan or a map is a copy of nature from the perspective of a bird in flight, and one must therefore copy everything as exactly as possible'.[13] This concern with accuracy, and this sensitivity towards the landscape had its counterpart in the aesthetic imperatives of design. The programme for a bridge competition held in 1791 reflects a serious concern to match the design to its site:

> Before projecting, it is essential that one examine carefully the sites and proportion of the bridge to the objects which surround it: in a wealthy and beautiful town, one builds a light bridge, enhanced by the beauties and agreements of architecture to which this kind of structure is susceptible; in the countryside or in a small town, a light and simple bridge; amidst the mountains, a heavy and bold bridge. If one seeks to build a bridge next to a substantial rock, it is essential that its construction combine a stability recognised and demonstrated by calculation, a boldness and a form as astonishing and remarkable as the mass which is adjacent to it.[14]

For the engineer, the territory was a kind of garden in which the design was located, much like a manufacture in a park (Fig. 79). Lightness and boldness were qualities which belong as much to the repertoire of the lover of gardens as to the vocabulary of the planner.

I shall use the map competitions of the Ecole des Ponts in order to elicit some of the principles governing the use of this vocabulary. Since they underwent a ceaseless transformation, these principles were actually akin to processes; the country was a shifting reality which could only be dominated through a constant effort of adaptation.

If one inspects the maps of the Ecole des Ponts et Chaussées, one is struck first of all by the concern to establish the axes, and thus to mark out the site in an imperious manner. In the eighteenth century, roads were generally straight. This desire for legibility found expression in a complete geometry, based upon translations, rotations and symmetries. Roads and canals were therefore organised in terms of various motifs, such as squares, circles, stars or goose paws.

Mindful of the advice of classical authors regarding the variety which ought always to be applied to designs, which would otherwise be dry and soulless, the pupils at the Ponts et Chaussées took care to introduce some dissymetry into their axial compositions, and to soften the repetition of identical elements by means of skilfully distributed modulations.

This geometry was subsequently deployed in a network. Being situated at the crossing of important axes, the services answered each other across space. This space peopled with signals was likened to the corps of the Ponts et Chaussées which had just taken possession of it. While design remained the prerogative of the isolated engineer, planning was the concern of a collective being, the engineers, who were organised and inspired by a common ambition to transform the world.

The interaction between these two levels of involvement took the form of a series of games of scale. Planning was to be reflected in each of its operations. Being separated by often considerable distances, the engineers' designs were so many microcosms which, although independent, were nonetheless ruled by the same laws as was the plan as a whole. Yet this correspondence between the whole and its parts differed fairly evidently in its principles from cartographic reduction. It consisted rather in a shared stock of ends and means. Intervening locally or in the territory amounted to the same thing; the actions governing the lay-out of a portion of a road or the constitution of the general network of traffic paths were fundamentally identical.

This latter aspect is especially clear if one considers the motif of the garden in the French style (see Fig. 80), which recurs to an almost obsessive degree in one map after another. The country, with its uneven topography, could never be mistaken for one of the productions of Le Nôtre. Yet the park of Versailles or its imitations do nonetheless provide us with an image of the engineer's work, with the network of their avenues opening out in the form of roads, and with their ornamental lakes which might turn into canals or estuaries. Through this process of dissemination, the garden was transformed into a grammar of planning.

Likewise, the towns which were scattered across the maps crystallised other possibilities, based, for example, upon the extension, sticking on or addition of new elements, which would be grafted on to the existing urban structure (see Fig. 81). Since it was often fortified, the town no more resembled the country than it did a garden in the French style. Its identification with a repertoire of planning procedures nonetheless anticipated its imminent assimilation.

The park of Versailles or the town thus constituted a representation of the

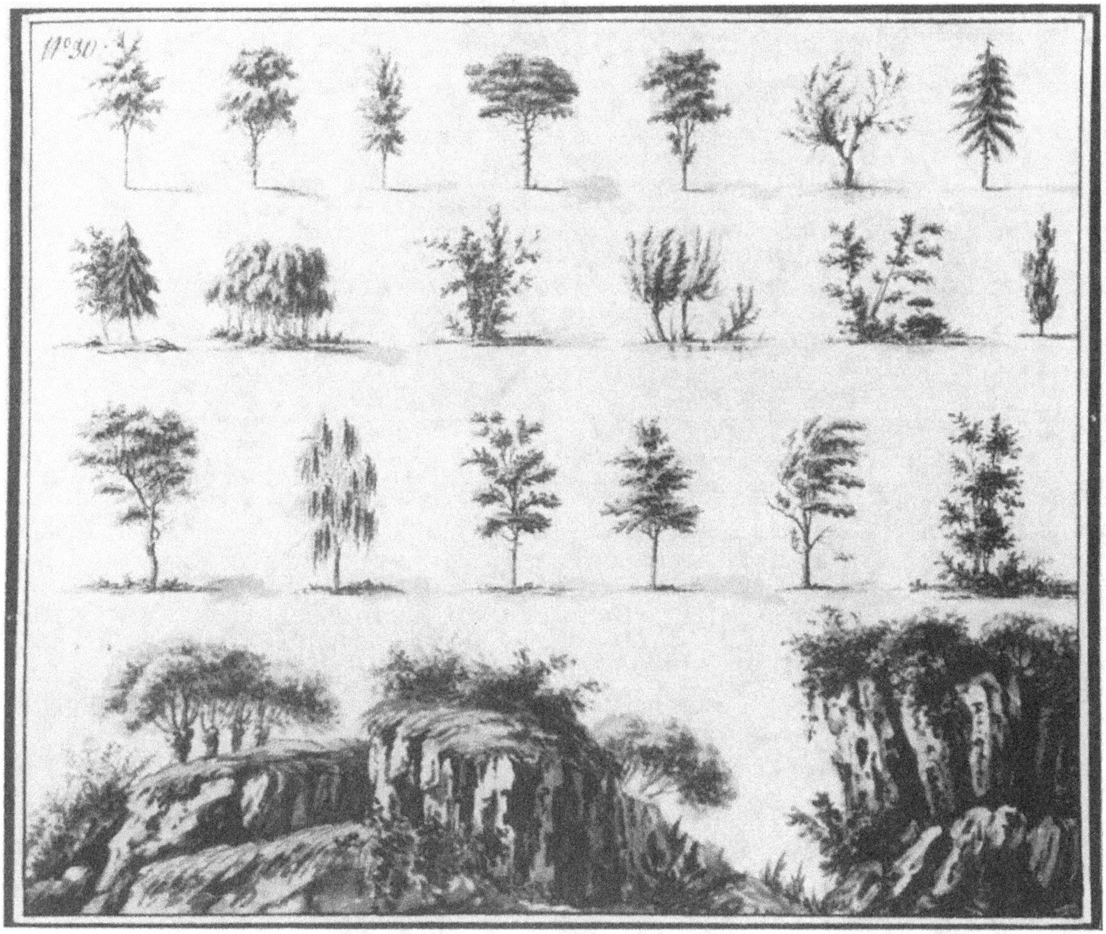

FIG.76 Outlines of trees for maps, 1793, ENPC, ink, watercolour; 18.5 × 26.5 cm. An apprenticeship in the drawing of maps was related to the study of nature.

planning of the land according to a scale other than that of cartography. To the reduction effected by the map were added infinite subdivisions of a new genre which saw the whole and the parts being exchanged on the plane of principles. This context helps us to understand the taste affected by the pupils of the Ponts et Chaussées for word play and symbols. Also in principle, the map and the word denoting it could stand in for each other, just like a plan and the various designs of which it was composed.

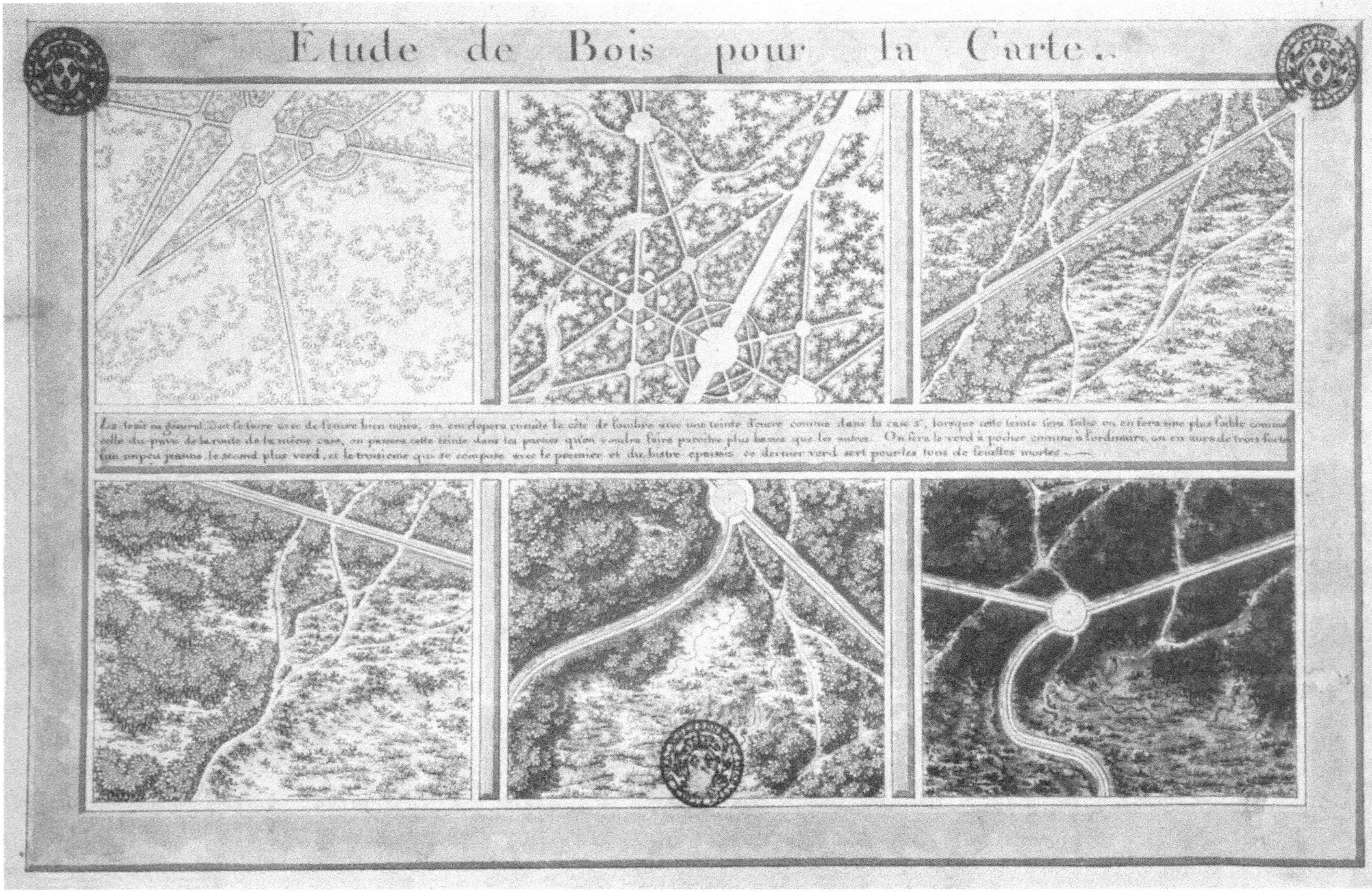

FIG. 77 'Study of woods for maps', 1793, ENPC, ink, wash, watercolour; 29 × 45.5 cm.

This to-and-fro between whole and parts, or between planning and one design or another, led to a kind of dematerialisation. Thus, whereas the garden in the French style had been an image of plenitude for Blondel, for the engineers it became a simple repertoire of means (Fig. 82). The town ceased to be a world turned in on itself, and came to seem a particularly dense zone for the network of services; it tended subsequently to be dematerialised, even if the lay-out of its fortifications protected it from total dissolution.

A further consequence was evident at the level of the territory. Distances seemed to contract or, rather, it now seemed possible to take in at a single glance the whole succession of topographic accidents and of designs, which seemed to be elements in a kind of journey in miniature. Traditional space was being replaced by a quantified universe. Such a replacement seemed highly improbable, if the fragmented and picturesque character of the map competitions were anything to go by, and yet the compositions of the pupils at the Ecole des Ponts et Chaussées present an accumulation of details, an excess which seems to require the irruption of figures designed to dispel the visual chaos. Within this confusion, the presence of great axes, the homology between the whole and the parts, occasioned a fictitious obliteration of distances which favoured the contemplation of a soon to be measurable system of relations.

The fictitious obliteration of distance also multiplied the contrasts by bringing a priori incompatible events together. Such rapprochements were common currency in the maps of the pupils at the Ponts et Chaussées (see Figs. 83 and 84). A sea port in a half-moon shape interrupts an indented landscape; a canal and a road cross distorted landscapes at right angles. Crossroads, irregular forests and deserts coexist with the reasonable architecture of the engineers. The latter actually composed their maps much as the gardeners had done, deriving pleasure from combining classical parterres with views of a more rustic reality. Marivaux's 'je ne sais quoi' is not very far removed from this spirit of 'oddity' in conformity with the wishes of the Abbé Laugier. Licence and a degree of confusion were necessary, just as order and clarity were in the exposition of great principles.

Finally, the countryside which features on the maps of the pupils of the Ponts et Chaussées was traversed by flaws and fractures. Mountains everywhere block the rush of the water courses, engendering a dust of islands and islets. Swamps insinuate themselves between bands of land with incomprehensible forms. Faced with this fragmentation, one wonders what meaning the engineer's action might still have.

The upheavals in the relief were reflected in the geometrical conflicts engendered by the documents thrown in bulk across the main drawing. Their jumble seems to condemn in advance any attempt to establish an order and a measure in the chaotic universe of the map.

Nevertheless, it was this situation of conflict which gave the engineer's approach all its meaning. The latter's task had been, as we have seen, to sew together the tattered shreds of a territory. His 'genius' was employed in reestablishing 'useful communications' between places separated by a relief which had seemed at first to be uncrossable. In their anarchic representation of territorial reality, the map competitions thus reflected one of the crucial aims of planning.

In concrete terms, the engineer was primarily concerned with physical communication. He was supposed to link up provinces by means of roads and canals, to throw bridges or to build ports. In order to avoid arbitrariness, the engineer also had to proceed rationally, on the basis of a general plan for the development of the transport infrastructure. The territory involved in this plan then acceded to a programmatic dimension. A scheme for overall planning was applied to filling its gaps, to overcoming its fragmentation, in order to restore its ideal unity. Given the sheer scale of the task, the engineer's struggle and sufferings sometimes took on epic connotations, as we can glean from the testimony of a pupil at the Ecole des Ponts et Chaussées:

> after having used theory to examine the progress of the infinite worlds subjected to his gaze, after having fixed his return upon one or other hemisphere, after having considered ways of combating the elements, of making them serve his purpose and of bending them to suit his designs, the scientist leaves his study and the thousands of arms subjected to his orders are so many levers which he uses to level the mountains, join up the seas and render the uninhabited regions fertile. Important works that have been undertaken, sometimes in burning sands, sometimes in pestilential marshlands or beneath a treacherous sea, require his presence and often expose him to illnesses occasioned by the fetid air, by the dampness of the water or by the heat of that star which he was able to fix and measure in thought, but whose burning activity he could only with difficulty bear.[15]

The confrontation between engineer and territory was, as we have seen, violent. There is something vertiginous about the unleashed elements, as there is also about the idea of thousands of arms subjected to a single will. It is also worth noting in passing the analogy here between the passivity of the worker and the blind force of nature. Could the engineer's relationship with the

FIG.78 Kirn, map study, 1792, ENPC, ink; 31 × 45.5 cm. Realism and the use of conventions are intermingled, both in the representation of the vegetation and in that of the relief.

workers not be likened to the struggle that the former waged against a hostile universe?

Although it is impossible to decide upon the latter point, we are now in a better position to understand the link between country and garden. Like the garden, the country had to be wholly mastered, and then moulded to man's advantage. The engineer was supposed to 'level the mountains, join up the seas and render the uninhabited mountains fertile'. In the eighteenth century, this image was still well in advance of the reality, and yet it was not long before engineers were to be fashioning large-scale landscapes, from the *département* of the Landes to the Po Valley, from reforestation serving to fix the dunes to the building of dikes and irrigation works.[16]

Country and garden were also linked in a more instrumental sense, inasmuch as the tools of the engineer were also those of the gardener. The terraces of gardens had their counterpart in the engineer's cuts and fills;[17]

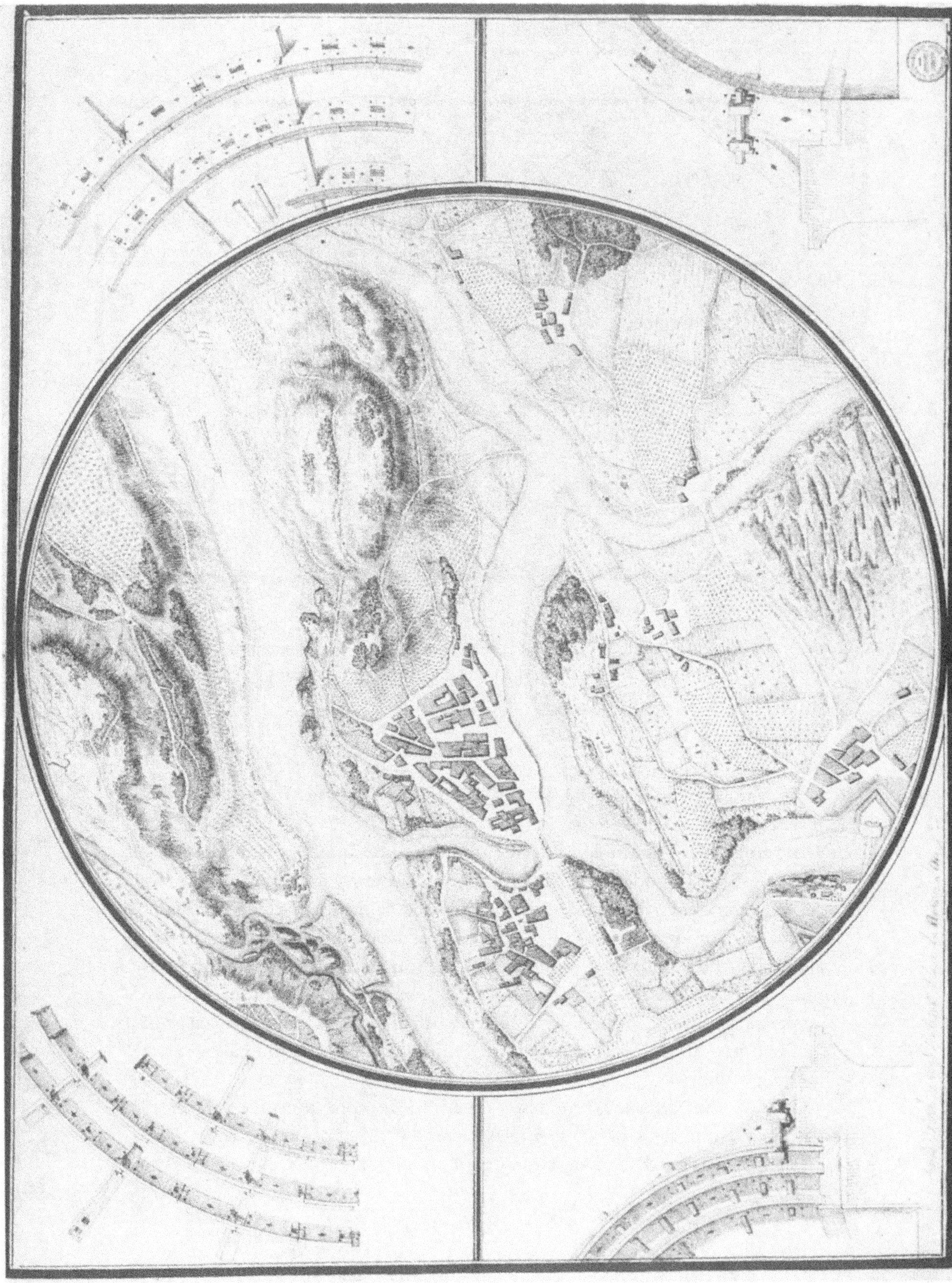

hydraulic works,[18] major road lay-outs and their retinue of plantations,[19] constituted a repertoire that was common both to planning and to the art of gardening. In a world in which metal was still rare, stone, earth, water and vegetation were at the basis of all possible combinations.

However, to judge by the competitions at the Ecole des Ponts et Chaussées, the country of the engineers no longer had much to do with the garden in the French style. The crucial features of such a territory, namely geometry, networks, contrasts and fragmentation, were no longer derived from the art of architecture. A new kind of space was emerging, one whose geometry seemed to float with no visible points of support or limits, and whose networks seemed to exclude any notion of a centre. In combining order and disorder, the territory could well constitute the most faithful expression of the contradictions in the engineers' system. There was no longer any unitary composition, but rather fragments which interlocked and which were articulated with each other like apparatuses crossed by flows of all kinds. I am thinking here of Piranesi's *Campo Marzio*, and of his strange machines in the antique mode,[20] but also of the Abbé Laugier's forest, which was an image of the uncertain future of the town. Was the countryside in any case not a town in disguise? Was it not already more urban than the fortified cities which it transformed into poles for its development? The country shared the town's excess, and its swarming detail, an invasive reality which sought to condense itself in clear and definitive formulae. On the maps composed by the Ponts et Chaussées, this excess took on the form of a crisis. It was in fact representation that was in a state of crisis, for it had both to control space by describing it wholly, and to explain its organisational logic. These were mutually contradictory requirements, since representational overbidding led to an irrevocable confusion.[21]

This crisis was not to be resolved until the 1800s. On the eve of the Revolution, it was nonetheless accompanied by a profound renewal of aesthetics, and one which involved both the territory as a whole and design.

FIG. 79 Debret, map study, Year II, ENPC, ink; 50 × 375 cm. The presence of Delorme's trusses in the corners of the composition seem to illustrate the shift from mapping to building.

FROM TERRITORY TO DESIGN: THE SUBLIME AND THE PICTURESQUE

This aesthetic renewal arose out of the contradiction evident in the engineer's combat with the elements which, as we have seen, involved 'burning sands', 'pestilential marshes' and 'treacherous seas'. As a man of the eighteenth century, the engineer was in fact natural, as was the countryside which he confronted. His 'genius' wrote under the dictation of nature, and it was this same nature which he had to try at all times to transform. In more prosaic terms, one wonders how the punctilious work done on maps could be reconciled with the often brutal decisions required in planning.

Engineers, however, refused to face up to this contradiction. As far as they were concerned, joining up the countryside did not involve using any irremediable violence upon nature. The engineer's work did not so much destroy nature as reveal it. In its very depths, nature thus remained close to the reasonable spirit which acted in its name. One could identify as its principle a rigorous geometry, which ruled over man's impulses and the behaviour of matter alike.

The bridge served to illustrate this process of reconciliation. Its primary function consisted in overcoming the obstacle of ravine or river, and thus in authorising a crossing which the site had seemed to refuse. A bridge thus seemed to solve a problem by imposing continuity upon a fractured terrain. But the structure was also inscribed in the immemorial practice of men who used to cross by a ford not far from that same spot, and who had marked out a road by taking into account the guiding lines of the relief. The bridge was thus a singular monument marking the presence of a fault; but it also expressed a regularity, a permanence, much like the road whose trajectory was determined by the overriding logic of both the territory and the needs of society.

There was thus a convergence between what the territory revealed and what the design required. Nonetheless, this approach often involved some conflict, and this would have some, at any rate superficial, consequences for

FIG. 80 Map for a competition at the end of the eighteenth century, ENPC, ink, wash, watercolour; 45.5 × 56 cm. There is a striking resemblance between the French-style garden at the bottom of the drawing and the one which features on Brulé's map study (Fig. 75). The Ponts et Chaussées maps were often composed using stereotyped elements, and this gave them an undeniably combinatory quality.

CARTE D'ETUDE
PAR
MILHAUSEN

the design. The latter was in fact always located at the precise point at which nature reversed its own forces against it, those of the engineer as it happened, who had to struggle and to suffer.

This reversal is reminiscent of the aesthetics of the sublime,[22] which was given its fullest expression by Burke in 1757, in his *Philosophical enquiry into the origin of our ideas of the sublime and the beautiful*. Burke's work probably did not exert any direct influence upon French engineers of the Age of Enlightenment.[23] The *Enquiry* was anyway more concerned with poetry than with architecture or planning; indeed, it excluded painting from the sphere of the sublime, although at the same time painting enabled one to exalt bridges, since through their boldness they defied the abyss. Together with cartography and design, the pupils of the Ponts et Chaussées were also involved in painting, especially because of the landscape drawing competitions.[24] Yet Burke's book belongs to a line of thought which, in the French context, may be traced back to the end of the seventeenth century, and to the famous polemic between Boileau and Huet,[25] and also to the beginning of the eighteenth century, and to the Abbé Dubos' *Réflexions critiques sur la poésie et la peinture*, which anticipated certain aspects of the *Enquiry*.[26] The notion of *ut pictura poesis*, so much cherished by the French theoreticians, means that the question of the sublime was also addressed in the domain of the plastic arts, as in poetry.[27]

Disregarding for the moment the question of sources and of influences, one should note that the notion of the sublime always turned upon that of a divided nature, which was subject to contradictory and violent currents. Within this contrariety, the sublime was established independently of any notion of the beautiful. According to Burke, one could regard as sublime 'whatever is fitted in any sort to excite the ideas of pain, and danger'.[28] Since the bridge was born of the suffering of the engineer and his workmen, and since it courted danger in braving the abyss and the impetuosity of the river, it was regarded as sublime (see Fig. 85). Being both breach and continuity,

FIG. 81 P.-C. Lesage, map study, 1775, ENPC, ink, wash, watercolour; 37.5 × 52.5 cm. There was plainly a polemical intent in placing a French-style garden and a system of fortifications opposite one another, on the diagonal of the composition. For the administration of the Ponts et Chaussées, fortification was but one of many forms of territorial construction, and this claim could not but offend their rivals, the engineers of the Génie.

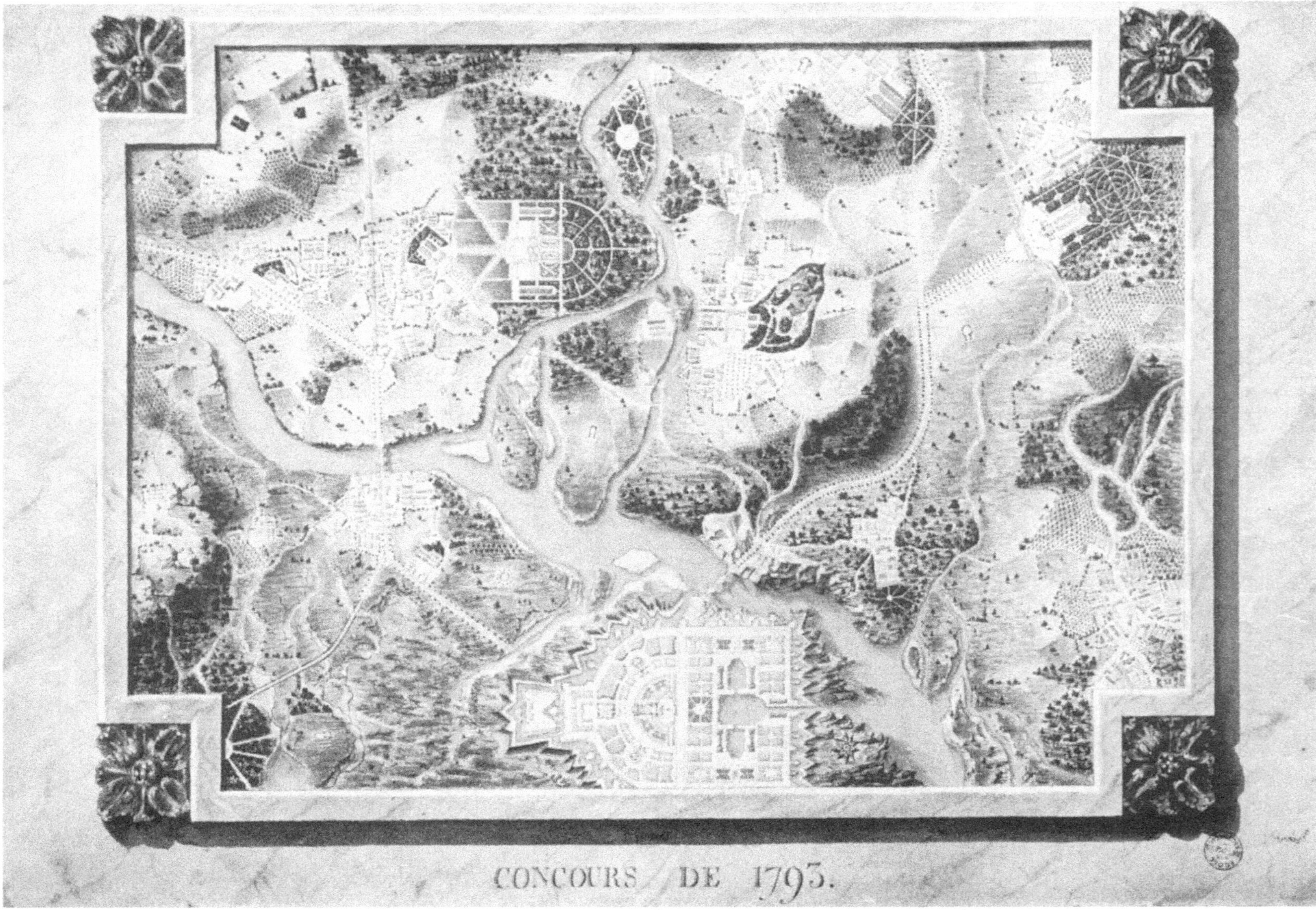

FIG. 82 Dubois, map for the 1793 competition, ENPC, ink, wash, watercolour; 44 × 58 cm. From top to bottom of the composition, the French-style garden and the model arsenal, the paroxystic features of the plan, seem to echo each other.

accident and necessary development, the bridge helped to express a duality and a conflict: the duality of the engineer, who was inspired by nature in order to combat it, but also a conflict between the ideal landscape revealed by planning, and the real landscape, which was riven with faults and fractures.

In more concrete terms, an intuition of the sublime may be discerned even in the detail of the engineer's work. The expression of structural conflict, the unadorned nature of the facings, the quest for violent contrasts of light and shade, were all, for example, part of Perronet's vocabulary. More generally, the building of substantial works often assumed a heroic and sublime character. Did it not in fact represent a lasting act of defiance towards gravity, a defiance which used nature in order to transgress its most universal law? It was at the very moment when Newton's conceptions had triumphed that an approach to construction was developed which sought to distort his teaching. Seen from this angle, the art of building stems from a basic contradiction. All construction is by its very essence paradoxical, since the scaffolded forms, those 'burdens' of stone (to use Patte's expression), were liable to collapse at any moment, and therefore required, if the building were to endure, the application of an unceasing violence.[29] The structural primitivism which the Abbé Laugier so cherished, the conflict-ridden expression of structure to which Perronet was prepared to make sacrifices, thus converged with a kind of cruelty of the builder, a cruelty of man towards nature, or a cruelty of nature towards itself. To build was to torture matter by subjecting it to the demands of a dangerous reason. It was this very danger which Patte, because of his allegiance in this respect to the *convenances* of a waning classicism,[30] had rejected.

Being essentially paradoxical, the exploitation of effects borrowed from the register of the sublime may now be more easily understood, if one bears in mind that they were directed as much at the building of the work as at the completed object. It was the various stages of the construction which were the most dangerous; the design was only sublime in so far as it required risky works which it then in some sense crowned. The best example of this is undoubtedly the Eddystone Lighthouse, which was built by the English engineer Smeaton between 1756 and 1759, and which was the wonder of Europe.[31] Lesage featured it in his *Recueil de divers mémoires, extraits de la bibliothèque des Ponts et Chaussées, à l'usage des élèves ingénieurs*.[32] The lighthouse's foundations were upon a rocky reef and, after countless mishaps, some of them resulting in death, we see it represented in an engraving as if upon the point of being submerged by breakers,[33] a polished cone, with a form of virtually machinelike

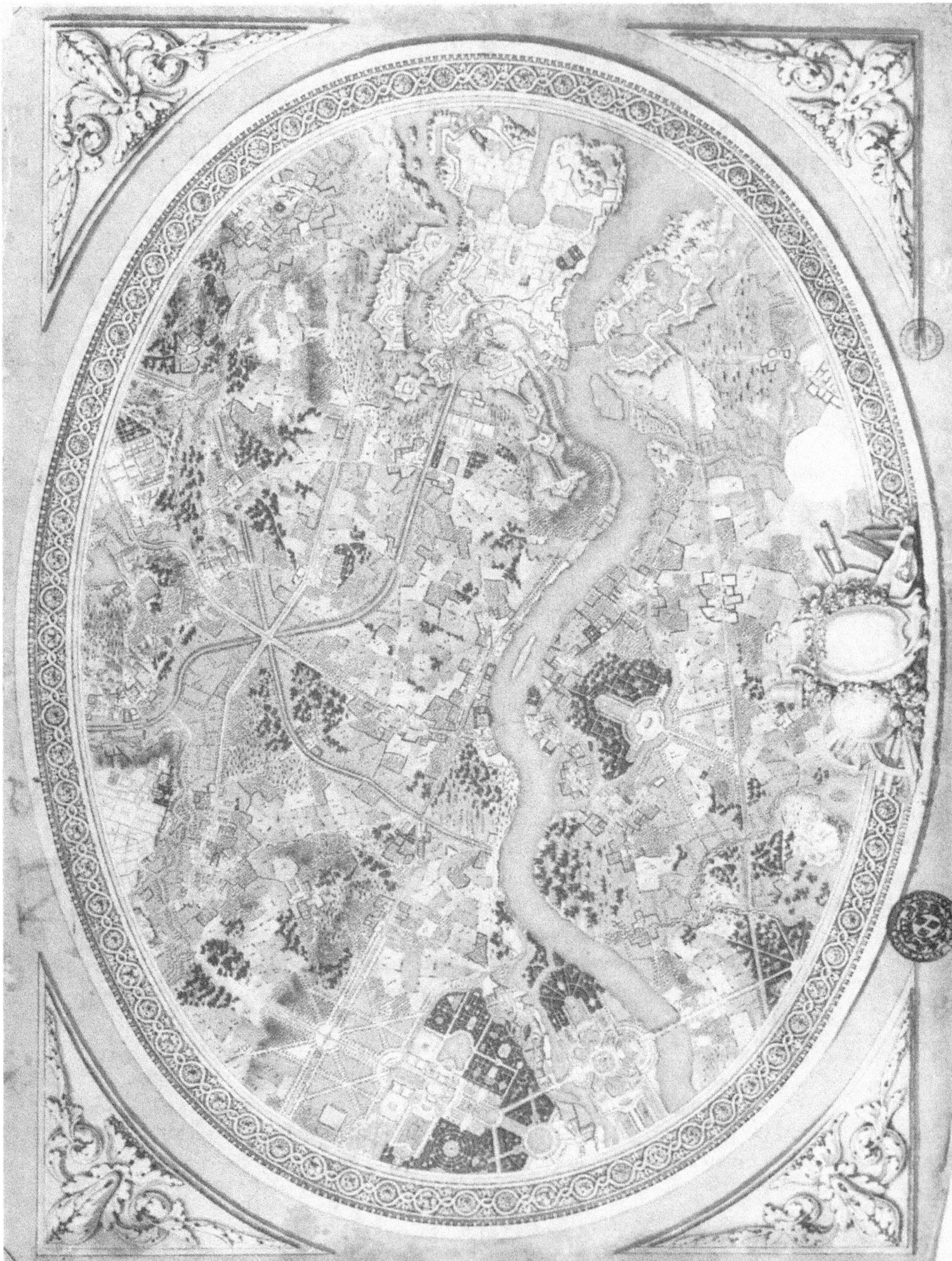

abstraction engaged, according to the engineer's conception and realisation, in a struggle against a 'treacherous sea' (Fig. 86). In Perronet or Smeaton, a new aesthetic had thus emerged, one of conflict and of process.

In his *A philosophical enquiry into the origin of our ideas of the sublime and the beautiful*, Burke also observed that, if 'to the sublime in building, greatness of dimension seems requisite', 'designs that are vast only by their dimensions, are always the sign of a common and low imagination. No work of art can be great, but as it deceives; to be otherwise is the prerogative of nature.'[34] Thus a bridge strikes the imagination through its defiance of gravity and of the impetuousness of the water, rather than by its actual dimensions. Like the elements, the arch of the bridge eludes human measurement; its real greatness consists rather in the performance that it effects, uniting what had seemed forever separated, playing with the void which it so perilously bestrides. It is in such terms that one should interpret Perronet's astonishing design in particular, where he envisaged the construction of great arches of stone 'of a two hundred, three hundred, four hundred and even five hundred foot span'.[35]

The principles governing the picturesque (see Fig. 87), the other aspect of the aesthetics of the sublime, were given by William Gilpin in his *Three essays on picturesque beauty*, which was published towards the end of the eighteenth century.[36] Etymologically, the picturesque is intimately linked to painting, and one is tempted to discern in it the sublime as represented and as if tamed by representation.[37] Indeed, a bridge, through the sublimity of its origin, became picturesque when painting captured it, and it thereby lost its anguished character. On the eve of the Revolution, bridges frequently appeared in paintings. In mountain scenery, one might see a wooden or stone construction jutting across a precipice, although it might also serve to punctuate an undulating plain, or be a sombre arch through which one would view a town half-concealed by ivy and other creepers. The picturesque thrived on contrasts; on a canvas, ruins served to temper the exuberance of nature. The bridge, with the contradictions which it expressed, became an easily exploitable subject and the pretext for countless stories. The maps of the pupils of the Ponts et Chaussées thus took on a narrative character.

The sublime and the picturesque were by now intimately linked, just as the

FIG. 83 Map for a competition, end of the eighteenth century, ENPC, ink, wash, watercolour; 35.5 × 49 cm.

design proved to be by turns necessary and contingent. This curious alliance in fact prefigured an exit from the system of the beautiful, such as the architects had codified it. Being outside of this system, the sublime and the picturesque enabled one to transgress the rules of *convenance*. One could not negotiate with the sublime, since it imposed itself with all the crushing obviousness of natural phenomena. In its apparent gratuitousness, the picturesque also conferred a degree of freedom upon the engineer, who was now able to develop the most improbable forms. The bridge was essential in so far as it completed the landscape, but contingent on the plane of its representation. It was then just a single step from this representation to the register of its architectural incarnation. The design could break the prevailing rules without therefore losing its natural character, and the necessity which that implied. Thus, once the bridge was assessed in terms of defiance and performance, one saw the progressive abandonment of the traditional requirements of stability. With the element of danger and suffering integral to the building of a bridge, the aesthetics of the sublime also paved the way for a structural model which turned conception and execution into so many stages of a single process.[38]

There would have been something paralysing in such a situation of exacerbated conflict, had not the picturesque served, as may readily be understood, to render the design less dramatic. In practice, it was a matter of being free to dispose the figures of planning across landscape. Freed from the yoke of *convenance*, the engineer was then able to imagine new conventions, which were no longer based upon a doctrine of imitation, but upon a radical redefinition of the relationship between the natural and the artificial. For, through the sublime and the picturesque, the genius of the engineer could imitate by inventing. Art and nature were violently opposed, as if they might then be recombined more effectively at the conclusion of a long process of unveiling.

Finally, this unitary conception rested upon ethics. As Kant was to show in his *Observations on the sentiment of the sublime and the beautiful* and in his *Critique of judgement*, the sublime resided primarily in the mind of the spectator, who had to have an 'aptitude for the sentiment of (practical) ideas, that is to say, for the moral sentiment'.[39] As far as the engineers were concerned, this aptitude could be discerned in their concern with social utility. It was their disinterested goodness which enabled them to struggle for the improvement of communications by all the means in their power.

Yet this power was to prove very weak in practice, since an engineer's

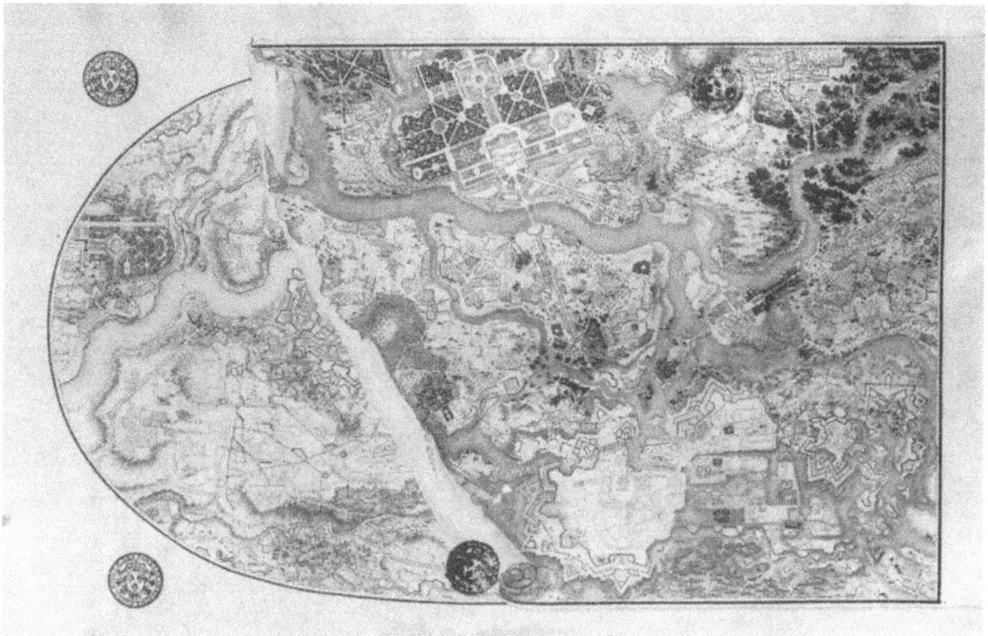

FIG. 84 Map for a competition, end of the eighteenth century, ENPC, ink, wash, watercolour; 27 × 43.5 cm.

designs often still referred to the principles of *convenance.* In other terms, the freedom which the engineer might enjoy with respect to the countryside itself posed a problem, in so far as it only took effect upon very particular occasions, such as the map competitions or in especially bold realisations. If, through the mediation of the picturesque, the bridge or structure was only *this* bridge or *this* structure, why should one concern oneself with the territory, why should one describe it so minutely while being preoccupied with the landscape and with the insertion of the design in its site, a preoccupation which was even reflected in the pastoral pictures done for the drawing competitions at the Ecole des Ponts et Chaussées? Was it a question of reconciling planning with nature, as if that reconciliation were not operative from the outset, and as if it were a question of distinct entities? It is at any rate hard to understand the risk entailed by this distinction, since every combination was a priori admissible. One had then to admit that the plan of the territory involved some sort of necessity, even before planning set out to perfect it. This necessity implied

FIG. 85 M. Cormier, project for a stone bridge for the 1791 competition, second prize, ENPC, ink, wash, watercolour; 92 × 252.5 cm. Whereas the background of snowclad mountains testifies to the sublime character of the work, the process of construction is still in evidence at the bottom of the drawing.

both a respect for the constraints of the site and the possibility of their harmonious transformation.

A PRODUCTIVE COUNTRYSIDE

This necessity had originally become apparent on the building-site. The pick and shovel were its essential tools, and the engineer was therefore obliged to take the slightest irregularities in the terrain into account. The apprenticeship of a pupil at the Ecole des Ponts et Chaussées entailed mastering problems to do with earthworks. The engineer's knowledge was particularly expressed by long columns of figures measuring the volume of the embankments and cuttings which the design required. The obstacles of the relief thus constituted an inescapable reality; the land constituted the natural limit to the engineer's action.

Another factor concerns the shortcomings of scientific knowledge, shortcomings which we have already emphasised several times. In the absence of any methods enabling one to achieve a genuine mastery of the design, the land may be compared to an instrument of control every bit as 'technical' as calculation.

With its claims to be natural, the engineer's practice in fact accorded a great deal of importance to built form, even if the process of conception and of execution served to temper its architectural character. Roads, bridges and canals were also monuments to technology. The aesthetics of the sublime also guaranteed the necessary link between design and territory. In the engineers' eyes, the latter then took the form of an instrument of verification, a verification which was a posteriori and which saw beauty, stability and utility harmonise around the design and its insertion into the site. Stretched arches were beautiful, stable and useful, and the landscape around a road was equally so. Preoccupations with territory or with landscape involved a persistent encounter with technology, with the economy and with society. Thus, the trees which were planted on the edges of canals, and which achieved such a striking effect, also helped to fix the earth and to combat deforestation, and so played a part in the solution of serious problems.[40] The territory was the natural soil in which both experiment and calculation were rooted. Furthermore, this rootedness was never so much in evidence as in the harmony of the design with its site.

In the eighteenth century, the engineers lived in a world of forms which all possessed some technological meaning. It is in this sense that one may talk of fundamental issues being involved in the debate over the nature of stability. In the wake of the architectural theoreticians, Prony accounted for the bulging of the Doric column in terms of resistance to buckling.[41] In order to resolve problems of statics, such as the width to be given to the piers of an arch, recourse was often had to an experimental geometry based upon regulatory lay-outs and scientific intuition.[42] The calculation of the angle to be given to the lock-mouth mixed mathematical reasoning with the interpolation of known results.[43] Form was the product of knowledge through necessity having been empirically observed. It can be compared to the expectation of a calculation of probabilities whose modalities exceed the narrow framework of mathematical physics. In this perspective, the territory represented a subsumption of all possible forms, and an assertion that there was a degree of unity in the world, in which both calculation and the action of the engineer had their place, so long as they conformed to preexisting logics. If the bridge was gradually turned into a process which later crystallised into a structural model, this process and this model still needed an image to guarantee their coherence. The territory answered the engineer, in a dialogue which entailed the existence of two protagonists serving as the reflections of a single reality.

This reality was also that of a technological system that was still largely classical. Tools and machines were still generally made of wood, and energy was hydraulic or derived from the wind. In this context, one can readily understand the importance of Bélidor's *Architecture hydraulique*, the bedside book of the engineers, which was devoted to locks, dikes, windmills, suction pumps and forcing pumps, and to a whole series of devices which were still fundamentally artisanal in nature. In spite of the increasing emphasis upon roads in the eighteenth century, waterways still enjoyed a privileged position. For the engineers at the Ponts et Chaussées, water and wind were still the determining factors. Forests were also crucial, since they provided a material which was basic to machines of all kinds. The countryside was of course nothing other than the place at which all of these elements converged. A forest was a potential warship or merchant vessel,[44] and all contouring therefore represented a possible source of energy. Symbolically, but also very concretely in many cases, the relief could be read as a geography of energy sources, a map in which the contours involved an inscription in filigree of the harshness of the earthworks, the lay-out of a canal, together with the possibility of a waterfall or of an undershot wheel.[45]

Thus, by summarising all of the possibilities offered to the engineer, the

FIG. 86 P.-C. Lesage, *Recueil de divers mémoires* (1806–10), Smeaton's Eddystone lighthouse.

territory in some sense gave rise to planning as the realisation of his deepest aspirations. Mountains and plains, rivers and ravines summoned up roads, canals and bridges, which would serve to overcome the obstacles of the relief. The territory provided the work of the engineer with an ultimate aim, and it elicited a number of instruments of verification and control, but its impact did not end there, for it was also the site of the products of nature. The territory enabled man to dispose of such products freely.

The engineers' thinking was close to that of the physiocrats and of agronomists such as Duhamel du Monceau or Rozier, and they therefore

FIG. 87 View of 'the temple of Jupiter Serapis' at Pozzuoli for the landscape drawing competition of 1784, ENPC, ink; 34 × 45 cm. This picturesque view is an interpretation of Louis-Jean Desprez's watercolours, executed in 1780 and reproduced as etchings by Francesco Piranesi.

thought of the products of nature as being primarily derived from agriculture, which they regarded as the source of all wealth.[46] This wealth had to be set in motion, however, and Quesnay, in his famous *Tableau économique* (1758), written a year after his article 'Grain' in the *Encyclopédie*, had laid the basis for an economic analysis in terms of flows.[47] Works devoted to the circulation of wheat proliferated. In this context, bridges and highways could be justified in terms of the need to exchange basic foodstuffs. It was because the land produced harvests that it subsequently gave rise to the notion of planning.

Nonetheless, the products of nature transcended the framework of agriculture. The land was a storehouse of materials of every kind, foremost among them those used in building. Nature in its abundance provided stone, wood, gypsum and iron. When competitions were held at the Ecole des Ponts et Chaussées, the pupils' syllabuses tended to specify the characteristics of the stone to be used, together with the site where it had been quarried. Estimates frequently mentioned transportation of materials by road or by waterway; designs took into account the cost of this transportation when calculating expenditure. As well as being an instrument for checking whether a particular approach was technologically appropriate or not, the territory was also involved at the level of concrete decisions. When building a bridge, for example, one had to take into account what materials were available, and one would tailor the structure to the type of stone found locally.

On an altogether different scale, the territory might be compared to an exhaustive catalogue of the products of nature and of art. Plans obviously enjoyed a favoured place among the products of art. Among the products of nature, special mention should be made of iron ore. Roads reflected a concern for the earth, to the detriment of the world of water.[48] This interest converged with a fascination with mines (see Fig. 88), and with the idea of a soil concealing unsuspected riches. In the second half of the eighteenth century, geology gradually emerged, as much through an interest in fossils as because of the hypothesis of a central fire, which had been criticised by, for example, Buffon, in his *Epoques de la nature* (1779). Guettard's *Atlas et descriptions minéralogiques*, which enjoyed the support of Lavoisier, appeared in 1780.[49] Scholars' discoveries were putting paid to the notion that the earth's age tallied with what the Bible had said. At the same time, the engineers were discovering the depths of the soil. The soil with which they were concerned was obviously that of the foundations, a crucial technological problem, but they readily produced cross-sections in which the stratification of the various alluvial layers could be read. The competitions of the Ecole des Ponts et

Chaussées often involved the use of such cross-sections, which served to identify the kinds of terrain which were to be crossed.[50] Nonetheless, the mineralogy of the engineers of the Ponts et Chaussées in the Age of Enlightenment remained rudimentary. In contrast to Werner's classification,[51] the engineers' vocabulary still turned upon terms such as peat, earths and rocks, although new factors, associated with the exploitation of the subsoil, were clearly at stake in their work. This exploitation was reflected in the proliferation of courses in mineralogy, such as the one offered by Sage and attended by pupils at the Ponts et Chaussées, and in the creation, in 1783, of the Ecole des Mines, which was modelled upon the Bergsakademie of Fribourg.[52]

One can discern in both agriculture and mining a new concern with the compiling of an inventory of space, which seemed both to increase its productive capacities and to render it more dense. With their numerous details, the maps of the Ponts et Chaussées constitute an excellent illustration of this phenomenon. In so doing, they reflect a significant shift of emphasis. As the English economist Steuart put it in his *An enquiry into the principles of political economy* (1767), the development of production and of domestic trade ought, once a particular stage has been reached, to take precedence over foreign trade.[53] More intense economic activity, the promotion of liberal policies and the planning of traffic routes together helped to promote this interiorisation of growth, and other, highly diverse ventures worked to the same end. Where travel had once paraded the attractions of the mysterious and exotic, voyages now seemed to uncover horizons that were strangely familiar. To visit far-flung places was to undergo a kind of initiation, which allowed one to identify the reforms which needed to be implemented at home. This is at any rate what Diderot had learnt from the discoveries of Bougainville. In a more historical genre, this also accounts for the success of the Abbé Barthelemy's *Voyage du jeune Anarcharsis en Grèce.* The theme of the voyage also featured in the maps of the Ponts et Chaussées; place-names often had literary connotations, and one might come across both the river Eldorado and the château of Thunder-ten-tronckh from Voltaire's *Candide.*[54]

The country's necessity could in short be read at several different levels. By contrast with the town, the country was first of all natural, occasioning the planning which served to perfect it. Secondly, it was a tool for both the conception and the execution of the design, which helped to unify these two moments. Finally, it represented a catalogue of products of every kind, the necessary conditions for economic progress, which might either appear on its

FIG. 88 *Encyclopédie*, cross-section of a mine.

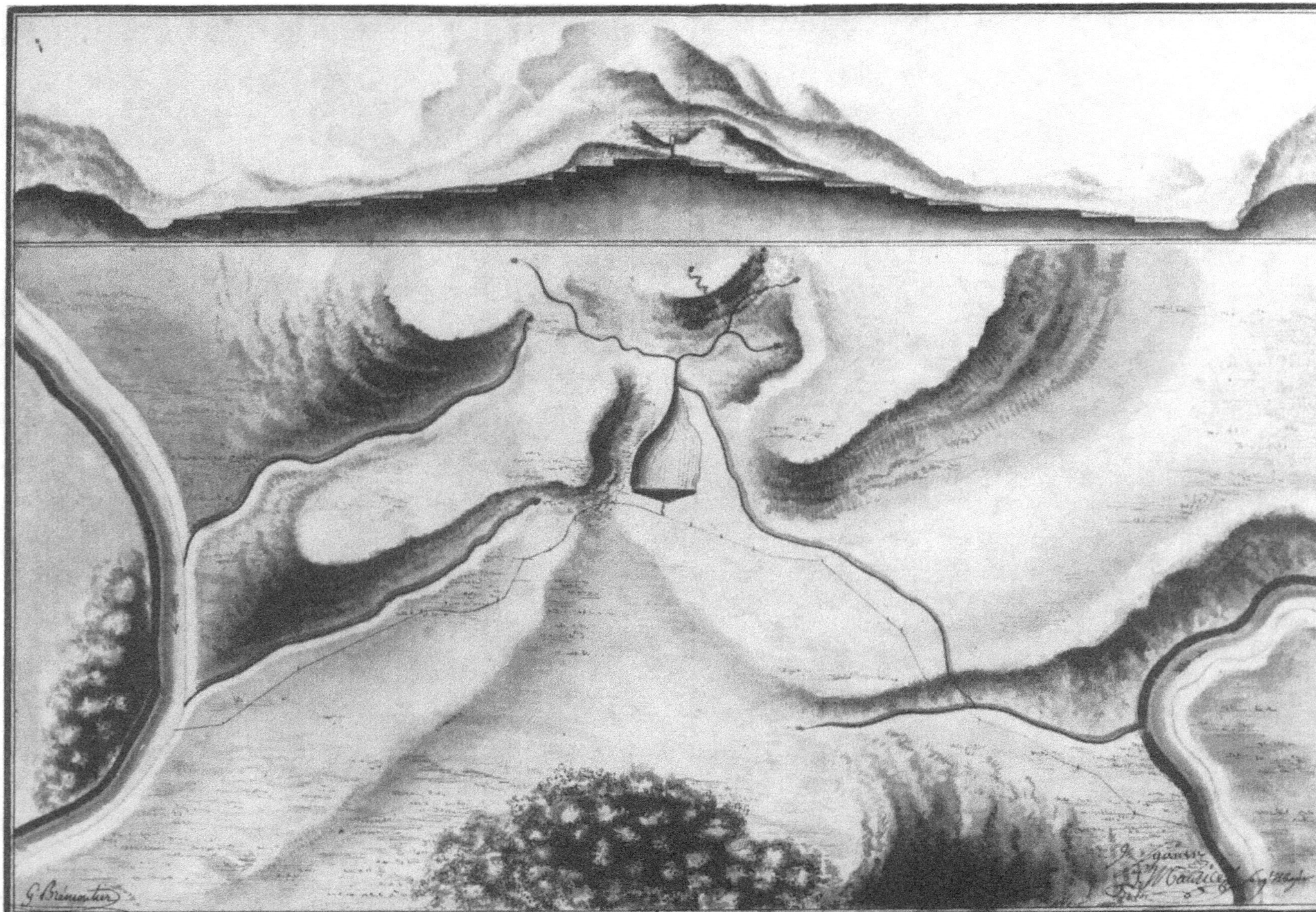

FIG. 89 G. Brémontier, canal project at summit level, drawing exercise set at the Ecole Polytechnique, private collection, ink, wash, watercolour.

surface or be concealed in its depths. This productive countryside could still be represented visually. The map, in the engineers' view, could be compared with an encyclopaedic venture. Nothing could be more telling in this respect than the observation an eighteenth-century author made regarding hydrographic maps. After he had described a model for a hydrographic map which represented relief 'using the variations of light and shade', this author remarked: 'A hydrographic map of this kind would serve as a model for maps which indicate the location of elements and the products of nature and of art. It would present a series as extensive as it was interesting of all the branches of natural history'.[55]

In spite of the engineers' confidence in traditional means of representation, the golden age of maps was drawing to a close. In the 1800s a dryer, more abstract cartography, often executed in black and white, began to replace the heightened realism of the previous period.[56] This shift is readily perceptible at the Ecole Polytechnique, where, in the early years of the nineteenth century, drawing methods were standardised. Nothing could be more unlike the map competitions of the Ecole des Ponts et Chaussées than the exercise, representing a canal at its summit level, which featured in the pupil Brémontier's portfolio in 1806 (Fig. 89).[57] No more than a dozen years separated this design from the works of the pupils of the Ponts et Chaussées, yet the underlying spirit of the composition was completely different, its analysis reminding one far more of Monge's descriptive geometry of a 'copy of nature seen from the point of view of a bird in flight'.

This was because, faced with the flood of information regarding 'the products of nature and of art', a scrupulous attention to pictorial detail proved inadequate. The available information had to be distributed in terms of a new grid; figures, graphs and graphic techniques now had to be combined. New methods for tackling territorial reality began to emerge, an excellent illustration of which would be the design which Prony conceived in 1800–11 for the draining of the Pontine Marshes (Fig. 90). Ever since the publication of his *Nouvelle lecture hydraulique*, since the *Recherches physico-mathématiques sur la théorie des eaux courantes* (1804), Prony had presented himself as an engineer who specialised in problems concerning water.[58] In his *Recherches*, which drew upon works by Chézy and Dubuat,[59] he had given a polynomial expression of the resistance to the flow in open-air drains and in pipes, at the same time as in his expression of the speed of the fluid he was anticipating the use of non-dimensional coefficients, whose hydraulics would subsequently play a now familiar part.[60] Prony used these results in 1811 to calculate the gradient and the cross-section of the canals he was proposing to dig. The science and art of the engineer thus converged in a precise application. A whole series of other data were also involved in the design, including the general surveying of the region, geographical and geological insights and measures of average rainfall. In the elaboration of his drainage plan, Prony isolated and listed the problems, much as he had always done. He thus took into consideration two chief classes of water, namely, running water and upthrow water. Each of these would be subjected to a separate treatment, and two kinds of drain would also be distinguished, being organised around a central canal, which would serve as the lynchpin of the design.[61] This latter also involved a number of bridges and culverts designed to relieve pressure of traffic (see Fig. 91). Being standardised, these constructions no longer had anything of the sublime, nor even of the picturesque, about them.[62] The books which Prony published in 1818 and 1822 sum up his whole approach: 'Des Marais Pontins' or the territory tamed by the planner's calculations.[63]

A STATE TERRITORY

Behind the territory, and behind the action of the Ponts et Chaussées, whether in open countryside or on the outskirts of towns, we of course find the state, which both inspired and coordinated the work done in its name. So far in this chapter we have spoken very little of the state, to which the engineers were subject by means of a hierarchical administrative organisation. If the territory was the concern of the corps of the Ponts et Chaussées, a decentralised organism, the structure of this corps constituted a kind of ideal architecture of territorial reality.

This architecture depended upon the ranking of material and human resources. Under the *Ancien Régime*, the holding of enquiries was commonplace within the administration of the Ponts et Chaussées; this practice continued and even intensified during the early stages of the Revolution.[64] Thus, in 1792–3, Roland, who was Minister of the Interior, entrusted the engineers of the Ponts et Chaussées with an enquiry into the state of agriculture, the roads, the manufactures, the *départements*' mines and quarries. The minister wrote as follows: 'The building of roads and the distribution of canals should from now on be calculated in terms of enlightened, administrative and benevolent views, for the more effective circulation of foodstuffs and commodities, the most prompt communication of assistance of all kinds, the most

FIG. 90 G. Riche de Prony, *Atlas des Marais Pontins* (1822), map of the marshlands.

rapid development of industry, and the abundance and wealth of society'.[65] In spite of the revolutionary rhetoric and the use of the phrase 'from now on', which was meant to affirm the birth of a new age of reason, the minister was really doing little more than reiterate ideas to which the engineers already subscribed. A number of new ideas were nevertheless grafted on to the requirement that agriculture be developed by means of the roads. First of all, one should note the use of the verb 'calculate', which served to bind economic considerations more closely than had been the case in the past to planning, introducing in addition a deterministic dimension to this relationship. But the most important aspect was undoubtedly the potential broadening of the engineers' involvement in industry. As a former inspector of manufactures,[66] Roland had proposed a short time before that the training of inspectors take place at the Ecole des Ponts et Chaussées; as Jacques Guillerme has shown, this was in effect to anticipate the production engineer, who in fact only made his appearance in the course of the nineteenth century.[67] Meanwhile, planning, agriculture and industry were already expected to dissolve into an overall policy, which was to be 'administrative and benevolent', and for which the engineer would be the spokesman. The state sought to take charge of the countryside by planning all the activities which took place there. Admittedly, this programme was not yet to be implemented; Roland was to be guillotined during the Terror, and a fair number of his proposals were buried with him. But the minister Neufchâteau's enquiries in the Years V and VI would revive the plan to rationalise the territory, although on this occasion they were addressed to the newly created prefects.[68] At the same time, the Revolution may be said to have made it possible to establish the modern instruments of state domination: statistics, which was gradually taking shape,[69] inheriting the mantle of the 'political arithmetic' of the classical age,[70] and above all taxation, with the land tax introduced in 1790, the *patents* in 1791 and the contribution on buildings in 1798.[71] These convergent measures enabled the territory to become increasingly instrumentalised. It became still more homogeneous with the abolition of the provinces and of local particularisms. Indeed, a new space arose out of the debris of the *Ancien Régime*. Being the heir of men such as Sully and Colbert, who had been aided by their clerks and their engineers, it also gave rise to fresh contradictions.

For this space was in the last analysis as contradictory as the one which it had replaced. Fundamentally, the contradiction derived from the conflict which arose between the collective mandate of the state and the class which it represented, between a planners' Utopia, tinged with social resolutions, and

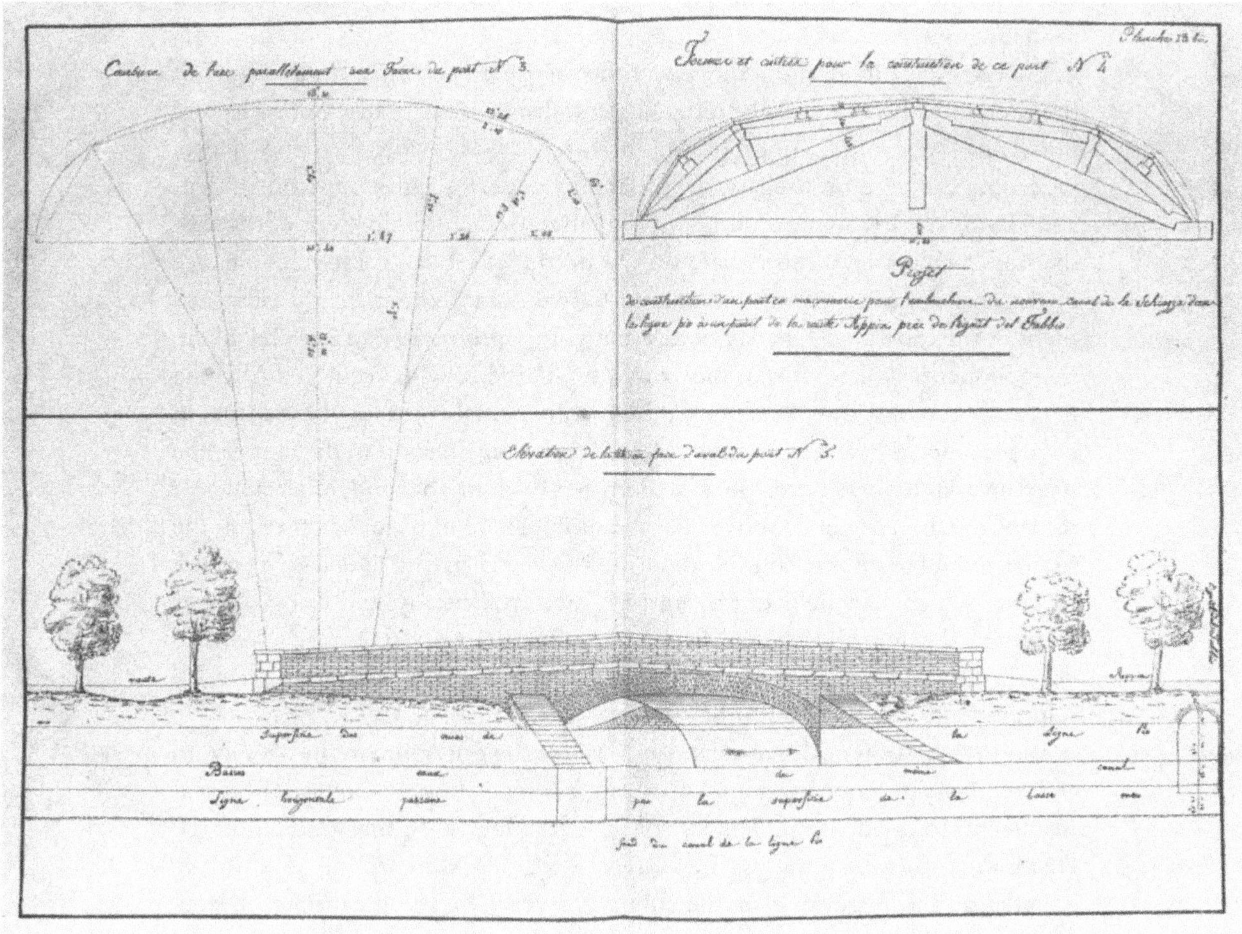

FIG. 91 G. Riche de Prony, bridge project. A standardised setting-out method is shown in the upper left-hand part of the picture.

private initiative. Under the Revolution, legislators would barely be conscious of such contradictions, since they believed that they were working for the good of humanity and in accordance with the 'general will'. One would have to wait until the nineteenth century for all the consequences of the work of the revolutionaries to become evident, such as the essential ambiguity of the relationship between the bourgeoisie and the state which it was appropriating without, however, ever managing wholly to subordinate it to its own interests.

Energies were focussed in the meantime on other problems, beginning with the alternative between continuing the centralising work of the monarchy and pursuing a decentralisation, desire for which was evident in the debate preceding the vote on the laws of 19 January and 18 August 1791 on Bridges and Highways. Involving arguments which were often reminiscent of those of the Le Chapelier law, this debate turned on the question of the maintenance or the abandonment of a single administrative management for the development of the territory. The day before the vote on these laws was to be taken, the opponents of the corps of the Ponts et Chaussées, who accused it of being unwieldy and of being characterised by a caste mentality, seemed for a time to have prevailed. However, owing to Mirabeau's contribution to the debate, the structure set up by Trudaine and Perronet was in the end reinstated, for reasons of efficiency and, above all, economy. The end of jurisdictions by the Estates would also serve to increase its influence, by giving it access to the last bastions which had held out against it, such as Brittany and Languedoc.[72] Although the Revolution had been inspired in its early stages by a wish to decentralise, in development, as in so many other fields, it resulted in a wholly unprecedented concentration of powers.

But the question of decentralisation was not the only one on the agenda. In order to manage a homogeneous territory, one had also, paradoxical though it may seem, to divide it up. The law of 22 September 1789 thus resulted in the creation of the *départements*; these new divisions were to be definitively adopted at the beginning of the following year. In the meantime, a large number of different systems were proposed in the Assembly. What was to be the basis for the projected division? Some counted on the virtues of geometry. Thus, Thouret,[73] and later Lhote, argued for a rigorous compartmentalisation on the basis of what were virtually squares. Mirabeau, on the other hand, defended a more balanced conception.[74] One should, he stressed, take the relief into account, for what was the point of uniting administratively what nature had kept radically separate? The legislator ought also to take into account population, its modes of grouping, whether in towns or in villages, its traditions and its activities. Once one began to ponder these different elements, the importance of field research became only too apparent. There was also nothing surprising in the fact that the engineers played a very prominent role in this area. As Director of the Cadastral Survey of 1791–8,[75] Prony in particular was called upon to make a statement regarding the 'plan proposed by Citizen Lhote for a new division of the territory of France'. Prony spoke as follows:

> Anyone who has given the slightest thought to this matter knows perfectly well that no plan of this nature could ever be based solely upon geometrical considerations . . . nature has imposed other conditions, to which geometry ought to be wholly subordinate, and which depend upon mountain ranges, upon the whereabouts of water, whether running or stagnant, upon the position of communes which constitute the chief centres for the consumption of wealth, whether territorial or industrial, of the nature of the different kinds of product yielded by the soil, on the surface or from its depths, and of population.[76]

At the end of the eighteenth century, geometry thus came up against one of its limits, namely, a territory which development could certainly knit together, and which could be unified politically, but which was nevertheless resistant to any narrowly spatial approach. Setting aside questions of lay-out, the division of the territory, together with the preliminary surveying of it, required new tools. Areas, populations, ratios and flows would soon make it possible to describe a specific space in terms of abstract data. The economy would also play an increasingly important role.[77] In the first half of the nineteenth century, ground rent loomed large in the debate over development; the territory of the engineer was no longer only a 'copy of nature', which might with patience be improved, nor even an architecture of reason, which administrative reforms might enable one to put into practice. Opposed forces began now to clash over the territory, as it became a kind of frontier at which the chief equilibria of the period could be deciphered, namely, those between the state and private initiative,[78] and between development and design.

In order to bring this chapter to a close, I want now to emphasise the paradoxical character of the embryonic territory which was contained in the Utopian maps of the Ecole des Ponts et Chaussées, and which the Revolution and the Empire were definitively to establish. It was the fragmentation of the territory which led, as we have seen, to the involvement of the engineer and of political agencies, and both were designed to remedy its shortcomings by reunifying it. But this reunification was essentially transitory, and was always balanced by the contrary tendency; the territory was alternately unified and divided. It was unified in terms of a nature which simply subsisted and was ever fertile, but which gave rise to obstacles in the relief which came between men and hampered the free play of supply and demand. This accident-ridden territory was replaced by that of the developer, who filled in ravines and levelled mountains, just as politics abolished obsolete privileges and compartmentalisations. Yet the territory which had been restored in this way to its unitary dimension could not help but be divided up again, in the name of the technical, economic and administrative treatment which was imposed upon it. Seen in these terms, the territory seemed to be in a state of perpetual transformation, under the impact of a force which was by turns positive and negative. Nature and natural obstacles, development and administrative divisions were thus both opposed and combined, just like the state, the society and the individuals of which it was composed.

For what this indefinite progress was soon to reflect was indeed the other movement which, in the name of progress, animated state, society and individuals. If calculation in fact had come increasingly to constitute the sole genuine measure of the territory, the latter, once it was tamed and traversed by the former, was assimilated to the conquests of the human mind, becoming the site of battles won against nature. The first great Utopias of the nineteenth century, Saint-Simonianism in particular, were also expressed in territorial terms, and were systematically in favour of canals and railways.[79]

The engineers of the late eighteenth century had not yet arrived at this point, but we are now in a position to understand just how important the territory was to them. As an image of society, the territory replaced the image which the architecture of a man such as Blondel had provided, and which had been buttressed by the spirit of *convenance*. Since it delineated a territory that was productive, this image made it possible to escape the ponderousness of urban space. While the town in Patte's conception remained immobile, in spite of his propensity to push hospitals and cemeteries into the outskirts, the existence of the territory was confused with the alternation in its phases of division and reunion. This rhythm was also one of dilation, under the impact of the accumulation of obstacles, and of the contraction which rapid means of communication established. It was in this sense that the territory was limitless, and it was from this point of view that it was increasingly to penetrate the town, fashioning it in its own likeness. The rational city of the *Mémoires sur les objets les plus importants de l'architecture* remained confined to the inside of the regular perimeter of its boulevards. The limitless town, on the other hand, was to oscillate between the recapture of its centre by means of sequences of services and the indefinite extension of its periphery. As the matrix of this twofold movement, the territory at the end of the eighteenth century constituted more of a promise of things in store than a precisely defined programme.

Its uncertainties correspond fairly exactly with the discourse of the engineers, with its superimpositions of order and disorder, of contintinuities and discontinuities, a mixture in which the technicians' ideology of the following century was forged.

10

FROM THE REVOLUTIONARY 'GENIUS' TO NEO-CLASSICISM

THE CHALLENGE TO THE CLASSICAL INHERITANCE

'I believe that a major change in architectural taste has only been perceptible in France since the death of Blondel', declared Mathurin Crucy, a pupil of Boullée and a former holder of the *prix de Rome*.[1] This change of course coincided with the emergence of revolutionary architecture, which presaged neo-classicism. But the abandonment of the concepts of the *Grand Siècle* and the consequent adoption of a contrasted geometry, based upon simple elements such as the cube and the sphere, had been a gradual affair, beginning around the 1750s, as the 'Greek taste' of men such as Soufflot or Contant d'Ivry took the place of rococo. Blondel's critiques of fantasies in a *rocaille* style, a number of his designs, such as his 'Palace with a frontage of sixty-six *toises*', were a part, as we saw above, of this tendency towards a hardening of forms.[2] Furthermore, in the work of De Wailly and Ledoux, one still encounters classical composition procedures.[3] The continuity of architectural currents in the second half of the eighteenth century should warn us against entertaining any too clear-cut opposition between conservatives and innovators.

Nonetheless, the death of Blondel in 1774 was a watershed. Now that the teaching posts which he had held at the Ecole des Arts and subsequently at the Academy had fallen vacant, it became possible for others to account in theoretical terms for the practical transformations which were already under way. Architectural theory, such as Blondel had conceived of it, now seemed to be threatened on all sides.

These transformations were as much concerned with the definition of architecture as a discipline as with the forms of knowledge which were involved in it or with the relationship between itself and society. We have already stressed how small a part nature played in Blondel's theory. However, nature was a universal point of reference, in philosophy and in the sciences, in social theory and physical experiment.[4] A fair number of Blondel's pupils now saw his architecture as one which, though endowed with rules and founded upon prestigious examples, was without a proper theoretical foundation. Even the rules had now to be treated with caution, since voyages and the emergent science of archaeology threatened to overturn them. Soufflot and Dumont visited Paestum in 1750, and emphasised the discrepancy between the squat Doric of the temples of southern Italy and the proportions which the classical authors had specified.

Leroy clinched the argument in 1758, with his *Ruines des plus beaux monuments de la Grèce*, which depicted ordinances which no longer corresponded with those stipulated by architectural theory.[5] The Orient was soon to take the place of Greece in the imagination of writers and artists, with the publication of Volney's *Les ruines* (1791), and of the engravings done by Cassas of Baalbek and Palmyra (Figs. 92–4).[6] Theory was also affected by research into problems of construction. The art of building was asserting its independence from architecture, so that, where Blondel had seen a genuine distance between theory and practice, a radical void was established, which the engineers filled in their own fashion. A yet more damaging factor was the emphasis on social utility in the writings of both physiocrats and technicians, a preoccupation at odds with the magnificence usually associated with the idea of architecture. The massive Doric of the temples at Paestum and the rhetoric of primary needs each served as sources of new information to which the architects were obliged to respond.

Blondel's response, as we have seen, was to reconcile architecture with society by means of a language referring both to the systematic designation of principles and of the parts of the art and to that of ranks and conditions. The *Cours d'architecture* was in this respect a failure. The correspondence which Blondel so much desired seemed to be forever eluding him, as it subsided beneath heterogenous lists or irreconcilable registers, distortions which showed just how weak the foundations of his theoretical edifice were. It was these foundations which the architects were subsequently concerned to

FIG.92 C.-F. Volney, *Les ruines* (1791), frontispiece. The philosopher stirred to thought by the ruins of Palmyra.

FIG.93 L.-F. Cassas, *Voyage pittoresque de la Syrie, de la Phénicie, de la Palestine et de la Basse-Egypte* (1799 and subsequently), 'Mausoleums located at the head of the valley leading to Palmyra'. The alliance between the sublime and the picturesque reveals the renewal of aesthetic categories at the end of the eighteenth century.

renew, by paths that were often contradictory. Blondel's pupils were increasingly aware of the shortcomings of the classical vocabulary and of the impossibility of overcoming them. It then became a matter of urgency to free themselves from the rules of the past, so as to recover the truth of architecture. This truth would sometimes be sought in a primitive knowledge on the basis of which everything might be reconstructed – a rational geometrism and combinatory of sensation in Boullée and Ledoux, a symbolism in the case of

Viel de Saint-Maux[7] – sometimes a historicism which, in the wake of Piranesi, was based upon archaeology.[8] The majority of the architects of the 1770s and 1780s thus combined the invention of a new combinatory with a taste for Antiquity, for it was in this period that the excavations of Herculaneum and Pompeii were beginning.[9] It might seem inevitable that there should be a contradiction between new rules, as immutable as those which they had wished to replace, and the discovery of history, the exploration of the vicissitudes of taste, the ornamentalism of a man like Delafosse,[10] or the eclecticism of someone like Lequeu,[11] but, as Georges Teyssot has observed, this contradiction was dialectical in nature. Indeed, the opposition between the 'regularity' of Boullée and Ledoux and the 'irregularity' of the French Piranesians concealed one and the same problem, that of reconciling necessity and invention.[12]

This necessity now took the form of 'genius', a category which had replaced the 'talent' recognised by Blondel, and of which Mansart had been the most admirable example. In Boullée and Ledoux, 'genius creates', whereas talent simply serves to put its inspirations into practice, and 'the sentiment of the art positions'.[13] There was something much more momentous involved in this reversal of Blondel's classification than a mere question of vocabulary.

A NATURAL 'GENIUS'

The first issue to be addressed was one's attitude to nature, which was no longer the vague repertoire of forms and sensations to which Blondel had referred. Condillac and Rousseau had put it once and for all at the centre of reflection on man and society; the natural sciences hymned its fecundity. As Buffon wrote:

> Nature is the system of laws established by the Creator for the existence of things and for the succession of beings. Nature is by no means a thing, for if it were, it would be everything; nature is by no means a being, for such a being would be God; but one may conceive of it as an immense, living power, which embraces everything, which animates everything, and which, subordinated to that of the first Being, has only begun to act through his orders, and still only acts with his cooperation or consent.[14]

In the years immediately prior to the Revolution, this enthusiasm was nevertheless sometimes tempered by a dumb anguish, for nature transcended man.[15] Indeed, nature was a mysterious entity, which posed a threat to man, with the

FIG. 94 L.-F. Cassas, 'Mausoleum of Iamblichus'. The geometrical reconstruction of this monument is reminiscent of studies relating to revolutionary architecture.

aesthetics of the sublime being positioned at the very heart of this disproportion between the creature and the work of the creator, a work which was itself traversed by contradictory and violent currents. Writers and artists from Sade to Fuseli began to be haunted by the strangeness of nature and the cruelty of the unleashed forces of evil. Man was himself a being by turns opaque and luminous, a mixture of impulse and reasoning, the key to which was very difficult to find.[16]

Moreover, it could be the case that the apparent absurdity of the human condition fundamentally referred only to death, as if the creator were absent and had withdrawn from his work, leaving behind him but a caricature of a world. In 1790, Benjamin Constant recalled a meeting with an 'homme d'esprit' whose cosmology closely resembled his own:

> He claimed that God, that is to say, the author of ourselves and of our surroundings, had died before finishing his work; that he had had the most beautiful and huge plans for the world and the greatest possible means; that he had already set several such means to work, just as one puts up scaffolding for the purposes of building, and that, in the middle of his work, he had died; that everything at present was done for a purpose which no longer existed, and that we in particular, felt ourselves to be destined for something of which we could form no idea; we were like watches without any dial, and whose wheels, being endowed with intelligence, would go on turning until they were worn out, without knowing why and forever saying: since I am turning, I must have some purpose.[17]

This sombre picture, which represented nature as violent and cruel, as something which man tore in tearing himself, as a cyclical phenomenon made in the image of a broken watch, as an indefinite movement interrupted only by the fitful lucidity of the human mind, coexisted with the hope that there might be a radical regeneration. For nature was still the source of all benefits, dispensing its gifts without counting the cost, and indicating the path to be followed if one wished to tame it. The discourse of the French Revolution set out to be optimistic, as if it were a new era, resolving all the contradictions of the past. In the meantime, if the taste for secret societies, for mysteries and for revelations spread among the nobility and the bourgeoisie, it was doubtless a reflection of the difficulties of the period, which saw persistent structural blockages and a context of masked economic crisis apparent since the 1770s.[18] Masonic ideas were increasingly popular,[19] while Mesmerism served to crystallise a blind faith in the virtues of science, which the Revolution was soon to exploit.[20] Architects were not untouched by these developments.

There was a vogue for 'buried' architecture, which Boullée claimed to have invented,[21] but which also found expression in the less austere drawings of Desprez. These works were characterised by chiaroscuro, tombs and bones, a disturbing atmosphere and a poisonous symbolism (see Fig. 95). One should range under this same heading the walk taken by William Beckford in the company of Ledoux, in the course of which the author of *Vathek* discovered strange sites and an architect making confused and exalted observations.[22] On the other hand, the influence of Condillac and Rousseau upon Ledoux and Boullée, and the admiration of this latter for Newton, reflected the expectation of the 'calm reign of the laws' of nature. 'Revolutionary' architecture was thus subject to a number of conflicting tendencies.

Perhaps the most vivid account of this state of affairs is that given by Ledoux in the text in the *Architecture* accompanying the 'Elevation of the cemetery of the town of Chaux' (Fig. 96): 'You who make a bugbear out of your own ignorance in order to frighten your active faculties, open your eyes! Observe the sublime harmony created by the parts of this vast tableau: the six epochs known as days prepare the munificence of the heavens and the various developments enacted in the heart of nature.' While the engraving transcends the register of man's productions, in order to display the heavenly vault and the 'immense order' of the planets, the architect's pen then goes on to describe these 'developments enacted in the heart of nature' and these 'shadows which obscure the earth and weigh down the seasons with grief':

> The forests are stripped bare; the lakes, which were formerly filled with aquatic wealth, have dried up. Destructive blasts spew forth epidemics, hunger and all those evils which destroy empires: unflinching destiny would seem to have decreed their end. The earth is isolated; man recoils in terror from the sight of this dread vault, which is to cover the debris of humanity: the hideous *politique* familiarises himself with crime, heaps up poisons and daggers; everywhere there is vengeance, injustice and ambitious passion, and the immutable order is shaken; vices piled upon vices stifle the virtues; motive powers weigh upon the two poles; murderous commotions stir up the mud which spreads its contagion. Everything here is confused, nature is shaken; men as moral beings no longer exist.[23]

This apocalyptic description was followed by the announcement of a new world, born of chaos, and then by the praise of the sea, which keeps the fiercest hatreds at a distance, thus avoiding the universal conflagration which the ambition and greed of men gathered in the same place would otherwise have been bound to produce. The text ends with an evocation of a society which

FIG.95 L.-J. Desprez, crypt and tombs, late eighteenth century, ENPC, ink, wash; 32 × 46.5 cm.

had 'improved the shores of this peaceful sea with thousands upon thousands of buildings', designed to increase the happiness of the century: monuments of charity, temples to the social virtues, agricultural ventures without which nature would indeed be nothing but a huge graveyard.

This strange meditation is in keeping with the whole book, which was published in 1804, with engravings that Ledoux had had commissioned from 1773, the year of his admission to the Academy. Between 1773 and 1776, he had also imagined his town of Chaux, a Utopian city organised around the saltworks whose project he had been ordered to design by the Ferme Générale.[24] During the Revolution, Ledoux had been starved of commissions, for he was suspected of royalist sympathies; imprisoned under the Terror, he was released during the Thermidorean reaction. Written in large part during this latter period, *L'architecture* is redolent of Thermidor, its author sometimes assuming a prophetic and disillusioned tone, and thus deliberately obscuring his purpose, whether out of prudence or in order to invest it with a dramatic dimension.

Like many other passages in the book, the commentary on the 'Elevation of the cemetery of Chaux' may be read on several different levels, whether as a narrative of origin, as a political fable or as a plea in favour of architecture, the arts or trade. But the crucial point is to recognise in it the presence of a nature which is in perpetual movement,[25] by turns terrible and benign, perhaps even cyclical in its behaviour, if one thinks of that immutable order of time bringing about the ruin of empires and their ineluctable succession.[26] As he proceeds, the architect even goes so far as to identify himself with the Creator: 'the Architect is there surrounded by whirlwinds, and by clouds which dispute with him the preeminence of the heavens'. In the same way, the cemetery at Chaux (Fig. 97) is compared to the earth in which it is embedded, a hollow sphere from which there radiate galleries ending in funerary chambers. The excessive quality of this design cannot help but remind one of Boullée's designs, and of the famous 'cenotaph for Newton' in particular, where the tomb is the sole material object, and the light supplies all the other effects. Boullée's writing may have been more sober, but his ambition was fundamentally the same: as master of both day and night, the architect was the 'metteur en œuvre' of a nature in movement: 'Architecture being the only art by means of which one can put nature to work, this unique advantage is proof of its sublimity.'[27]

This sublimity now concerned the artist as much as his productions. It was in this that his genius resided, a creative faculty which borrowed its means, such as light, from nature, but its principles also. According to Kant: 'Genius is the talent (natural endowment) which gives the rule to art. Since talent, as an innate productive faculty of the artist, belongs itself to nature, we may put it this way: Genius is the innate mental aptitude (*ingenium*) through which nature gives the rule to art.'[28]

The quest for the fundamental rules of art became a priority for the revolutionary architects. Imitating nature no longer meant taking for one's model its objects, from treetrunks to leaves, but rather aping its laws of creation.[29] It was in this sense that architecture was a creative art, like the nature whose fecundity it reproduced. Furthermore, since genius could be identified with nature, the principles of architecture would be precisely the determinations of the architect, the movement of his mind creating freely. Necessity and freedom of invention thereby converged, genius never imitating nature so faithfully as when it obeyed its own impulses. In the last analysis, this configuration within the field of architecture was simply the consequence of a far more general tendency. Having set out to explore nature, preoccupied with the moral and physical sciences, the eighteenth century gave rise to a subject who would both synthesise the knowledge so far acquired and at the same time define its limits. The trajectory which leads from Blondel to Boullée and Ledoux was a part of a line of development which culminated in Kant's critical idealism.[30] The 'raisonnement de l'art' espoused by an architect such as Blondel was replaced by a genius seeking to base itself upon a system of immutable laws, inscribed in the processes of both nature and the mind. It was in this sense that the harmony of the spheres was a prelude to the unleashing of the elements and of the sensibility manifest in the 'Elevation of the cemetery of Chaux'. But astronomy was of course only a metaphor; the real principles of architecture had still to be decided upon.

These principles were to preside over the inventions of genius, and they were primarily concerned with design. In his *Essai sur l'art*, written on the eve of the Revolution, Boullée had declared:

> What is architecture? Shall I follow Vitruvius in defining it as the art of building? No. This definition rests on a crude error. Vitruvius mistook the effect for the cause. One must conceive if one is to put something into effect. Our ancestors only built their huts when they had conceived the image of them. It is this production of the mind, this creation which constitutes architecture.[31]

To the movement of condensation of nature in a genius who reproduced its laws there corresponded that of the architecture which was organised around

FIG.96 C.-N. Ledoux, *L'Architecture* (1804), 'Elevation of the cemetery of the town of Chaux'.

the execution of the design. In Blondel's eyes, what was built constituted a crucial reference; it was the gap between itself and theory which made the critique of architectural creation possible. For Boullée and Ledoux, the future of architectural expression lay with design.

Architectural expression depended primarily upon sensation. In the course of the second half of the eighteenth century, Condillac came to wield a powerful influence. In 1780, Le Camus de Mézières' *Génie de l'architecture ou l'analogie de cet art avec nos sensations* was published. Even if this genius remained tentative, in confining itself to the Orders and to the planning of apartments,[32] a new direction was taken with the assertion of the preponderant role of sensation. As Condillac had written in his *Traité des sensations* (1754),[33] 'all of our knowledge derives from the senses'. Knowing nature amounted to knowing its impact upon us; it was also a way of knowing oneself. Boullée was undoubtedly the architect who pursued this line of enquiry the furthest. Thus, in his *Essai sur l'architecture*, in which he finally relinquished the doctrine of proportions,[34] he turned towards simple masses, seeking to identify 'the analogy [between them] and our own organisation'.[35] Regularity, symmetry and variety constituted new principles in accordance with our senses; the sphere was taken to represent the image of perfection, combining 'exact symmetry, the most perfect regularity and the greatest variety'.[36] Squares and triangles, and indeed, all the regular geometrical forms were seen as expressing so many laws drawn from nature.[37] Boullée hoped to establish a combinatory, or a grammar of sensation, the hypothesis being that there was a rational correspondence between geometry and the operation of the human mind. It was a matter of producing images: 'The art of producing images in archi tecture derives from the effect of masses, and it is this which constitutes its poetry';[38] the combinatory was thus flanked by a poetics.

Although it depended upon the eternal laws of nature, the poetics of architecture also consisted of its mode of appearance. To the grammar of simple and regular masses there was added a theory of character, which constituted the application of that grammar:[39]

> Let us direct our gazes at an object! The first sentiment we experience then obviously derives from the manner in which the object affects us. And I term 'character' the effect which arises from this object and causes some impression or other in us. Investing a work with character entails employing appropriately all the means which will cause us to experience only those sensations which should result from the subject in hand.[40]

In more concrete terms, it was a matter of deploying the elements of a composition in such a way as to produce a grandiose or a picturesque effect, by combining masses, by playing with light, by always drawing one's inspiration from the changing spectacle of nature, from the severe coldness of winter to the 'correct and pure' perfection of summer. Character also affects the different kinds of building, so that sombre sadness is fitting for a court or a prison, while an enchanting spectacle would suit fairs or places of pleasure such as *vauxhalls*. But one should avoid the slavish imitation of nature; the architect should rather select whatever means serve to arouse and hold the attention of the spectator. The concept of character also occasioned the introduction of a notion of economy. As Le Camus de Mézières had already written, 'the more

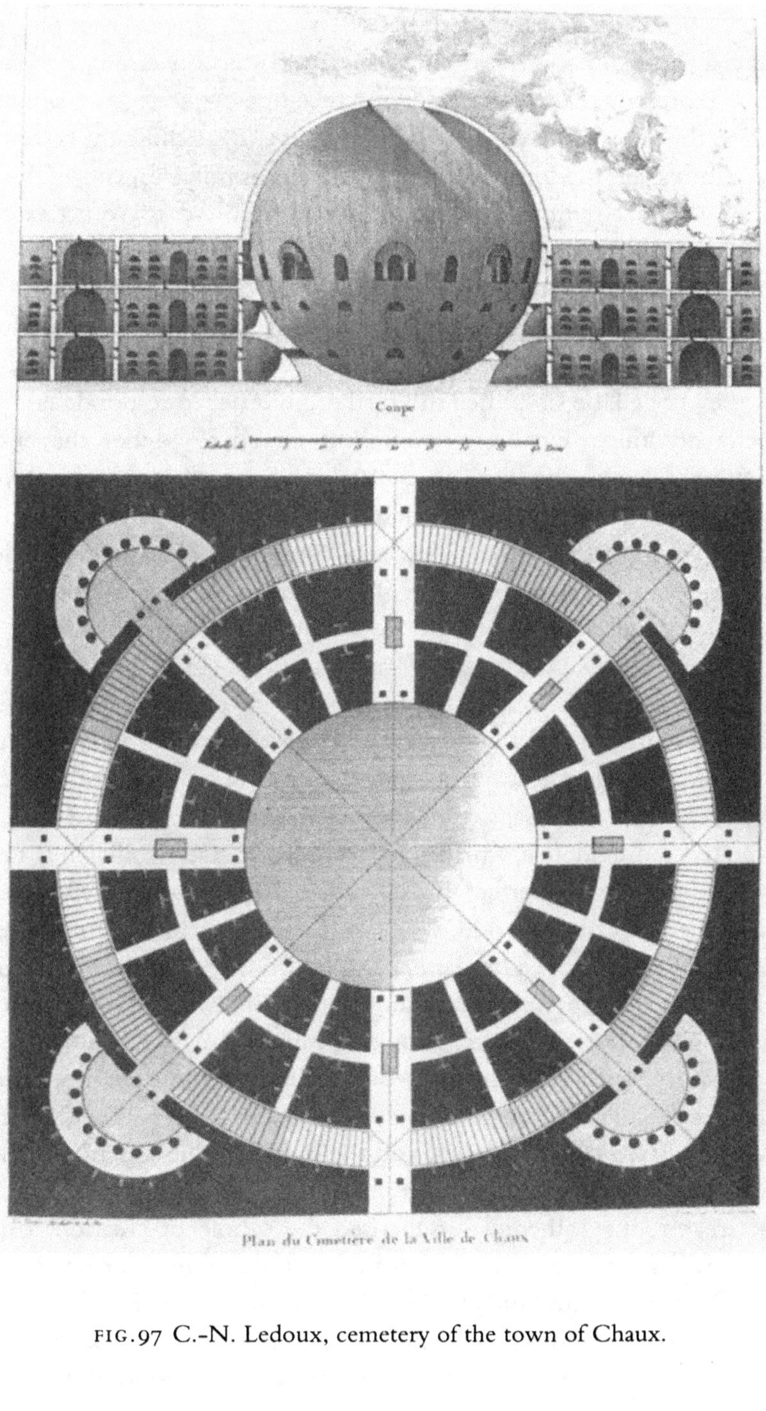

FIG.97 C.-N. Ledoux, cemetery of the town of Chaux.

closely I have examined the matter, the more I have found that each object has its own peculiar character, and that often a single line or a simple contour are sufficient to express it'.[41] As far as the latter was concerned, Servandoni's opera sets represented an unparalleled instance of a living *œuvre*.[42] Boullée also referred to the art of painting, observing that five to six different colours were sufficient to render 'the immense range of tones, or nuances, and all the effects of nature'.[43] Indeed, his theory of masses constituted to some extent the equivalent of this basic palette, with the characters being compared to the mixtures or harmonies that were subsequently obtained. The proliferating lists we find in Blondel were thereby avoided, while at the same time the necessary link between architecture and society was preserved, a link which would, they hoped, be simplified still further, in order to respond better to the exigencies of utility. Finally, a symbolism served to render the effects of architecture still more precise. Thus, Viel de Saint-Maux, in his *Lettres sur l'architecture des anciens* (1787), considered the order to be akin to 'a poem on fecundity'; 'one lists in it the first stones, or votive altars, from which the columns or simulacra were derived, stones which were the mothers of the sciences and of the arts, since they bore the first hieroglyphs or representative signs, to which we owe the origin of painting and of language'.[44] The message is clear, since the symbolic register enables one to go back to the origins of painting and of language. Boullée was also concerned with an expression prior to all division or classification, a synthetic formulation of the fit between architecture, nature and usage. Once revealed, this reading seized hold of and reinterpreted the evidence which had preceded it. De Wailly thus held the circle to be not only the perfect form, but also the measure of a man installing himself in the heart of the world, while for Ledoux it was 'the immense circle of human affections'.[45] The latter also believed that the pyramid was reminiscent of the tapering flame and of the soul which rises up to dwell among the immortals.[46] Finally, Boullée's cathedral evoked the three kingdoms of nature, with the mineral being evoked by the base supporting the building, the vegetable domain being referred to by the ranks of columns in the nave compared to quincunxes, and the spiritual sphere being invoked by the cupola, an airy temple which overhung the altar. As a building dedicated to the Creator, the cathedral summarised the whole of creation, from the inanimate rocks to man, the sovereign animal. Light, which 'produces effects', could well represent this God at whose behest nature works.[47]

At the end of the eighteenth century, a 'speaking' architecture thus made its appearance, an architecture intended to make the design speak prior to language and to combine with society through the expression of an essence as primitive as the needs which this latter experienced. This immediately calls to mind research undertaken around this same period, from Rousseau's *Essai sur l'origine des langues*[48] to Court de Gebelin's *Monde primitif analysé avec le monde moderne*,[49] in which the quest for the original language resulted in an awareness of the limits of articulation, and came up against the existence of an irreducible and founding cry.[50] In a certain sense, revolutionary architecture stems in a direct line from this cry, a perfect imitation of nature or a total invention.

REVOLUTIONARY DESIGN

Although we are now better equipped to grasp the scope of the genius, the 'exalted rationalism' which Aldo Rossi has attributed to Boullée[51] but which is also to be found in Ledoux, we still have to assess its impact upon the spatial plane. Since Emil Kaufmann's pioneering studies, the characteristics of revolutionary architecture have been the object of numerous works. There is probably no need to linger here over the various elements involved: a contrasted geometry, based upon simple forms, 'alphabetic letters from the best authors' (as Ledoux put it),[52] and the autonomy of elements freely associating instead of being subjected to a single continuous hierarchy. The juxtaposition of parts, the 'Pavillon system' which Kaufmann has observed, replaced the classical mode of composition. Instead of rehearsing this analysis in any detail, it may be more appropriate here to investigate the new procedures which it presupposed, both in terms of the architect's way of working and in terms of public concepts of design. I shall try to consider these procedures on three distinct and complementary levels, namely that of architectural form in the strict sense, that of the town (a theme present in the work of Boullée and Ledoux) and, finally, that of architectural detail, an element whose status was profoundly modified, as structural doctrines evolved.

TEXTS, IMAGES, FORMS IN MOVEMENT

Ledoux's engravings, together with Boullée's drawings, epitomise the resources available to revolutionary architecture. Only some of these engravings featured in the *Architecture* of 1804, since the projected publication in four volumes was not completed before the author's death.[53] In 1847, however,

Daniel Ramée was to publish all of the engraved work, although he failed to reproduce the text of 1804, deeming it obscure and too removed 'from the subject which it treats or the drawings which it explains'.[54] This distancing from the text is itself suggestive. In spite of a number of somewhat ponderous passages, Ledoux's writing was not without its virtues.[55] Its evocative qualities would only be matched by the prose of Le Corbusier and, indeed, it is only in this respect that Emil Kaufmann's genealogy[56] is wholly justified, for the same apologetic ambition and the same lyricism animates the author of the salt farm of Arc-et-Senans and that of the radiant City.[57] Yet it was not merely a craving for immortality that accounts for the architect's recourse to writing. For, in Ledoux, the text has a new function which Blondel had only glimpsed when he declared that, before convincing, 'one must begin by stirring the emotions'.[58] Like his former teacher, Ledoux naturally set out to convince the reader also, and, as Anthony Vidler has quite rightly emphasised,[59] the *Architecture* is markedly didactic in intention.

This didactic purpose clashed, however, with the forms of the past. Ledoux therefore waxed ironical over Blondel, the master 'circumscribed within the five Orders',[60] and nowhere in his work will one find a systematic account of proportions or a dissertation upon the principle of *convenance*. There is no trace of the 'raisonnement de l'art' which the professor of the Academy had valued so highly; he insisted, rather, upon the immediate seduction of architecture: 'If you wish to become an architect, begin by being a painter.'[61] There was then but a single step from painting to engraving. According to Boullée, the poetry of architecture consisted in producing images, and Ledoux was in agreement with him.

But it is difficult to understand precisely what role writing was supposed to play. If it was a question of arousing emotion, engraving ought to have sufficed; if one had to convince, one could have resorted to patient reasoning and exposition of principles. Yet no principle was directly formulated.

This paradox turns on the nature of the image. For Boullée and Ledoux, there was no image except in the mind of a spectator, the same spectator as was frequently addressed in the *Architecture*.[62] The book began, moreover, with the discoveries of a traveller arriving in the vicinity of the ideal town of Chaux, a traveller whose attention was in turn attracted by the bridge over the Loüe, a decorated barn, the house of the directors of the Loüe or the *canal de graduation*.[63] It was not by chance that the narrator was now on the move, discovering designs from a distance, drawing nearer to them, and circling around them as if the better to take possession of them. To this movement in space there corresponded the succession of sensations. Architecture was the art of producing images or that of linking up impressions, which although initially fugitive, accumulate and interconnect, so as to form a single complex and composite entity, in which the genuine character of the building could be discerned. The function of the text then becomes plain, for, if architecture was image and traverse, the physical impossibility of actually traversing an engraving meant that a supplement was necessary. To be more precise, the text took the reader on a walk, suggesting reveries and emotions to him, and recreating artificially the sequence of sensations which had originally constituted the design.

As Ledoux's book proceeded, the traveller disappeared, being subsumed by the figure of the architect, who now addressed the reader directly, as if he too were discovering the designs which he had himself composed. But this was because the traverse of the spectator was to a degree symmetrical with the work of the artist, both of them being based upon the immutable laws of nature. And since 'the simplest designs take on the colouring of the soul which conceives of them',[64] architecture was confused with its mode of appearance, an indefinite movement comparable to that of the universe:

> Architecture resembles those kindly stars which shed light on the world. Like them, it has its vicissitudes and its phases; like them, it experiences those shocks which confound the elements; thus, its light slips away, as theirs does, behind the thickness of the clouds, and, victorious over the shadows, it launches itself forth once more, still more brilliant and more beautiful than ever. It is surrounded by dazzling satellites which reflect its splendour and are sustained along with it in the heart of the immensity, by its imposing equilibrium.[65]

These vicissitudes and these phases could nevertheless refer only to the history of architecture. A few lines before this passage, Ledoux in fact declared that he had opted for a chronological order for expounding his designs, but also for demonstrating the progress which the art had known since the 'marvels of the Hebrews' and the 'ashes of Ilium'.[66] Although one might at first suppose that this programme bore a strong resemblance to that of Blondel, it was in fact based upon a wholly different set of assumptions. If Ledoux chose to refer, as his former teacher had done, to an Egyptian weightiness subsequently lightened by the Greeks, this reference seemed to entail no a priori condemnation. Indeed, the massive character of Egyptian buildings was to him exemplary, by contrast with the often cramped creations of the modern period. Ledoux associated early Antiquity with simplicity and freedom, untouched by any erudite affectation. Where Blondel, with an eye on Mansart, Lemercier and Le Vau, would barely cover more than a century, Ledoux bestrode whole millennia, since it was his wish to reconcile first principles with contemporary usage.[67] There was in actual fact nothing historical about this ambition. For, in spite of its allusions to the 'ruin of empires', the *Architecture* made a *tabula rasa* of the past. The possibility was even entertained, as the obsession with the figure of the circle implies, that history was actually circular. This was because the design absorbed history, subsuming it within the development of sensation. All that finally survived was a presentiment of it, together with the evocation of an immemorial symbolic register. The history of architecture was reduced to a bundle of significations which were, like rays of light passing through a lens, refocussed by the design. The probable circularity of the universe occasioned a poetics of vibration; in its movement of unveiling, architecture reflected the successive decompositions and recompositions of the atoms.

This movement now took place between extreme terms, oppositions which structured the practice of architecture, and which enabled one to give an account of it. To begin with, there were effects of contrast between the building and its site. Blondel used to surround a dwelling with gardens conceived in its image, which led one gradually from the apartments to the countryside. Ledoux, on the other hand, put his designs in abrupt contact with the landscape. The cubes and spheres which featured in the *Architecture* sprang up from meadows and thickets as if they were artificial objects, their artificiality referring to a purified expression of nature. Sometimes nature might even cut across the design, as if to emphasise more strongly the opposition and the close affinity between nature and the architectural object, as was the case with the waterfall of the house of the directors of the Loüe (Fig. 98),[68] or the river of Meilland, which passed below the Ecole Rurale.[69] In fact, architecture dominated the countryside, and surveyed it, as was borne out by the frequently repeated motif of the belvedere, the observatory or the mirador, which stood above the house of the director of the saltworks[70] or above the House of Education.[71] When commissioned to build the barrier-gateways around Paris, between 1784 and 1787, the new customs wall for the Farmers-General (dubbed 'propylaea' in the *Architecture*),[72] Ledoux invoked a space that was squared by an administrative and accountable gaze.

This surveillance did not rule out, however, the effect of surprise, but it was that of the spectator who discovers the site. This was another source of surprise than that of the comparison of elevation and plan. To the symbolic

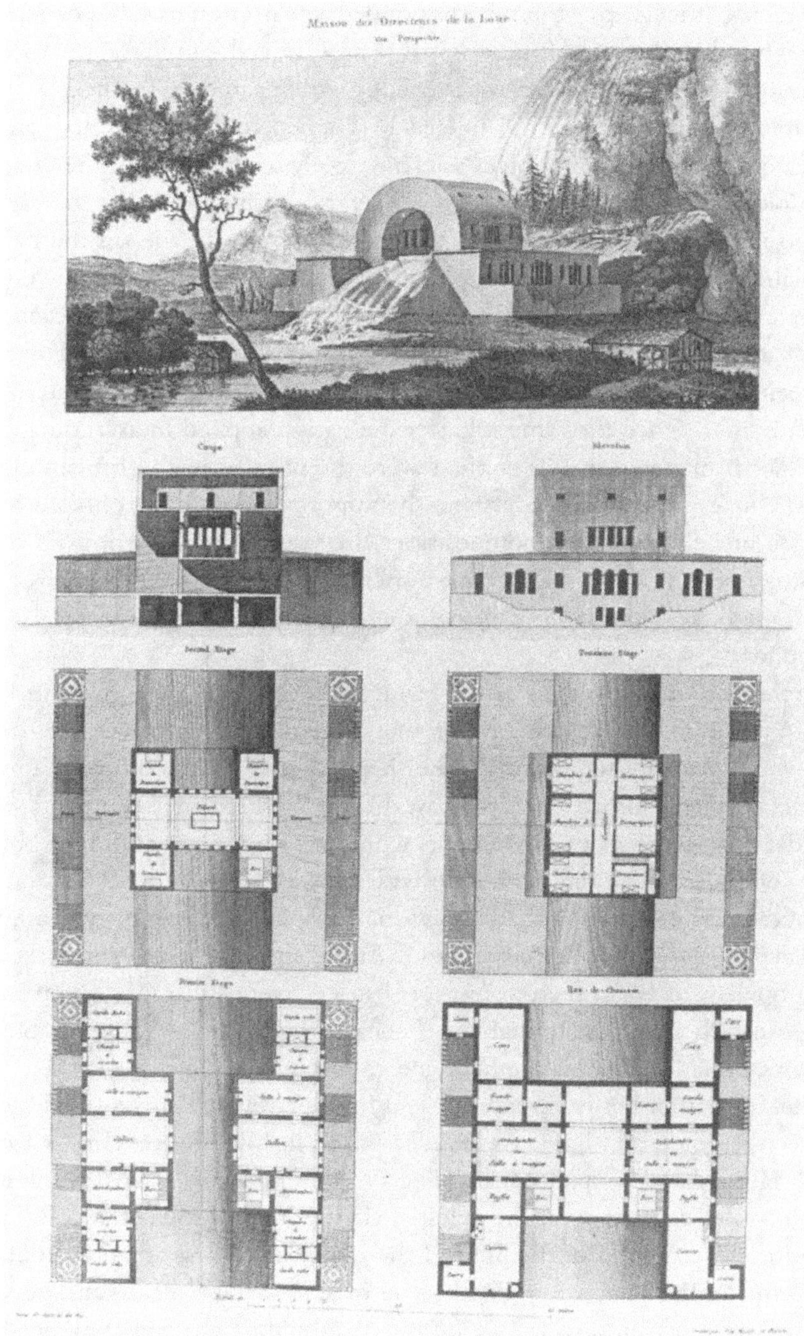

FIG.98 C.-N. Ledoux, 'House of the directors of the Loüe'.

volumetry of the hoops[73] there corresponded a projection based upon squares, in which only a shadow of the original form survived. Similarly, the cylinder of the house of the directors of the Loüe disappeared from the plan. The principle of 'autonomy' could indeed be applied to the different representations of the architectural object referring solely to the unity of the design as it unfolded. We have come a long way from the principles of classical composition, where one was able to reconstitute the façade on the basis of the plans and the information supplied by a module. Although they were not systematic, these paradoxical effects nevertheless bore witness to a breach with the ideal of *convenance*, such as Blondel might have applied to the relationship between exterior and interior. One could argue that exterior and interior no longer existed, since the same logic of the design applied indiscriminately to both. What mattered in this context were the interstices, the fringes and the limits. One can therefore understand the importance of the part played by the staircase in Ledoux's great compositions, from the director's house[74] to the *palais de justice* in Aix-en-Provence.[75] Being an essentially ambiguous place, which might be turned into a chapel if need be,[76] the staircase marked a kind of boundary.

There were many other ways of handling limits, from the simple masonry wall to those porches which gave an unexpected depth to the façade. Architects were generally tending to make their porches more complex. Thus, if Ledoux corrected his façades, doing away with those mannered ressauts which Blondel had described as 'movements which are sometimes authorised by the graces of the art',[77] he replaced them with another kind of movement, arising out of the play of light and shade between the columns of a peristyle, as in the *pavillon Guimard*[78] or at Louveciennes.[79] At the entrance to the saltworks or at the propylaea of Paris, these transitional spaces were still more complicated, with a grotto being hollowed out beneath the peristyle, or with bossage columns struggling against the light and casting violent shadows.[80]

The opposition between exterior and interior was in fact replaced by that between darkness and light. As Boullée wrote in his *Essai sur l'art*,[81] 'I make light'. His cathedral exploited the effects of the 'mysterious day' of the nave, which was in contrast with the celestial clarity of the dome. It was Boullée's ambition to seize hold of the light: 'I therefore say to myself, and I confess it with some pride: your art will make you master of these means and you too, you will have occasion to say *fiat lux*; at your bidding a temple will be ablaze with light or it will be naught else but the dwelling-place of shadows.'[82] The cenotaph for Newton was the most complete fulfilment of this programme, for it achieved an alternation of day and night within a perfect sphere, which symbolised both earth and the celestial vault. Ledoux's compositions were never so radical, although his church at Chaux presented a striking contrast between the altar, which was to be lit by a lantern, and the nave, which was to have blind walls plunged in semi-darkness.[83] Ledoux, like Boullée, accorded great importance to light and shade, which also enabled one to counter the dissolving effects of distance: 'We know that all masses diminish and taper away as we get further away from them; they lose their proportions.'[84] Revolutionary architecture would employ the light/shade dyad in order to assert its domination over nature; the design would function as a powerhouse of energy, transforming day into its opposite, decomposing it and analysing it.[85]

Although architecture constituted a nodal point for the organisation of space, and although it had, if it was to recall the eternal principles of nature, to be distinguished from the sweetness of undulating landscapes, one still had to isolate a new method of composition which took this imperative into account. In Ledoux's work, the introduction of such a method was long obscured by the extraordinary virtuosity of an architect whose artifices ranged from the diagonal plan and the polygonal interlockings of the hôtel de Montmorency[86] to the clever partition concealing the ellipse of Mademoiselle Guimard's theatre in a corner of the plot.[87] It was only with the propylaea of Paris that his thinking about space became organised in a systematic manner. This systematic approach diverged from the research undertaken by Blondel into matters of planning. Ledoux's problem was in fact no longer that of counting and of the analysis of the functions and of the rooms of a dwelling; his buildings testify in this respect to a complete mastery of the knowledge of his time. Subsequently, his thinking about design was to be much more concerned with composition, even if the *Architecture* marked the emergence of the new programme of the house.[88] We have already emphasised the role of regular forms; one should now note the importance of the square, which featured in the majority of Ledoux's compositions. He would often divide up this square, into three bays or into nine compartments, inscribe a circle within it, or hollow out its centre or its corners. The theme of hollowing out often featured, for example, in the decorated barn with which the *Architecture* opens or in the various country houses surrounding the ideal town of Chaux. The tensions which hollowing out created suggested the notion of an 'autonomous' development of form, from the centre towards the periphery or vice versa. Other arrangements fostered this same impression of movement, whether the centre was emphasised by means of a circular space on the basis of which the wings were organised, as was the case with la Cénobie,[89] or whether a multiplication of the square was involved, with other squares being stuck on or added, sometimes generating a regular, squared space, as at the market at Chaux,[90] sometimes a more massive composition, after the model of the *palais de justice* at Aix-en-Provence.[91] These methods might involve a repetition or a proliferation; in either case, they would be controlled by a square serving both as a unit of the composition and as a regulatory lay-out, and they would refer to a single process, with the form expanding or contracting. Finally, one should note the appearance of diagonals, which were appended to the traditional axes of the composition; architecture from now on expanded in directions which eluded the single register of frontal figuration.

These various techniques of composition led to a new concept of design. Exhaustive analysis of the geometrical possibilities of the circle or of the square in fact resulted in a typological approach, in the sense that these possibilities became the object of a systematic investigation, followed by a classification which cut across that of function and use. In the work of Ledoux, architectural production began to be polarised in terms of services and habitation, with the traditional opposition between the monumental and the vernacular being subsumed within the public/private dyad.

On the one hand, one therefore had the lawcourts, the prisons, the markets, together with the church and the cemetery, while on the other there were the houses of artisans and of employees, of woodcutters, and even the dwelling of the poor man, depicted in one famous engraving as being beneath a tree (Fig. 99).[92] We have already noted the unintentional irony involved in representing the poor man exposed to the bad weather and viewed by the Gods of Olympus with complete indifference. The commentary, however, stressed the need to shelter the destitute man, whose virtues were hymned in Rousseauist vein: 'Consider these winds, which are folded one into the other, are condensed and develop in order to assemble the magnificence of the sky with the benefits of the earth . . . Well! this vast universe which so astonishes you is the poor man's house.'[93] Nature was the quintessential dwelling-place of the poor man, and of the generic man who 'occupies but a little space', and who 'does not at the same time fill the immense void of the universe'.[94] Whether it was a question of housing a wealthy merchant or a humble journeyman, architecture isolated and partitioned, while at the same time remaining permeable to the breath of the universal spirit, that of nature and of the architect. But at the same time as the unity of the architectural work was

FIG. 99 C.-N. Ledoux, 'The poor man's shelter'.

affirmed, the poor man's tree perhaps referred to a radical discontinuity, for it was no longer a matter of programmes which merely had to be stripped down in order to pass from a palace to a hut. Architecture may still have been universal in its inspiration, but the solutions which it advanced were divided into zones of distinct pertinence. The typology was also characterised by the simultaneous affirmation of a continuity and of discontinuities, by the use of an immutable geometry and by the definition of programmes which could not be deduced from each other through successive differences. The expansion of a form no longer proceeded by, so to speak, addition or subtraction, for such movement from now on was achieved by means of multiplication or division.

However, one should not exaggerate the scope of the typological investigation. The agreement between programme, form and function was not yet based upon a rational arithmetic, or even upon one that claimed to be such. The combinatory of sensation was distinct from the classificatory spirit of Durand, whose simplifications inaugurated an abstract space, together with the analytic formulae studied by his pupils at the Polytechnique. In the *Architecture*, character presided over the encounter between design and its utilitarian determinations: 'Character, being based upon need of all kinds, is all the more varied for being also adapted to the *convenances*.'[95] There is nothing fortuitous about the reference to classical principles here, for their abandonment did not occur overnight, and many traces of Blondel's teaching survived on the blotting-pad. In the meantime, the typological approach often produced its own symbolic systems. The reader will perhaps be acquainted with the phallic form of the Oikema,[96] a 'Greek monument' designed for amorous revels at which 'Hymen and Amor would conclude a treaty serving to purify public morals'.[97] Although I cannot wholly accept Serge Conard's interpretation,[98] it may be that Ledoux's plans sometimes functioned as a kind of writing, referring to the mysteries of the universe and of life. The revolutionary architects' concern with investigation and with typological rationalisation was here tempered by a poetics descended from the hermeneutic tradition of Father Kircher and from Warburton's theories regarding the hieroglyphic language of the Egyptians.

THE TOWN, SOCIAL REFORM AND PLANNING

The town was not immune to such symbolic interpretations either. Thus, in order to complete the semi-ellipse of the saltworks of Arc et Senans, Ledoux conceived the nucleus of an ideal settlement whose lay-out evoked the course of the sun (see Fig. 100).[99] He thereby became the first architect to organise a town explicitly around a productive function. Whereas Patte, in his *Mémoires*, had merely pointed to the presence of factories on the periphery of the rational city, the *Architecture* specified the installations that would be needed for the making of cakes of salt, from the *bâtiment du graduation* to the workshops and the workers' housing. Approaching the kinds of conception which would subsequently result in the construction of arsenals by the engineers,[100] Ledoux proposed to rationalise this production by means of a rigorous spatialisation. Given this perspective, the idea of surveillance, which was facilitated by the elliptical emphasis, played a crucial role, the gaze of the director being directed

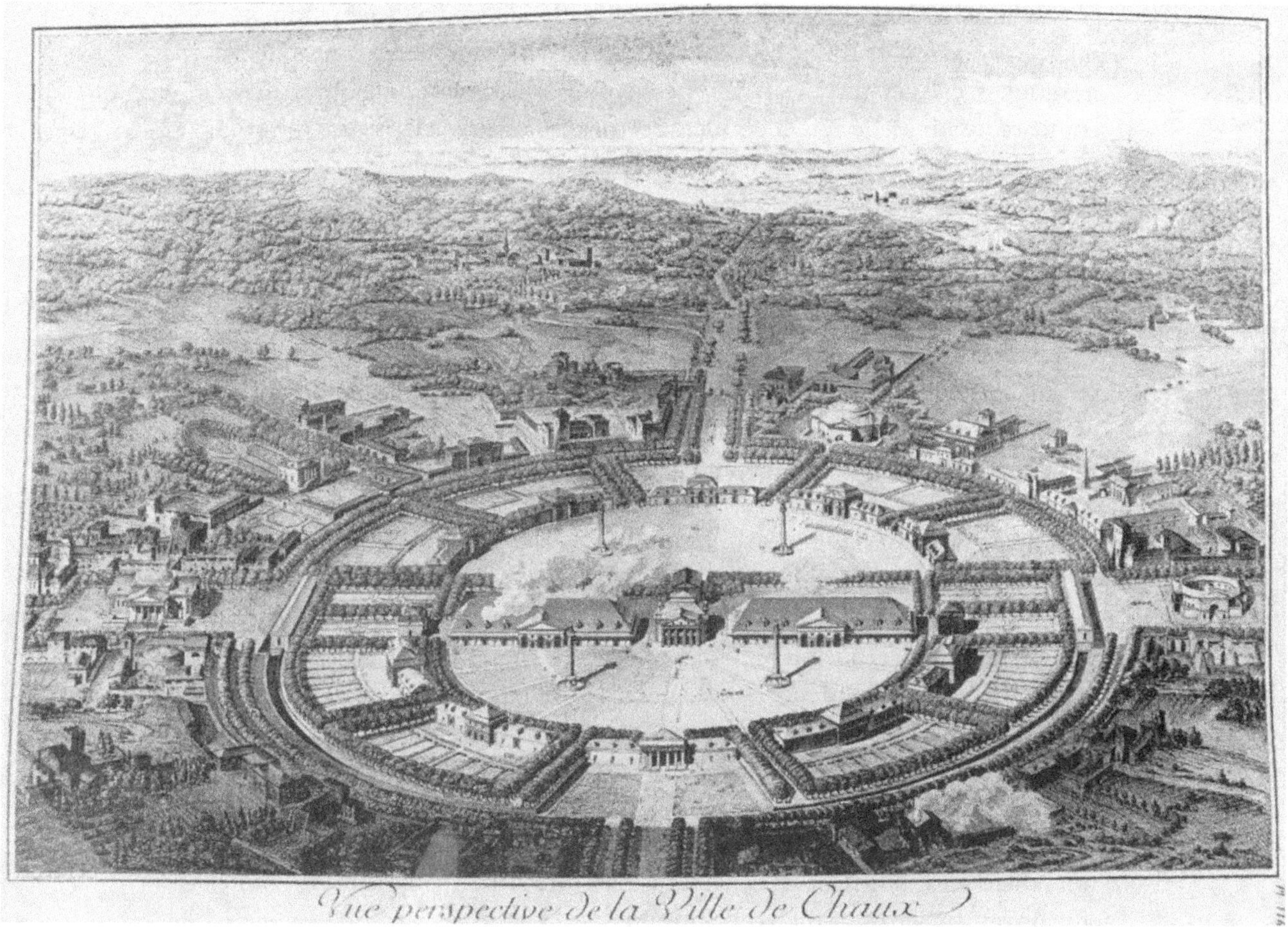

FIG. 100 C.-N. Ledoux, 'Perspective view of the town of Chaux'.

in turn at the various stages of the production. But this surveillance would have proved wholly ineffectual if there had been no consensus. The saltworks, and the town which served to complete it, operated through the 'general will', the director's calculations and the workers' actions alike being motivated by a shared intention to behave well. The factory and its environs were the framework for a perfectly adjusted social representation; though symbols of the universe, the ellipse and the circle also referred to the notion of a community that was transparent.[101]

Nevertheless, this transparency presupposed a collective celebration. Consider, for example, the semicircle of steps, which are contemplated by a motionless eye, in the theatre at Besançon.[102] This motionless eye may be compared with the gaze which the director of the saltworks directed from his house. But the theatre was above all else a site of sociability; the real play unfolded in the auditorium, which was filled by a brilliant gathering of people out to win admiration. For Boullée and Ledoux, the theatre served to exalt the beauty of women; through their finery, the whole of society was staged.[103] The eye in the Besançon design therefore looked towards the public. Likewise, at the saltworks, work became a festival, done under the enlightened surveillance of the director; architecture had been given the function not only of rationalising such work, but also of turning it into a celebration. Ledoux's thinking here took on a social and a political character. The Revolution was later to try, as is well known, to represent itself in the ceremonies of the Champ de Mars.[104] This line of thought was well developed in Boullée's *Essai sur l'art*, and it culminated in his Colosseum design, which was to 'announce the happiness of the nation'.[105] Ledoux's republican sentiments were undoubtedly more qualified, and a passage in the *Architecture* in fact waxed ironical regarding the desire for change;[106] I therefore cannot see my way to accepting the account which Bernard Stoloff has given,[107] which suggests that Ledoux had suddenly discovered the virtues of a people who had been violently opposed to the propylaea of Paris,[108] and who had later thrown him into prison. Being clearly Rousseauist in inspiration, the town of Chaux was nonetheless a product of its time, and therefore participated almost unintentionally in the ideals of the Revolution.

The *Architecture* generally maintained a discreet silence regarding the revolutionary events, where it did not employ veiled allusions, but it was nonetheless reformist in intent. An architect should seek to improve social relations. It was no longer simply a matter of reflecting the hierarchy of conditions and customs; the wish to transform society – though it was a question of transformation rather than of revolution – had replaced the ideal of *convenance*. Since it celebrated the civic and the familial virtues, and the love of sciences and of the arts, Ledoux's reformism was primarily concerned with collective morality. In crystallising the spectators' emotions, architecture served the cause of social regeneration, as advanced by programmes such as the Panarétéon, which was a 'school for morality, where the duties of man are taught',[109] or the House of Union, which was decorated with fasces (Fig. 101).[110] These designs were usually crowned with belvederes and had a

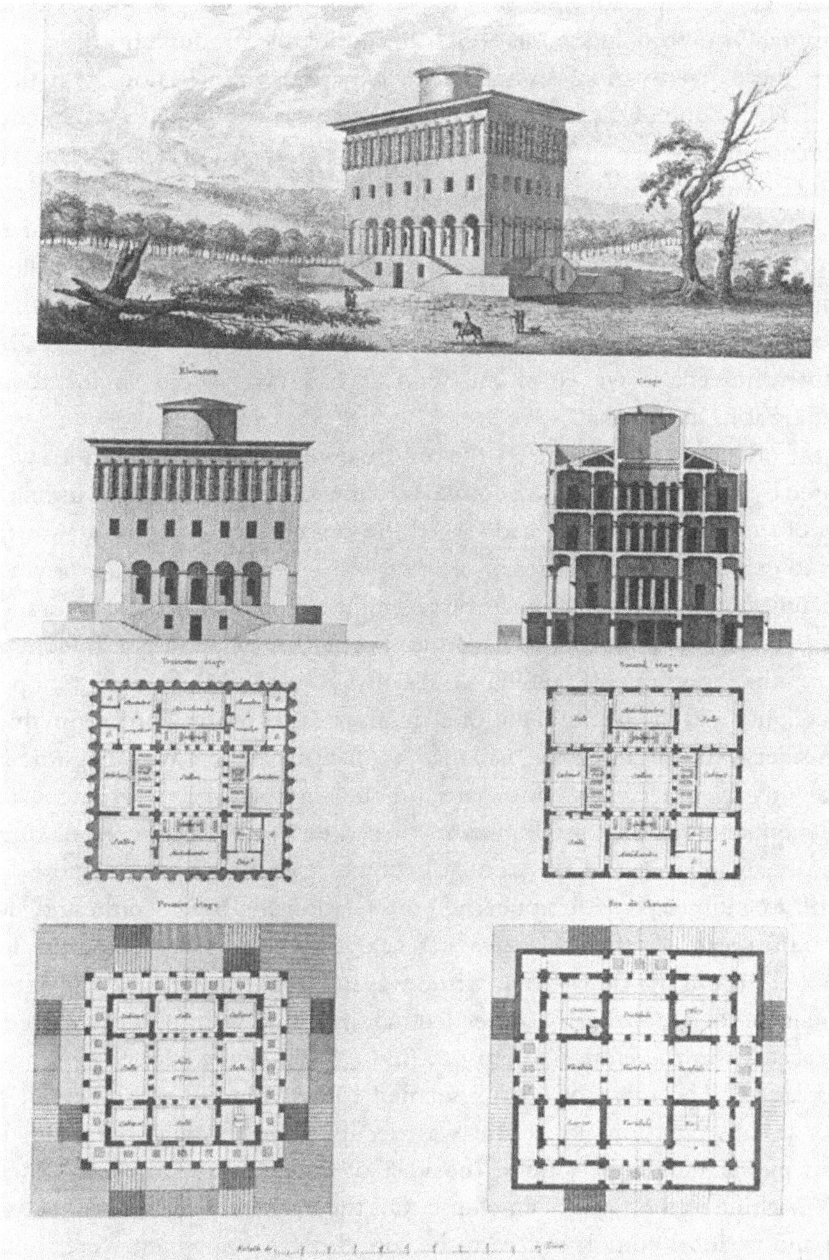

FIG. 101 C.-N. Ledoux, 'Assembly house'. 'Every authority is weak if it is not assembled.' The cubic form and the fasces decorating the upper storey of the building suggested the strength which devolves from social harmony. The whole and the ornamental details 'speak' with one and the same voice.

generally cubic shape, for 'the form of a cube is the symbol of immutability' and 'morality must include in its annals an immutable monument'.[111]

But there were more mundane celebrations also. Alongside these 'immutable monuments', architecture had also to contribute to the satisfaction of more immediate needs. The town of Chaux provided a pretext for this, and Ledoux constructed a market, some public baths, a church and a cemetery there. This notion of services was also evident in Boullée, as the Library, Colosseum and Museum described in his *Essai* serve to show. In Boullée's drawings, the town was also concentrated in the public buildings, which seemed to absorb it, subsequently rising up on a terrain clear of any other structure. Indeed, it was as if the services had taken hold of the town, subjecting it to its designs.

At the other end of the scale lay the private dwelling, whose variations were explored by both Boullée and Ledoux. If architecture turned its back upon the pomp of the planning which had served the cause of the nobility, it was the better to exploit the possibilities of geometry. But practitioners were now less preoccupied with such combinatory possibilities than with the polarisation between services and dwellings. As Bruno Fortier has emphasised: 'In contrast with . . . the "monument", which almost always implied the manifestation of a sovereignty and the deployment of a language (a theoretical problem which the Academy of Architecture had had to face from the very beginning), services involved a confrontation around the interplay of government and population, and this on a predominantly urban terrain.'[112] This confrontation was especially apparent in the case of Ledoux. If one were to transpose this issue on to another plane, it concerned the relationship between the individual and the group. As the *Architecture* states, 'every authority is weak if it is not assembled'.[113] On the other hand, virtue was in essence natural and the entertainments of society had early succeeded in corrupting it. In order to preserve the peace, it was therefore necessary to find a middle way between assembly and isolation. The town of Chaux seemed to provide the rudiments of an answer to this question. Its centre was occupied by the saltworks and the various public buildings exalting the idea of collectivity and composing a single machine in the form of an ellipse. On the periphery of this productive, social and political complex were to be found dwellings which were sometimes isolated and sometimes grouped in twos or in fours, thus providing a whole range of possibilities of association, including the Cénobie, where sixteen families lived and worked. This range of residences referred back to the centre, from which it derived its existence and its legitimacy; at the same time it made various degrees of withdrawal possible.[114] To the design's movement of expansion and contraction there corresponded a similar process in the town, which was diffused and disseminated through a perfectly adjusted mineral nucleus. About the same period, the *idéologues*, Tracy and Sieyès, developed a theory of national representation by which they sought to reconcile individual liberty with the existence of a structured community.[115] If there was indeed a Utopian dimension to Ledoux's work, it should probably be identified in his concern to make architecture into the privileged means of expression of collective life. The adjective 'attractive' often featured in his writing, a reference both to Newton and also no doubt to the reflections of those contemporary chemists who were enquiring into the affinities between different substances. This was because architecture, as the trustee of the laws of nature, was supposed to be attractive in the same sense as the sun which caused the planets to move. It was supposed to inspire the kind of love which set in motion the celestial bodies in medieval theology, impressing upon them a circular trajectory around a motionless and radiant God – with the caveat that there was no longer any genuine sovereignty, the universe being reduced to a succession of contradictory impulses. One of the major differences between Blondel and Ledoux consisted precisely in this absence, from which revolutionary architecture would draw all the appropriate conclusions. For if there were no longer any sovereign principle except in a people that was by turns assembled and divided, there was no longer any centre except a geometrical and arbitrary one, which was liable to be abandoned at any moment. The classical hierarchy no longer worked, and one therefore came upon a neutral, isotropic space, in which the role of the architect was to make forces converge or diverge, to manage interactions, and to handle thresholds, stages and limits. There was therefore a considerable risk that architecture might become a simple ordering of factors which were radically alien to it. Ledoux thought to overcome this obstacle by referring to nature, but it was soon to crop up again, with architectural work being already reduced in the case of Durand to the putting into a rational form of a predefined programme.

We need now to consider the preoccupation with planning which was evident in the approaches of both Ledoux (see Fig. 102) and Boullée, and which was linked to the new concept of space which was then emerging. In the *Architecture*, the settlement of Chaux was justified by the presence of axes of traffic, of agricultural wealth and of salt deposits.[116] The town then seized control of the country, by means of the roads which served it, thanks to a network of habitations and of minor services – foresters' and woodcutters' houses, workshops – which represented an intention to exploit the territory systematically. Boullée, on the other hand, was fascinated by Bonaparte's expedition to Egypt: 'This great General and the *savants* who have followed him shall teach the whole world the art of shaping a great settlement.'[117] He recommended the elaboration of a 'general plan', which would decide upon the allotment of the soil, and which would determine where towns, monuments, canals and ports should be located. 'Finally', wrote Boullée, 'I would envisage this great plan as resembling the tree of science.'[118] The work of the architect should thus transcend the stage of a particular organisation so as to link up again with the great objectives of the nation, which might be compared with the 'plan of the universe formed by the Creator'.[119] If it were a question of spreading the poetry of architecture, this poetry was never anything else but the ultimate expression of the needs of society. The revolutionary architects can thus be seen to have followed in the footsteps of Blondel, who had already extended planning to gardens, and then to the town. But while the latter restricted himself to spaces ordered by a notion of *convenance* inherited from the classical age, Boullée and Ledoux, like the engineers, sought to address society as a whole. As far as they were concerned, it was the architect, who might be compared in this respect to the Creator or to the General who had been victorious in the Campaign of Egypt, who should inspire and coordinate the initiatives of each. This led to a significant modification in the scope of design. If we pass now from the question of typological investigation to that of architectural detail, we may gain a better sense of the kind of transformations which were occurring.

ARCHITECTURE AND MASONRY: THE FUNCTION OF DETAIL

In the classical age, architectural detail was, as we have seen, one of the chief preoccupations of an architect, since it referred both to his own good taste and to the existence of a consensus between the latter and the workmen.[120] It was at the same time beset by apparently contradictory tendencies, since the theoreticians' attempt at codification conflicted with the necessary freedom of the artist. This freedom thus entered into conflict with the desire to rationalise production. Vignola's prestige ultimately reflected the need for a reconciliation between the rules of beauty and the quest for simplicity permitting the execution of the details by the workforce. Since it raised questions having to do with optics, detail also represented a field of enquiry which was both

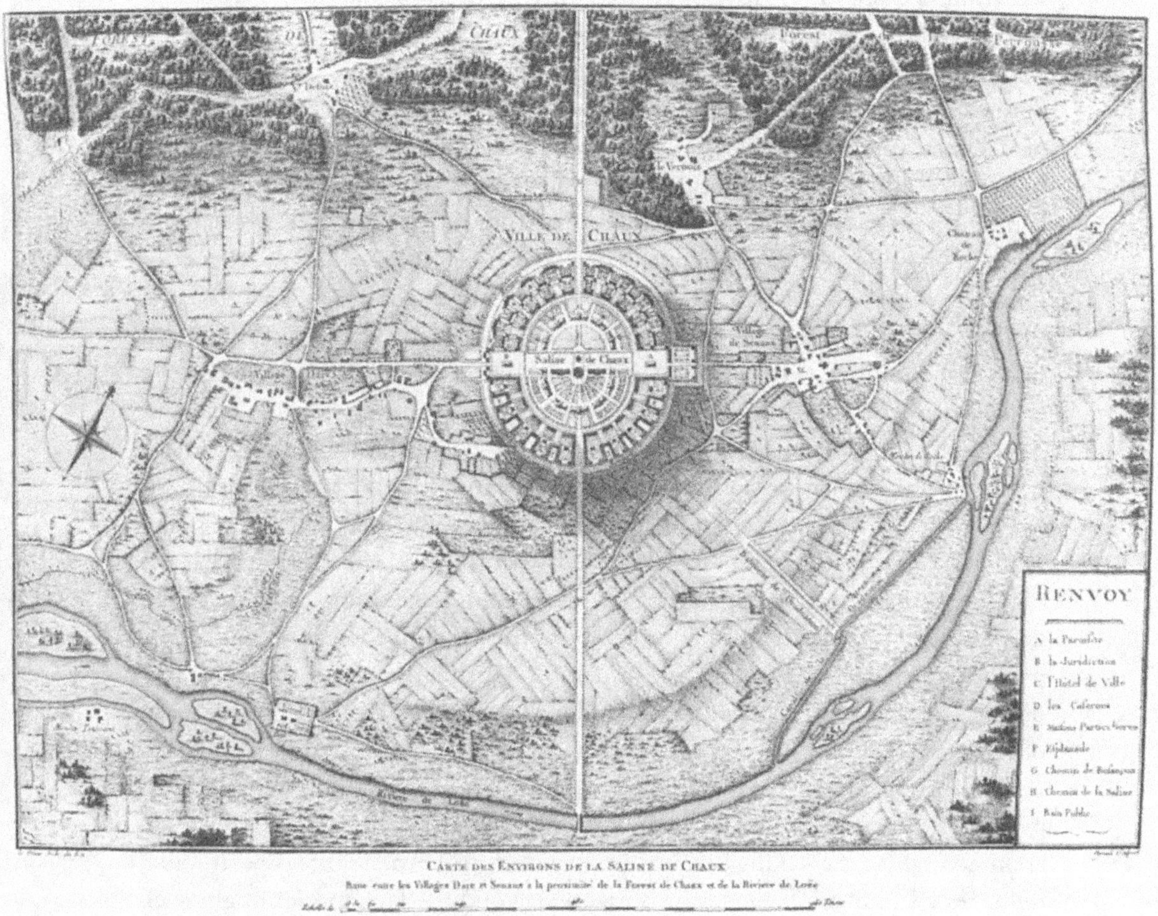

FIG. 102 C.-N. Ledoux, ‘Map of the surroundings of the Chaux saltworks’. In his concern to exploit natural resources, Ledoux relates to the reformism of the Enlightenment. The map resembles those drawn by engineers. However, the country was here centred upon the town and, as a consequence, upon architecture.

scientific and social. In its link with visual perception and with custom, it proved to be profoundly experimental.

In the course of the eighteenth century, the experimental aspect of detail diminished. As it became dissociated from structural considerations, detail first of all lost its role as a principle in the processes of rationalisation. It was, moreover, very difficult to distinguish between the art of sectioning and that of handling a given material; the notion of structure subsequently grew clearer as it became distinguished from matters of decoration. More generally, the art of building, which detail had previously kept on the margins of architectural theory, now became one of its explicit components. The reintegration of structural questions within theory should itself be related to the generalisation of design initiated by Blondel in his *Cours*, in terms of the relationship between decoration and distribution, and continued by Boullée and Ledoux through typological investigation and the emergence of the notion of composition.

Given this context, classical ornamentation was charged with being artificial. Once architecture was entrusted with the task of restoring a natural order within society, with construction reflecting the primary function of shelter allotted to design, the complexity of the classical Orders seemed to be irremediably estranged from the formulation of primary needs. It was a matter of setting out in search of origins, in order to reinterpret design and the status of its details.

The Abbé Laugier had already initiated this enquiry in the 1750s, with his primitive hut (Fig. 103), the paradigm of primary needs and of a purified architectural logic.[121] This hut was built with four corner posts, upon which there rested four other pieces of horizontal wood.[122] This model might well have seemed the first step on the way to a return to origins; Laugier derived from it a whole series of consequences concerning decoration. This source of inspiration had, however, given rise to buildings as complex as that of the church of Sainte-Geneviève or Perronet's great bridges, by a kind of cunning of structural reason which one could hardly have anticipated beforehand. With the four corner posts of his hut in mind, Laugier recommended by analogy the use of isolated columns supporting an entablature, especially in churches. The Greek temple constituted a fairly faithful transposition of this arrangement. But since one could no longer use monolithic lintels after the fashion of the Ancients, the lintel courses were to be cut up into voussoirs, thus creating horizontal or oblique thrusts which had not existed in the original model. One thus arrived at a system of abutments and counter-abutments similar to Gothic structure, hence the famous formula by which Soufflot proposed to reconcile

FIG. 103 M.-A. Laugier, *Essai sur l'architecture* (1753), frontispiece.

at Saint-Geneviève 'Gothic lightness with Greek elegance' (see Fig. 104), even at the cost of occasioning those 'indirect forces' which were later to be criticised by Charles-François Viel.[123] One thus passed almost without a transition from nature to the most absolute artifice. This was because Laugier was mistaken in believing that one could imitate nature by simple analogy. For Boullée and Ledoux, imitating nature no longer meant, as we have seen, aping its productions or the first products of human industry, which amounted to the same thing, but indeed going back to the beginning.

This beginning consisted of partitioning. Architecture sheltered; it had therefore to isolate man, while allowing him to assemble with other men if he so wished. The architect's task was therefore to deploy fullnesses and emptinesses, and to establish boundaries and thresholds. Design was first and foremost a matter of walling, as is borne out in Boullée and Ledoux by the ubiquitous theme of the wall, whether it be the blind wall of the Library described in the *Essai sur l'art*,[124] or the walls of the country houses in the *Architecture*.

This could not help but result in paradox. As Boullée observed, 'What is architecture? Should I follow Vitruvius and define it as the art of building? No. This definition is based upon a grave mistake. Vitruvius mistakes the effect for the cause.'[125] Ledoux likewise declared that 'architecture is to masonry what poetry is to *belles-lettres*'.[126] Although already present in the *Cours d'architecture*, this distinction between architecture and the art of building, and between architecture and masonry, took on its full meaning in the work of the revolutionary architects. But if design consisted of raising walls and partitions, the work of the architect strongly resembled that of the mason. Architecture essentially depended upon masonry, for all its claims to be radically distinct from it (see Fig. 105).

This tension between architecture and masonry could well cast a fresh light upon the replacement of classical ordinance by a form of decoration that was more assertive and increasingly repetitive and rigid. Although it had been constitutive of the building at the outset, the detail gradually won its independence; it 'came unstuck' in some way from the wall, so as to become a trace, an emblem of an architecture that was always on the point of disappearing. Through its nakedness, the wall was likened to the coherence of primary needs; decoration was supposed to mark the process of transformation undertaken with respect to design. In the *Cours d'architecture*, repetition was introduced as a provisional answer to the problem of the relationship between exterior and interior, but in Boullée and Ledoux, it was to become

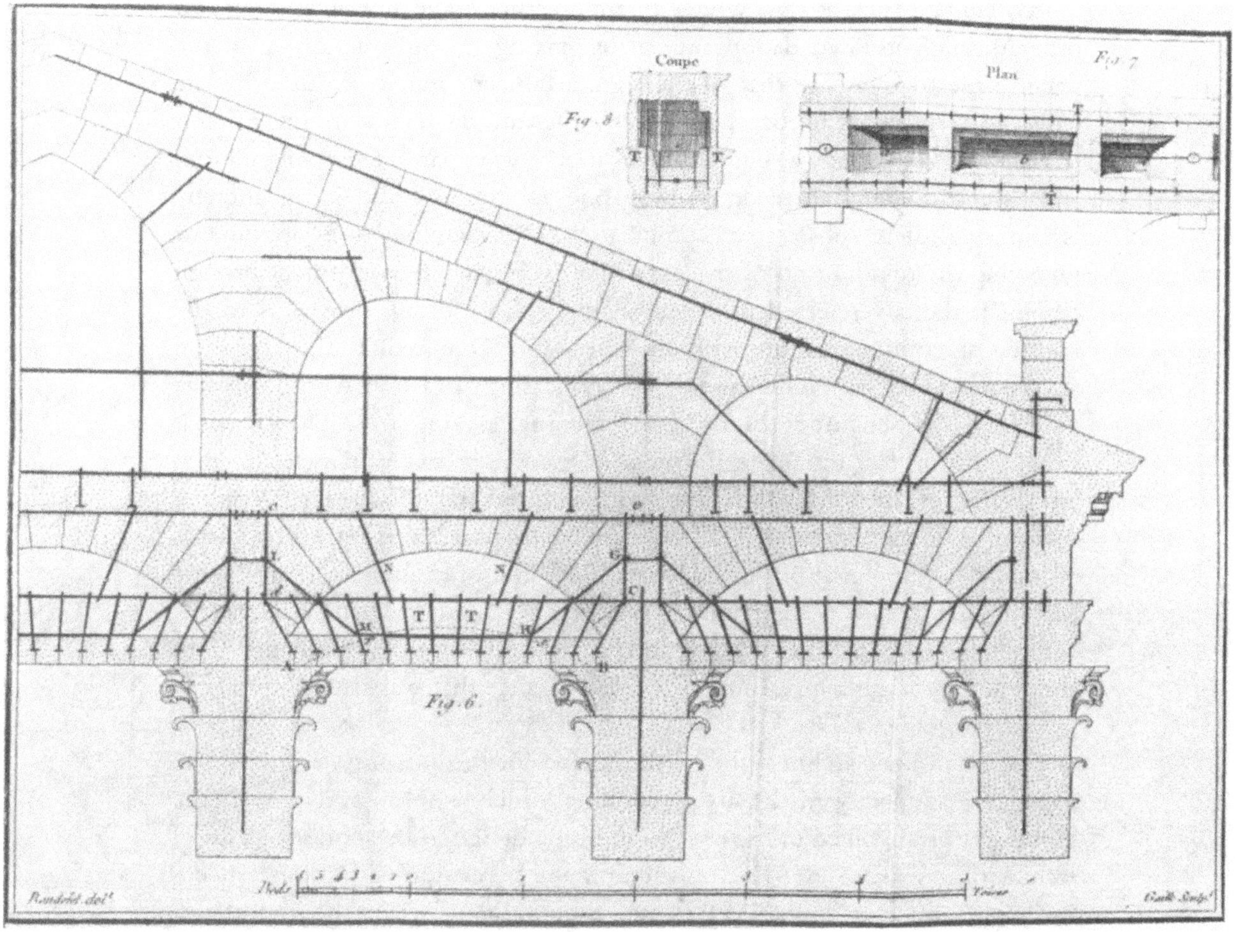

FIG. 104 A.-J.-B. Rondelet, 'Lintels of the French Panthéon'. The cross-section shows the iron reinforcement, which took the 'indirect forces' of the structure.

the index of a movement, just as the concentration of effects around a porch made it possible to emphasise the importance of the passage, and of transition. It was no longer a matter of arriving at a compromise between beauty and its realisation by workmen, but actually of signifying architecture, or of having it signified. Revolutionary designs also 'spoke' through the Atlases of Boullée's Library,[127] or through the petrified fountains which decorated the walls of the saltworks at Chaux.[128]

In a context such as this, where an architecture could only install its own order through its own disappearance, by staging it, the functioning of the detail came to resemble that of the sign, a concept much favoured by the *idéologues*, and soon to be theorised by Destutt de Tracy in his *Elémens d'idéologie.*[129] The sign was both arbitrary in its always particular configuration, and essential inasmuch as it enabled one to circumscribe ideas, thereby legitimising their subsequent combination. Was architecture not an art involving the dividing up of spaces, which were later reconstituted through design? It thereby referred to sensation, the source of all ideas, and to the register of communication with the spectator. One would thus have an architecture which signified itself by disappearing, and one whose message awakened the sensations of the person contemplating it. This was what Boullée had in fact intended, through his theory of simple masses, aimed at articulating geometry and sensation, with architectural knowledge becoming a sort of grammar which existed before verbal language, since the movements of the body, the sensations of impenetrability and of resistance are undoubtedly among the very first which we experience. Detail then became a commentary upon this process, an analysis of its development, both announcing and giving warning of a disappearance that was always just about to take place.

The detail kept architecture on the surface of the masonry. But can one speak of a simplicity in the art of building which would contrast with the sophistication involved in the execution of the design? The agonistic status of architecture was extended to construction, whose relation with primary needs was just as much in doubt. At the same time as they asserted the importance of walling, the revolutionary architects continued to be interested in structural research.[130] More generally, nascent neo-classicism explored the domain of technology with a renewed curiosity, studying both the boldest realisations and vernacular construction. There was an ever-increasing repertoire of such techniques, from the Delorme assembled frames, of which Legrand and Molinos developed a modern version in 1782–3 for covering the corn market,[131] to Saracen arches or the use of puddled clay. As Jean-Marie Pérouse de Montclos emphasises: 'Neo-classicism is a two-faced subject. The obverse presents architectural design, the imaginary world of Piranesi and the Utopia of Ledoux, while on the reverse there appear the technical studies of the architect-builders and the engineers.'[132] The simplicity of origins yielded to this proliferation of studies and research, which took on the appearance of a veritable explosion.

FIG. 105 A.-J.-B. Rondelet, 'Ruin of a wall built of small tufa stones, near to the Metella tower at Rome'. As in the case of Piranesi, archaeology and ideas on the basic elements of architecture and of construction are combined here. Given the desire to raise monuments which would defy the centuries, would architecture be reduced, in the last analysis, to masonry?

Boullée and Ledoux never really managed to choose between columns and structures which, paradoxically enough, took their inspiration from the primitive hut, and walls which enabled one to circumscribe primary needs, although it was not an issue that could be easily evaded. Did this dilemma perhaps express their distance from the origin, which became a ceaseless to-and-fro between column and wall?

On the theoretical plane, this debate was subsequently resolved, although to the detriment of the primitive hut, which was criticised by Durand in his *Précis des leçons d'architecture données à l'Ecole Polytechnique.*[133] Rondelet's *Art de bâtir* also expounded the actual building methods used by primitive peoples, and emphasised their diversity, which was linked to differences in climate and to the use of local materials.[134] Rondelet would thereby point to the meaning of the neo-classical answer, which was tied to the refusal of a unique model of supposedly universal validity. Taking their cue from the experiments and voyages of the Enlightenment, the neo-classical architects acknowledged the diversity of structural options, and the origin thereby lost the transparency and the explanatory character which Condillac and Rousseau had ascribed to it.

Boullée and Ledoux had themselves begun to acknowledge this diversity, even though they wished to believe in the existence of a nature ruled by the laws of sensation. Architectural stability was, as we have seen, already cracking under the weight of accumulated experience, through its contact with the figures of constructive prowess and risk which the realisations of Soufflot and Perronet served to illustrate.[135] The revolutionary architects' Utopia also consisted in wishing to reconstitute the exploded domain of the art of building, to control it through design, just as the latter was designed to restore the social equilibrium by formalising the to-and-fro between the collectivity and the isolated individual. Architectural form sought to seize hold of the alternative of column or wall, to reabsorb it by rendering explicit the tensions which it had generated. Between the supposed simplicity of primary needs and a technological flight forwards, was it possible to restore some kind of order and to recover some kind of stability?

The desire for such an order gave ornamentation a technical quality, and it became an instrument of persuasion, the sign of a unity regained within a nonetheless heterogeneous universe. One should emphasise in this respect the new affinity between architectural detail and what the engineers around the same period called the 'expedients' serving to perfect design. Being distinct from the basic principles of construction, the 'expedient' served to reinforce them. In a certain sense, detail very often functioned as a kind of 'supplement'[136] to construction, and one which was supposed to account for its meaning.

Within this technological sphere, there was however a split between structural and architectural detail, both of them referring to this same register of communication with the spectator. Since it was a matter of signifying the design, decoration would be able to function, if need be, in such a way as to contradict the structural 'truth'. Massive pilasters would thus punctuate a wall built of wood and plaster, thereby recomposing on the plane of appearance a wholly different technical reality. A 'direct' ornamentation would often be superimposed on a structural arrangement of the most 'indirect' kind.

Between structural truth and apparent solidity, between technical heterogeneity and the unity of the execution of the design, the detail thus became the instrument of a complex staging. Its nature as 'expedient' would subsequently triumph in the *Détails de construction d'une maison*, by Mandar (see Fig. 106), who was professor of architecture at the Ecole des Ponts et Chaussées from 1804 to 1820, and who presented a pavilion of his own invention, conceived at the beginning of the Revolution.[137] Lithographed in an edition of 250 copies in 1818, for the pupils at the Ponts et Chaussées only, the work was to be published in 1826, with a number of additions, under the title *Etudes d'architecture civile*. Mandar undertook the systematic deconstruction, for educational purposes, of a design (Figs. 107 and 108). The plates could be divided, roughly speaking, into separate series, with some being devoted to the basic elements of construction (structure of walls and floors, masonry and carpentry), others presenting a mass of details regarding assembly or decoration. Expedients, assemblages and ornamentation fundamentally participated in the leeway left to the designer, in so far as he followed the basic principles of construction and a logic of design for which Durand would give the most general formulation around the same period.[138] Whereas this logic referred to a notion of economy of forms and means, the detail generated a space of commentary and seduction, a seduction which could easily lapse into eclecticism. When adding a garden to his house in his *Etudes d'architecture civile* of 1826, Mandar thus filled it with Gothic or Moorish manufactures, with pagodas, and with ice-houses in the Egyptian style.[139]

This eclecticism was lacking in the *Essai sur l'art* or in the *Architecture*. Yet one can readily understand how it might have inspired Piranesi, Delafosse or Lequeu. On the eve of the Revolution, two attitudes were already possible.

FIG. 106 C.-F. Mandar, *Détails de construction d'une maison* (1818), perspective view. His *Etudes d'architecture civile* (1826) included a line engraving of this view, with some changes tending to make it less picturesque.

One might, on the one hand, espouse the purism of Boullée and Ledoux, who sought to rarefy and concentrate decoration the better to affirm its semantic link with composition. On the other hand, one might opt for an ornamentalism, which involved exploring, with a degree of prolixity, the new relationships possible between architecture and construction, architecture and principles of invention, architecture and history. Piranesi's staging of Roman construction or Boullée and Ledoux's bare walls basically reflected the same concern to develop an architectural expression which would resolve the various paradoxes we have just encountered. In either case, detail now had to be accurate, with an accuracy eluding the dimensional coordination of classicism.[140] Whether sign or expedient, architectural detail functioned as a kind of fragment of a machine. Purism and eclecticism culminated in an identical preoccupation with efficacy. The theme of repetition, the rigidity which was installed in frontages, were connected to this register of the machine which an absolute distance separated from its different fragments or mechanical parts. Only the functioning of these different parts, once the work had been effected, enabled one to re-establish an overall coherence, to recover the machine as an inorganic totality. The new status of an architectural detail proceeded from a somewhat similar schema. One is therefore bound to agree with Kaufmann when he sees 'autonomous' architecture as marking the end of an era characterised by the organicity of architecture and by the sensuality of the materials.[141] In neo-classicism, matter would be inert, for the same reason as the universe was regulated, according to Laplace, by the immutable laws of universal gravitation. 'Revolutionary' architecture had not yet arrived at this stage, but the abandonment of classical fluidity marks the possibility of an evolution. This evolution involved during the same period an obsession with archaeological statements, which were at last accurate, a recourse to moulding taken directly from ancient monuments and fixed with an absolute archaeological precision. Dufourny's collection, which was begun in the 1790s, contained a fair number of these mouldings, which are evidence for a new attitude towards the problems of dimensioning.[142] Now that it was unimpeachably perfect, the architectural detail could become repetitive, thus giving rise to a genuine industry at the beginning of the nineteenth century, providing architects with standardised elements, such as the terracottas which decorate the house realised by Durand.[143] The destiny of the detail could be seen as a progression from archaeology to the standard, with freedom of invention gradually being transformed into arbitrary norms, as Claude Perrault had anticipated.[144]

FIGS. 107, 108 C.-F. Mandar, details of the porch of the house, seen from below and in cross-section. Construction and ornamentation informed by the same rigorous, analytic spirit.

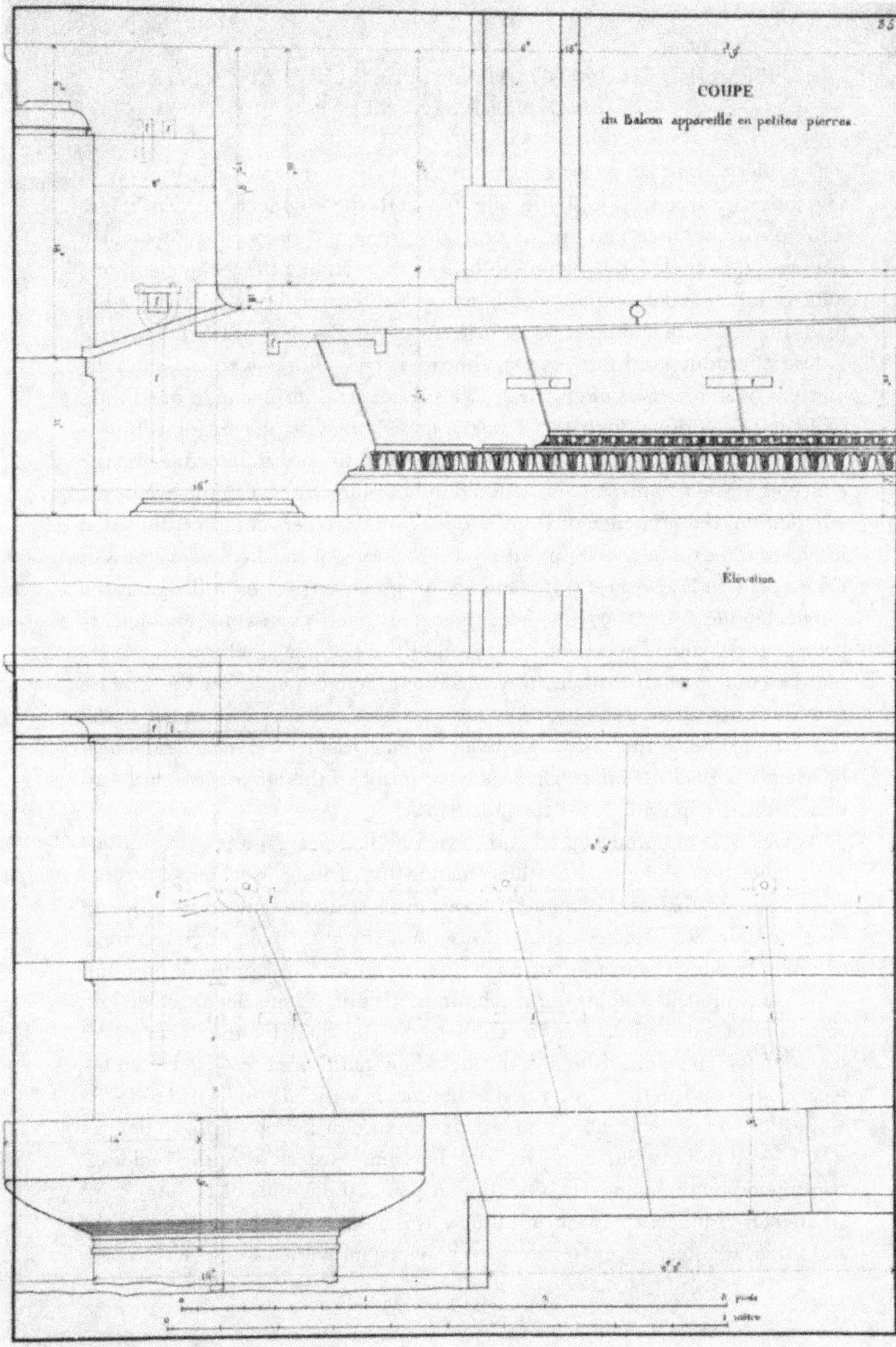
35
COUPE
du Balcon appareillé en petites pierres.
Élevation
pieds
mètre

REVOLUTIONARY ARCHITECTURE AND THE ENGINEERS' SYSTEM

The reader cannot fail to have been struck, at the end of this rapid survey of revolutionary architecture, by its affinities with the engineers' 'system'. This kinship may be discerned first of all in the architects' preoccupations which refer to a universal, active nature, in order to revert, like the technicians, to a logic of utility and economy. These demands for rationality were reflected in programmes, with public services in the case of Boullée and Ledoux, pre-industrial production in the case of the latter; civic works were as much the concern of architects as of engineers. This is not at all surprising, if one thinks of Ledoux's dealings with the *Ferme générale* and with the major administrations of the kingdom. The links between revolutionary architecture and the concepts of the technicians also worked in both directions. On the eve of the Revolution, the influence of Boullée upon the engineers could be discerned in concrete terms in the architecture competitions of the Ecole des Ponts et Chaussées (see Figs. 109–11).[145] Durand, his pupil, was to teach design at the Polytechnique from 1797 on. More generally, neo-classicism very quickly became the favoured mode of expression of the engineers, who were covering the country with markets, town-halls and prefectures under the Empire and the Restoration. Alongside Durand's treatises, which owed much to the research of Boullée and Ledoux, Bruyère's *Etudes relatives à l'art des constructions* or Mandar's *Etudes d'architecture civile* were proof of the engineers' loyalty to neo-classicism up until the 1820s and 1830s.

As well as being preoccupied with civic building programmes, revolutionary architecture also prepared this situation by making itself into a system comparable to that of the engineers both in its structure and in its mode of operation. It was Boullée and Ledoux's ambition to make architecture a natural art, which was accurate in its principles and its combinations, and which was modelled upon the machinic perfection of the details of design. Accuracy of calculation was therefore to be matched by the accuracy of sensation which architects might, through their genius, put into effect. Architects did not disdain the register of the sublime, in which Blondel had still seen a 'spirit of oddity'.[146] Boullée's buried architecture or Ledoux's prison designs may be said to be sublime in precisely the same sense as were the buildings constructed by the engineers. On another plane, the genius of revolutionary architecture found expression in numerous conflicts, whether between light and shade, architecture and territory, or between individual and society. These

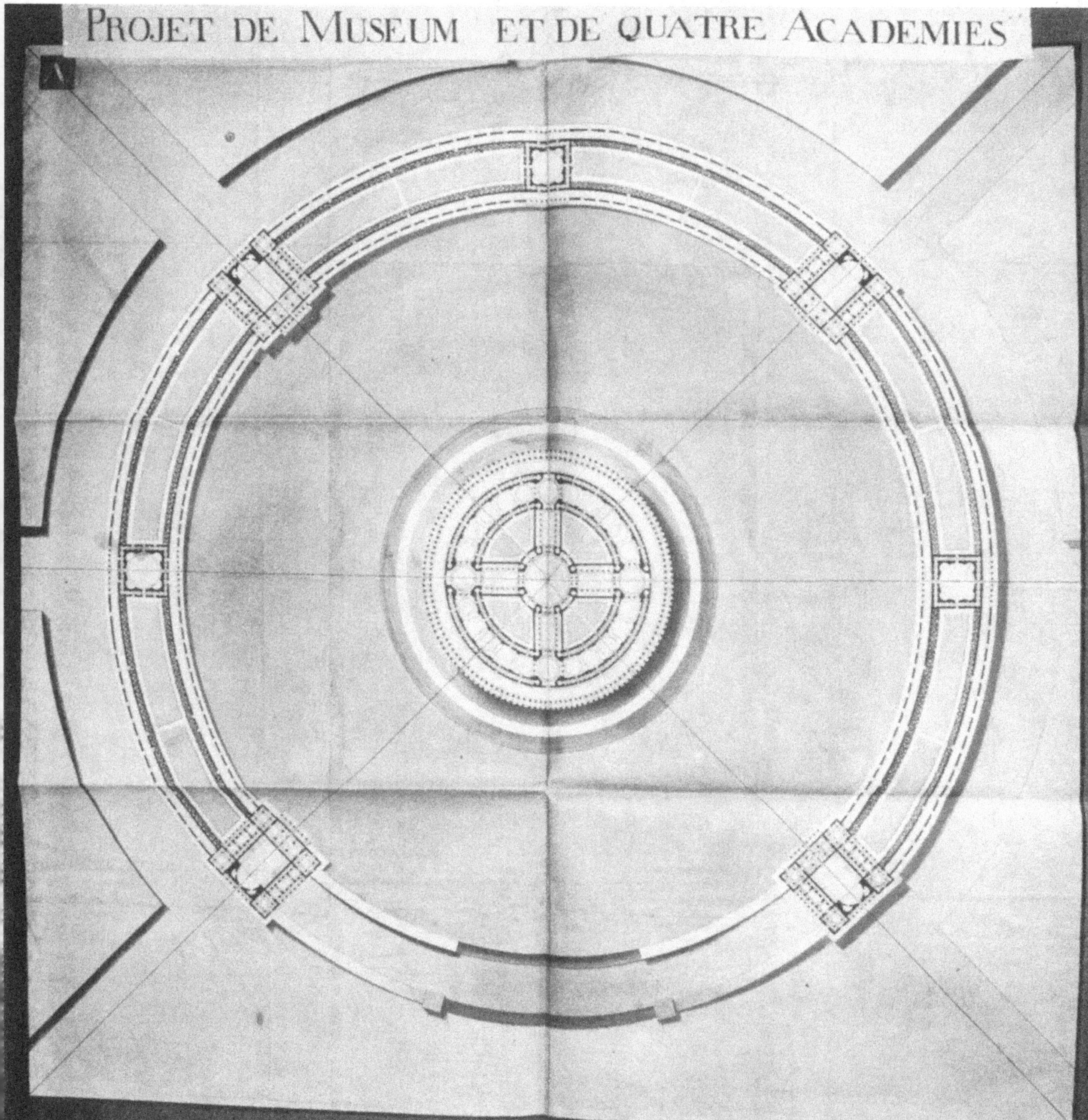

FIG. 109 N. Thouret du Breuil, project for a museum and for four academies for the 1786 competition, second prize, plan, ENPC, ink, wash, watercolour; 170 × 170 cm.

FIG. 110 N. Thouret du Breuil, project for a museum and for four academies, elevation, ENPC, ink, wash, watercolour; 26.5 × 310 cm. Boullée's influence is clearly in evidence in this project, which was done at the Ecole des Ponts et Chaussées.

FIG. 111 J.-L. Carbon, stock exchange project for the 1788 competition, second prize, ENPC, ink, wash; 62 × 138 cm. A presentation that was both sublime and picturesque informed this drawing, which is of a common sort, indicating late eighteenth-century notions of a proper service architecture. Thomas de Thomon, a former pupil of Ledoux, was to provide the most elaborate version of this type of project in his design for the St Petersburg Stock Exchange of 1801.

conflicts generated a new space of flows and movements, founded, like that of the engineers, upon the alliance of contraries and upon the notions of threshold and limit, a space in which was inaugurated a process of dematerialisation of the object, to the benefit of a regulated series of oppositions for which, through its form, it was supposed to account.

Boullée and Ledoux no longer believed that architectural theory should depend upon a slowly perfected judgement, in the sense in which Blondel had understood it. 'Raisonnement de l'art' was not replaced by principles which belonged immediately to nature. The traditional distinction between theory and practice was abolished, to be replaced by a creative subject who concentrated natural powers. Without any mediation, the idea from now on 'worked' architectural matter just as the mind of the engineer bent the world to his designs. Between the revolutionary designs and the notion of structural model, there were numerous *rapprochements*, beginning with the idea of process, the effect of surprise, between plan and elevation in the case of the architects, between visual stability and structural stability in the case of Perronet. Surprise was always produced by indirect means: the indirect day of Boullée and Ledoux had its counterpart in the 'indirect and powerful' forces which Viel attributed to the engineers. Revealing themselves in the immediacy of their effects, calculation and sensation often arrived at their destinations by circuitous paths.

It would be tempting to pursue such correspondences further, by developing, for example, the analogy that may hold between typological investigation, which was based upon procedures of free association between simple volumes, and the mineralogical curiosity of Haüy or Romé de l'Isle (see Fig. 112).[147] The alliance of contraries, the notion of flow, the importance accorded to the trajectory of the spectator and to the movement of the design represent, on the other hand, more tangible evidence, being always linked to the idea of a nature in perpetual development, thanks to privileged actors, whether architects or engineers were involved. In either case, there was the same curiosity as regards technology; through experimentation with calculation or with sensation, it was a question of triggering a process of reasonable self-organisation of nature, a process of which architecture and planning would constitute the transposition.

This transposition depended in fact upon the only method capable of reconciling calculation and sensation, namely, analysis, which could be applied as much to the domain of scientific discovery as to the understanding, as the *idéologues* tried to show. Thus, Destutt de Tracy, in his *Elémens d'idéologie*, set

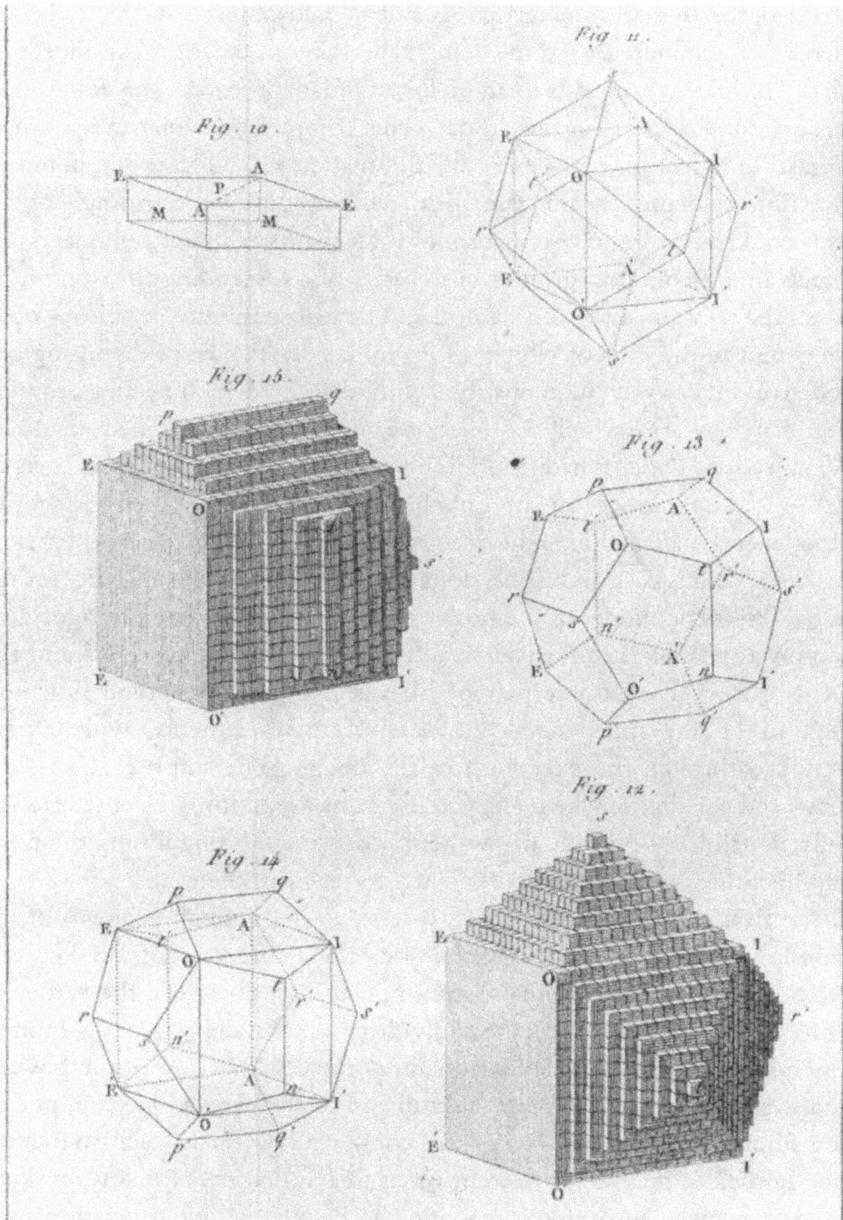

FIG. 112 R.-J. Haüy, *Traité élémentaire de physique* (1803), form of crystals. The regularity of the phenomenon of crystallisation seems here to justify the ambitions of an architecture conceived on the model of the laws of nature.

out to describe human intelligence 'as one would observe and describe the properties of a mineral, or some remarkable circumstance in the life of an animal',[148] while Cabanis was to argue for a similar approach in his *Rapports du physique et du moral de l'Homme* (1802). This complexity, which was born of the refusal of a dualism between mind and matter, would later be refuted by mechanism, in France at any rate. But, while awaiting the triumph of the concepts of Laplace or the systematism of someone like Durand, engineers and architects found common ground in the idea of transcending the opposition between the register of the mechanical and that of sensibility, by means of an active genius in the eyes of which, as Prony put it, 'the art of discussing and analysing is in no way incompatible with that of depicting and arousing emotion'.[149] This art – which was of course analysis – thus provided an answer to the objections Diderot had raised in his *Pensées sur l'interprétation de la nature* (1754).[150] It was no longer a matter of choosing between two such basic terms of experience as calculation and sensation, but rather of associating them.

For the revolutionary architects, design was therefore to be analytic, for the same reasons as the model of the bridge, which was based upon as rigorous a separation as possible between piers and roadway, between bearers and the elements borne. The counterpart of this separation of structural functions seemed to be the independence of architectural elements, which were recomposed through the execution of the design. The surprise effects, the contrasts and paradoxes, which characterised this execution, in fact referred to analysis, which had originally made them possible by distinguishing the poles from which they stemmed, before seeking to reabsorb them.

The analytic method was designed to resolve the difficult problem of the relationship between architecture and society. In the case of Boullée and Ledoux there emerged, as we have seen, new tensions between the particular and the general, and between the individual and the collective. As Hubert Damisch has emphasised, 'The idea of autonomy only has . . . meaning when related to that of equality',[151] but one still had to answer the question put by Rousseau and find a form, that of the contract, by which 'each individual uniting himself to the others . . . remains as free as before'.[152] It was this kind of freedom which the simple forms and the details, whose association produced architectural composition, were designed to embody. Like Blondel, Boullée and Ledoux were rationalists, but their rationalism, although they were still concerned to bring architecture into alignment with society, was based upon different premises. In Blondel's case, the claim that there was a specifically French manner reflected a desire to see architecture constitute the privileged expression of a collective consciousness akin to that 'spirit' which Montesquieu ascribed to the various nations. Revolutionary architecture, on the other hand, evoked *The social contract* rather than *The spirit of the laws*. It was no longer a matter of arriving at a compromise between the rules of beauty and social requirements, but rather of moulding the latter, of imposing a form upon them by depending upon nature. Boullée and Ledoux proposed in the last analysis to reveal to society its own legitimate ambitions – the surprise effect which a design provoked was never anything else but the recognition of a forgotten truth. In this context, the tableau of term-for-term correspondences between architecture and society was transformed into a process of unveiling. Discovery of a design became a discovery of oneself and of one's relationship to the collectivity.

If I speak of a self-discovery here, it is because Destutt de Tracy, in his *Elémens d'idéologie*, which was designed to introduce pupils at the Ecoles Centrales to the analytic method, reckoned that the sentiment of self, as distinct from the world, arose out of the inner resistance to movement which the thinking being experienced when faced with obstacles which it could not name, but which constituted it, and structured it in some sense, through their opposition to its will. Form would therefore be the locus of this opposition, 'the confines of the power' of man, as well as that which gradually organised him.[153] From this point of view, the architect who composed a design would merely be anticipating the sentiments which the spectator would subsequently experience. The composing of a design or the composing of an itinerary of sensation and of self-discovery would amount to basically the same thing. Form would be the absolute site for the constitution of the self and for the power of this self over others. The ultimate ambition involved in the execution of a design was thereby revealed. Rejecting in advance the possibility of a conflict between the 'general will', the will of the architect and that of the spectator, one would from now on regard the geometrical working of matter as at the same time a working upon oneself and upon others, in a perspective of reconciliation which strongly resembled the artifices of the discourse of the engineers. This could well be the most accurate description of the Utopia of Boullée and Ledoux.

Up until now we have only referred in passing to the notion of Utopia, under which heading a fair number of historians have ranged the creations of Boullée and Ledoux. One needs a clear understanding of the meaning of this term, which was coined by Thomas More, and which literally means the negation of place. A dictionary of philosophical terms defines Utopia as 'a plan or dream of a society and, by extension, a desirable future, but one which is regarded as chimerical'.[154] The *Essai sur l'art* and the *Architecture* are certainly not located in an indeterminate elsewhere. Boullée's town and services concerned the society of his own time, while Ledoux even went so far as to assign a specific site to his ideal city. What of the social dream of these architects? Was it a chimera? Unless the engineers are to be regarded as potential Utopians also, we have no good cause to see Boullée and Ledoux in this light either. One should also note that their designs could almost always be realised. Admittedly, the sheer scale of Boullée's services seems to rule out their transposition to the site, but, in his bid to explore the creative possibilities of the execution of the design, should we see this architect as lapsing into Utopias? The technology to which he referred was explicitly that of his own time; the dimensions of his designs expressed an indifference towards the rules of *convenance*, as well as a breach between services and the old town. If one excepts the cemetery of Chaux, and the wardens' house in the forest of Maupertuis, Ledoux's creations were still more reasonable. If there was, in the last analysis, a Utopian quality to the architecture of Boullée and Ledoux, it plainly had nothing to do with 'the dream of a society and, by extension, a desirable future', nor even with formal manipulation at one remove from structural realities.

What was in fact Utopian about revolutionary architecture was a desire to control society and its exchanges which found expression in exclusively spatial terms. Boullée's town and Ledoux's saltworks referred to a concept of architectural planning. But is it legitimate, at the end of the eighteenth century, to restrict one's enquiry to the register of *mise en forme*? Is space, fully invested in the name of a reason just then discovering the full extent of its powers, sufficient to account for that same reason? This is the question one should ask Boullée and Ledoux, and it is this that goes some way towards accounting for their failure. At the same time as the combinatory of sensation was being elaborated, the more lasting domination of calculation was, as we have seen, being prepared. Although revolutionary designs and the structural model of the engineers were both organised in terms of an analytic process, the scope of this process was not the same in either case.

This divergence may be illustrated in terms of the conception of bridges. The bridges proceeded on the basis of contradictions which had been overcome, especially as regards the contradiction between the direction of the passage across and the perpendicular course of the river. Perronet resolved this conflict by reducing the piers; Boullée and Ledoux, on the other hand, gave them a monumental character. Entrusted around 1787 with the task of improving the design of the pont Louis XVI, which had been conceived by Perronet, Boullée began by observing that: 'The civil engineers, entrusted with this aspect of architecture, had performed miracles in the scientific aspect, but the artistic aspect had eluded them. Bridges, in general, have not been characterised by beautiful decoration.'[155] Boullée wished to replace the narrow piers which the engineer had chosen with boats' hulls. This idea also featured in the *Architecture*, with Ledoux conceiving a bridge of the same kind in order to cross the Louë near to the town of Chaux (Fig. 113).[156] This solution, which was vested with symbolic significance since it derived from the primitive pontoon bridge,[157] allows one a better understanding of the polemical positions involved. By heightening the spatial contradictions, the architects were still referring to a form capable of tempering them. The use of stretched arches, on the other hand, produced a plan of coherence which was no longer spatial. It mattered little that the direction of the bridge and that of the river clashed; it became pointless to seek to reconcile them by giving the piers a boat-shaped form.

Likewise, while the country surrounding the town of Chaux referred to the ideal qualities of a regular urban structure, that of the engineers, through its apparent disorder, seemed rather to call for a description in terms of figures and graphs.[158] If the limits of the classical theory which Blondel had reinvoked arose from the impossibility of achieving a correspondence, through a systematic designation of the principles and parts of the art, between architecture and society, the limits of revolutionary architecture were inherent in architectural space itself.

These limits may now be apprehended by reverting to the notion of analysis, which was susceptible to very different interpretations. For the *idéologues*, this notion was defined in terms of an essentially discursive perspective, where it was a question of analysing language in its relation to thought.[159] The engineers, by contrast, were mainly concerned with mathematical calculation.[160] On the one hand, we have analysis of language, history of language or, what amounts to the same thing, history of man and of society, on the other, mathematical analysis, within the universal method which Condillac and his successors had sought to establish, and which was already occasioning new polarities, which the revolutionary architects had to take into account.

The revolutionary architects sought to make architecture 'speak' prior to language, while at the same time claiming to calculate, without having recourse to mathematics, the effects that this discourse would have on the spectator. But would an expression of this sort be possible? Was an architectural analysis viable? The task was an awkward one, unless architecture were to choose to give full expression to its contradictions by stating them. The loophole therefore lay in a poetics based upon the desire for a spontaneous speech prior to verbal language, and for a non-mathematical combinatory accuracy, both of which were altogether impossible.

This poetics was in the last analysis based upon the implicit recognition of the threatening rationalities of language and of the exact sciences, which constituted the real limits to the architectural approach. If there was a misconception involved in Boullée and Ledoux's proposed buildings, this misconception was undoubtedly inevitable. With its boat-shaped piers, the bridge was about to 'speak', whereas its flattened arches and its narrow roadway evoke the calculations of the technicians. The relationship of revolutionary architecture to *idéologie* and to the engineers' system, though very close at the time, turned out to be particularly ambiguous, since the design referred to analysis without ever deciding between the different interpretations which had generally been advanced of this term.

Nevertheless, this attitude supplies the premisses for a kind of submission, inasmuch as the limits of design were both stated and admitted. The hegemonic will of the revolutionary architects could only with great difficulty mask the demands for rationality which had been imposed upon them. The ambiguities of design were also those of an architectural discipline confronted with a more and more demanding programming,[161] and more generally with a will to anticipate which reflected major advances in technological rationality. The latter had formerly been articulated upon realised objects, which enjoyed the status of models, and which therefore served as supports for technological knowledge. Classical solidity had thus entailed rules of stability which had been asserted and confirmed from building to building. Its gradual effacement was to assume a symptomatic value. From this time on the emphasis would be upon conception and upon a relation between conception and realisation regulated by technical and economic laws which no longer had anything to do with an architectonics.[162] 'In order to realise', wrote Boullée, 'one must first conceive', and he thereby substituted a conceptual logic for the 'most excellent buildings of France' upon which Blondel had still based his doctrine. But, fundamentally, did this reversal not represent the transposition on to the theoretical plane of a *fait accompli*? Being caught between the language of programming, flanked by an emergent art history, and the salient calculations

FIG. 113 C.-N. Ledoux, 'Perspective view of Loüe bridge'.

of the engineers, could a specifically architectural space still claim to run society and its interchanges? For all that Ledoux concentrated history in an approach based upon design, or Boullée distinguished between this 'production of the mind' and the science of building, was not architectural space already becoming, along with architectural detail, a commentary upon decisions which had preceded it, and a form of seduction designed to facilitate the adoption of these selfsame decisions?

This commentary undoubtedly maintained a high degree of autonomy. If the engineers' services were on the other hand rationalised, a margin of liberty was recovered as regards individual dwellings. But the political reformism of Boullée and Ledoux testifies to a loftier ambition. Once it had been disavowed, one enters fully into the modern dramaturgy of the avant-gardes. Truth and lies, Utopias and betrayals constitute a new kind of theatre. With Boullée's enthusiasm and Ledoux's complaints, architecture entered into a cycle of exaltation and disillusionment which it would prove very hard ever to break.

FROM THE 'APPROXIMATE WORLD' TO THE 'UNIVERSE OF PRECISION'

Alexander Koyré's famous formula ' from the 'approximate world' to the 'universe of precision'[163] provides an accurate enough summary of the shifts which occurred in the closing years of the eighteenth century. These shifts were mainly concerned with science, whose efficacy increased considerably, thanks to advances in mathematical analysis, which gradually won its independence from the universal method advocated by the *idéologues*.

THE TRIUMPH OF MATHEMATICAL ANALYSIS

Classical science had produced a description of the solar system. In contemplating the stars and the planets, the *savants* derived from them the assurance that the world was run by a God who was a clockmaker, the mathematical rigour of whose designs could also be discerned in the purity of the figures and theorems of geometry. At the opposite end of the scale, in a more elementary form, arithmetic enabled one to describe reality in numerical terms. Between these fields of application, involving the sublimity of the stars in their courses and the prosaic quality of what could be counted, there extended an immense zone populated with phenomena whose nature remained uncertain. Classical thought was in this respect still the heir of Aristotle, and divided the universe into a supralunary and sublunary world.[164] Everyday reality continued to be that of the generation and corruption, along paths that were still obscure,[165] of beings, and of a permanent outflanking of man by nature.

However, in becoming experimental, science had drawn closer to technology,[166] as the mechanistic conceptions of Descartes or Mersenne testify.[167] But the machines of the classical age, the majority of which were made out of wood, participated in spite of everything in the incessant generation and corruption of beings.

As an 'approximate world', the classical world had not known precision, except in the domains of clockmaking and optics. The measures which were used, from fathom (*toise*) to foot to inch (*pouce* = thumb), were characterised by an anthropomorphism which ruled out any idea of standardisation. Precision was restricted to geometry, while the everyday productions of men were characterised by a constant negotiation between the purity of the ideal, the perfection of models and the necessary risks involved in execution.

The relationship between architecture and scientific and technical thought was necessarily very close. For was not architecture itself caught in a vice between, on the one hand, the rules of ideal beauty and, on the other, the realities of construction which, being artisanal in nature, were invariably approximate? In a world in which rationality and geometry also tended to be confused, architectural detail suggested a number of rewarding topics for reflection. What could be more telling in this respect than the interest displayed by a doctor such as Perrault, an Academician, in the theory of proportions?[168]

On another plane, this complicity between architecture and scientific and technical rationality was reflected in the definition of hydraulic architecture (see Fig. 114), namely, 'the art of building in water' bridges, highways, quays, dikes, aqueducts, locks and mills.[169] Considered as one of the branches of architecture, hydraulic architecture adopted for its own uses the systematic mode of exposition of the great classical treatises. In 'The art of directing, raising and handling water for the various needs of life',[170] Bélidor still placed himself within the 'Vitruvian paradigm', with the theory of Orders featuring in a chapter of his *Science des ingénieurs* which was in fact devoted to fortifications.[171] Likewise, Frézier's *Théorie et la pratique de la coupe des pierres et des bois* included a 'Critical and historical dissertation upon the Orders of architecture', which was published separately in 1769.[172]

In spite of Bélidor and Frézier's loyalty to the canons of classical architecture, the relationship between architecture, science and technology gradually loosened in the course of the Age of Enlightenment. This dissociation paralleled the advance of analysis, in the wake of the pioneering studies of Leibniz and Newton. In France, the Marquis de l'Hôpital provided an early synthesis of such arguments in his *Analyse des infiniment petits* (1696).[173] Taylor, Bernouilli and Euler were next to take up the challenge.[174]

Through analysis, mechanics was able to employ newer, more reliable methods also. Solid state mechanics was finally formalised, while Bernouilli, d'Alembert, Clairaut and Euler laid the foundation for the mechanics of perfect fluids.[175] Finally, 1788 saw the publication of Lagrange's *Mécanique analytique*, a work in which its author reduced the studies of his predecessors to some sort of order, thereby investing them with an overall coherence. With the *Mécanique analytique*, we have finally entered the age of rigorous scientific calculation. Lagrange was to take great pride in the fact that his work contained no illustrations whatsoever, since in it reasoning had for once ousted visual intuition.[176]

Lagrange's work thus represents an important stage in the history of science. Geometry advanced in the course of the eighteenth century, especially as regards the study of surfaces, but the works of Euler and Monge were from now on to be dependent upon analysis.[177] Visual intuition retreated before calculation in this area also.

In the technological domain, the balance-sheet was less one-sided. We have already emphasised several times the remaining obstacles; the Enlightenment was still classical in its struggle with a recalcitrant nature. From another angle, scientific idealisation engaged this struggle, as it moved from astronomy to the perfect solids and fluids. Analysis opened up fresh vistas; Newton's disciples tried to apply its methods to everyday reality, indeed, even to the structure of life itself. For the engineers, the sciences were still a 'torch' – the reproduction of models inherited from the past nevertheless coexisted with the expectation that new tools would emerge. The situation was anyway unblocked in some branches of concrete knowledge, as was the case with applied hydrodynamics, which arose out of the repeated experiments of Chézy, Dubuat, d'Alembert and Bossut.[178]

In many respects, eighteenth-century analysis proved to be very tentative. The concept of a function of a real variable, upon which Euler relied, did not yet have all of its generality; the definition of continuity and the notion of limit were still vague.[179] But analysis had already transcended 'Vitruvian' geometry. Through its capacity to manipulate proximities, thresholds, and limits, to approach the infinite by means of a succession of finite entities, and to characterise the continuous by means of the discontinuous, analysis already constituted a viable model for engineers seeking to combine continuities and discontinuities, whether it was a question of discoursing upon society, of rationalising the construction of buildings or of developing the countryside. The tensions which were expressed seemed thus to prefigure the recourse to calculation. It was this perspective which had chiefly been lacking in Boullée and Ledoux, who had persevered with a dynamics of architectural form, while the technicians were pioneering a dynamics of mathematical formalism.

I ought now to emphasise the startling acceleration evident in the final years

FIG. 114 B.-F. de Bélidor, *Architecture hydraulique* (1737–9, republished in 1782), frontispiece. Hydraulic architecture, which is defined as the 'art of directing, raising and managing waters for the various needs of life', is naturally concerned with the art of gardens.

of the century. In parallel to the invention of the metrical system, conditions governing the training and status of engineers changed radically. Being concerned with scientific and technical knowledge, the Revolution founded the Ecoles Centrales, where mathematics, physics and chemistry were allotted a by no means negligible place.[180] In 1794, the Ecole Polytechnique crowned the whole edifice, its function being to train future engineers by bringing them into contact with the most eminent men of science of the period (Lagrange, Laplace, Monge, Guyton and Morveau), and with engineers such as Prony, who were capable of consolidating the link between physico-mathematical theory and its application.[181]

The Commission's introduction of a metrical system and the courses at the Polytechnique together served to uncover the 'universe of precision', in which mathematical accuracy represented an ideal. This framework saw the emergence of a number of new themes, namely, the possibility of a 'particular and new' science of the engineer, a desire for technological control by means of mathematical modelling, a passage from invention to innovation, that is to say, from isolated discoveries to the gradual and regulated development of a system of theoretical and practical knowledge. The catalogue of techniques, and their static description, was reorganised in terms of new polarities. The transition was made to a more dynamic structure, in which the idea of process and that of progress played a predominant role.

FROM INVENTION TO INNOVATION; PRONY'S NEW HYDRAULIC ARCHITECTURE

The two volumes of Prony's *Nouvelle architecture hydraulique* mark a crucial stage in this process. Although the title of the work still seems to hark back to Bélidor's *Architecture hydraulique*, it is possible to discern some original tendencies within it. Thus, in the first volume, which was published in 1790, Prony set out to describe the means by which one might raise water (Fig. 115) under two different aspects: 'to know as regards machines and as regards motors' before specifying that 'the order of the description must be subordinated to their reciprocal filiation'.[182] In its treatment of so hackneyed a subject, the *Nouvelle architecture hydraulique* displayed a classificatory and genealogical spirit which seems to prefigure the *Plan d'instruction des élèves de l'école nationale des Ponts et Chaussées pour l'an* 7.[183] The exposition of the theoretical conclusions of d'Alembert and Lagrange also served to dispel Bélidor's confusion regarding the principles of statics and of dynamics.

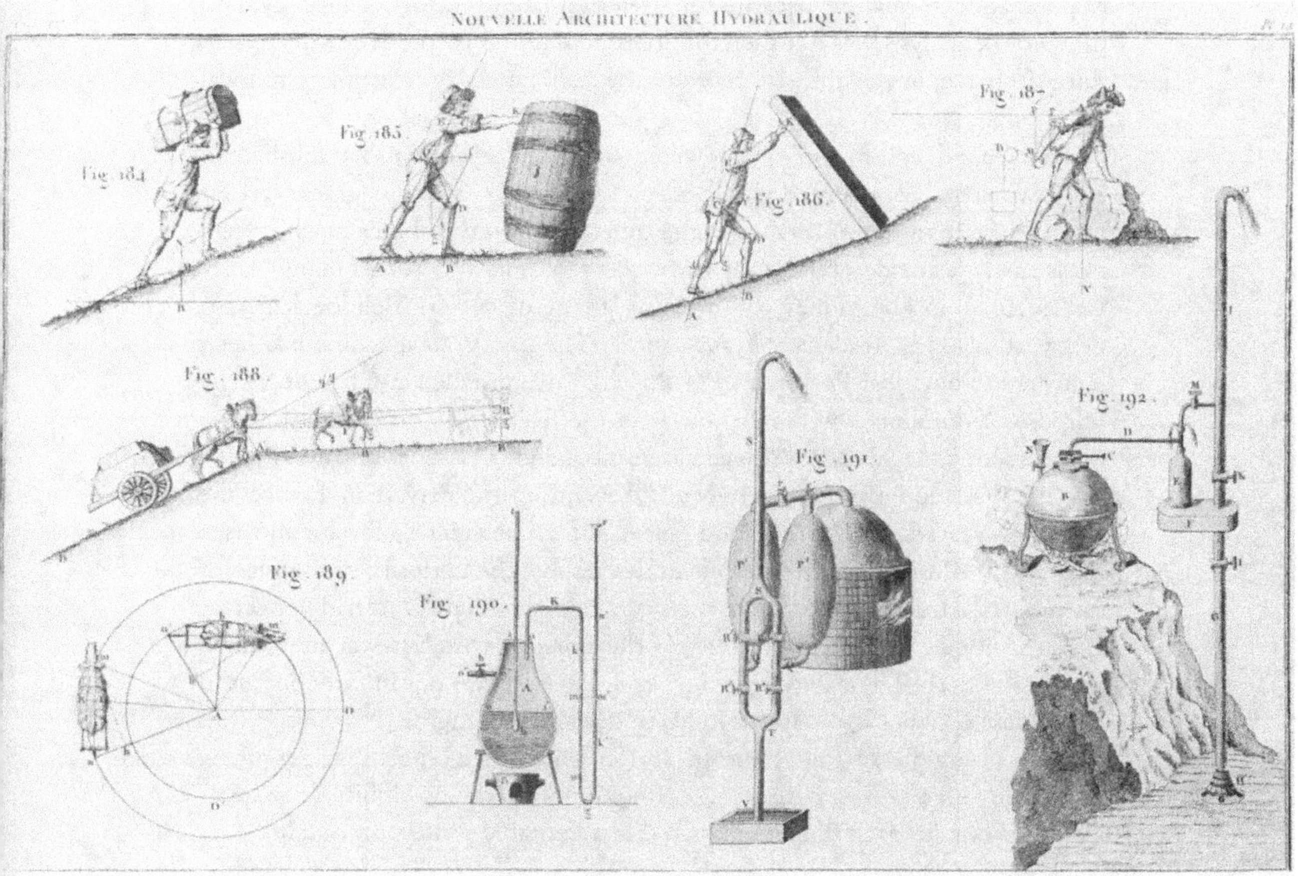

FIG. 115 G. Riche de Prony, *Nouvelle architecture hydraulique*, the various motors: animal, human, steam.

Mathematical formalism became the preferred language for technology, and the spirit of analysis was not therefore restricted to the methods of calculation, since it made it possible to reorder the full range of existing empirical knowledge.

The second volume of the *Nouvelle architecture hydraulique*, published in 1796, was largely concerned with heat-driven engines. Steam engines were far from being unknown in France at the time, since Newcomen's machine had made an appearance at Passy as early as 1726, and was subsequently fairly widely used in the mines for draining them of water; Bélidor had also described it in his *Architecture hydraulique*.[184] The first Watt machine had been delivered to engineer Périer in 1779–80 and, once installed at Chaillot for the purpose of pumping out water, it was in use from 1781 on (Fig. 116). The double-acting machine, developed as a consequence of a series of patents taken out by Watt and Boulton between 1782 and 1787, arrived in France the following year, thanks to the good graces of the chevalier de Bétancourt (see Fig. 117).[185] Prony's treatise reflects in this respect the curiosity of the men of science as regards the potential of the steam engine which Watt had perfected. But it is not so much this curiosity as the spirit in which it was met which interests us here. The *Nouvelle architecture hydraulique* was not in fact content to give a description of the various kinds of heat-driven engine. In addition, this work also completed the scientists' and engineers' investigations, supplying numerous charts and including some 'Enquiries into the possibilities of expansion of elastic fluids', which culminated in a veritable profession of faith:

> Physics has been enriched for around the last forty years by a large number of observations conducted with great care by learned and experienced men. This stock of knowledge is expanding daily, and is becoming more and more precious as the perfection of the new instruments renders our experiments more precise. The philosophical spirit has already seized hold of the proliferating number of facts furnished by observers; phenomena have been assembled, compared and classified; the language of a significant part of science has become analytic.[186]

The study of nature, according to Prony, entailed the explanation of effects, which 'consists of finding, in a given class of composite phenomena, the simple or primitive phenomena of which all the others are merely the various modifications or combinations',[187] and in the measurement of effects. Analysis of heat-driven engines belonged to this context. The shift from invention to innovation was precisely expressed by the emergence of this context, which was common both to the existing technology, to the

FIG. 116 G. Riche de Prony, steam engine at Chaillot.

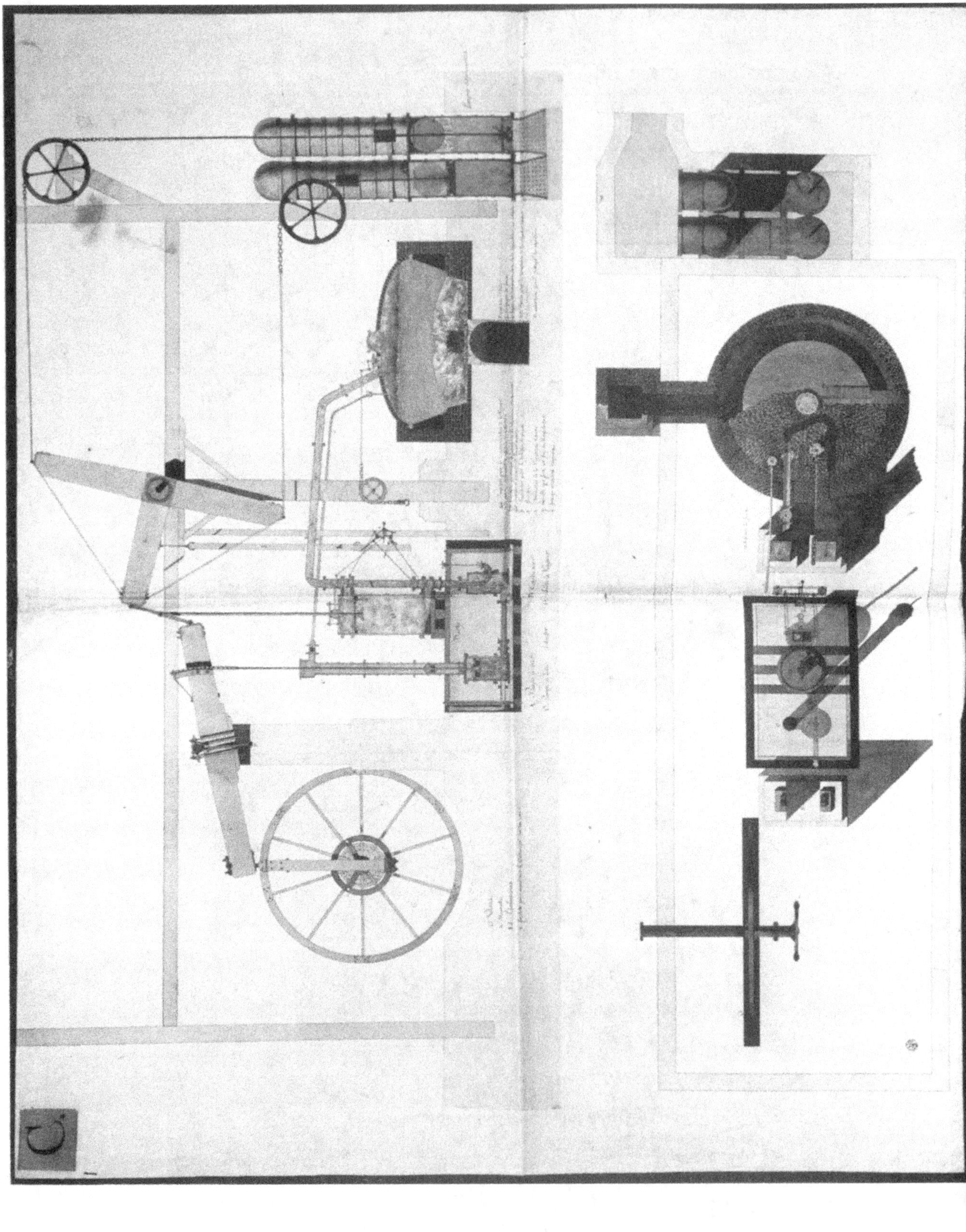

theoretical data and to the engineers' research. Prony's book in reality belongs to a long series of publications devoted to the subject of machines, from Lazare Carnot's *Essai sur les machines en général* to d'Hachette's *Traité élémentaire des machines*, publications which Sadi Carnot would effectively summarise in his *Réflexions sur la puissance motrice du feu* of 1824.[188]

The example of steam also allows us to emphasise the shortcomings of the system then being introduced. These shortcomings were primarily economic in nature. From 1810, Périer began to grow alarmed at the gap which was opening up in this area between France and England.[189] Towards 1830, France had barely 3,000 steam engines in operation, whereas surveys had given a tally of 15,000 for England.[190] The majority of the treatises devoted to the subject were, however, French.[191] In the early years of the nineteenth century, the engineers' supremacy was still linked to their planning of the territory. In spite of the involvement of the Société d'Encouragement pour l'Industrie Nationale, founded at the instigation of Chaptal, France was not to be genuinely industrialised until the July Monarchy, with the development of the railways.[192]

The Empire and the Restoration were characterised by an impressive development in the science of engineering. The example of Poinsot, Navier, Biot, Cauchy, Coriolis or Vicat testifies to the vitality of the research undertaken in domains ranging from mechanics to the strength of materials and to the use of hydraulic limes. In his homage to Navier, which was published in 1837 in the *Annales des Ponts et Chaussées*, Prony therefore felt able to write:

> The system of knowledge which constitutes the science of the engineer (architecture is assumed to be included within it) embraces at the present time, from the theoretical, experimental and practical point of view, virtually all of the various branches of the physical and mathematical sciences. This system has not always been anywhere near so extensive; but if we were to compare monuments executed at successive periods, and if we were to use as a standard of comparison not so much their size or the quantity of work invested in them but rather the degree of instruction which was necessary to draw up the designs and to supervise their execution, we would be able

FIG. 117 Dumousseau, project for a steam engine for the 1790 competition, first prize, ENPC, ink, wash, watercolour, gouache; 117 × 147.5 cm. The organisation of a steam engine competition at the Ecole des Ponts et Chaussées testifies to the engineers' interest in the new technology. The steam engine was, however, more an object of curiosity than an instrument of economic development.

to discern an evident correspondence between the scientific march of progress of the human mind and that of the art of construction. The forms of knowledge which Vitruvius judged to be necessary to an architect, and which he listed in his treatise, were probably not all possessed by the pyramid builders of Egypt, and yet they would now constitute but a mediocre form of instruction.[193]

Being included within the science of engineering, architecture was thus clearly subordinated to it. This subordination then had to be expressed in material terms. This in fact was the historical role of Durand, professor of architecture at the Polytechnique between 1797 and 1833.

JEAN-NICOLAS-LOUIS DURAND, OR ARCHITECTURE UNDER THE YOKE

Durand expounded his doctrine in his *Préçis des leçons d'architecture données à l'Ecole Polytechnique*, the first two volumes of which were published in 1802–5, and were frequently reprinted up until around 1840.[194] The success of this course may be judged by the influence enjoyed by the architect. His teaching was in effect a response to the administration's demand for engineers who were trained in architecture.

The *Précis* began by defining architecture as 'the art of composing and executing all public and private buildings',[195] by placing itself moreover under the sign of utility. *Convenance*[196] and economy would consequently be the essential qualities of architecture. The chief object of the work would consist in indicating to the pupils in 'the little time which they were able to devote to it',[197] being occupied elsewhere by studies in mathematics and physics, a rational method of design to be applied to every kind of building regardless.

This method could no longer depend upon the traditional distinction between decoration, planning and construction, a distinction which Durand described as 'vicious',[198] since it overlooked the to him fundamental notion of composition. The course at the Ecole Polytechnique was divided into three parts, the first being devoted to the elements of buildings, the second treating composition in general, and the last the analysis of buildings.

The elements of the buildings embraced the theory of Orders of architecture, which Durand undertook to reduce to 'a small number of general and fertile ideas'.[199] He criticised the complexity of classical rules, together with the theories of the Abbé Laugier on the primitive hut. As a disciple of Perrault, he saw beauty as arising in large part out of habit; like the author of the *Ordonnance*, he then proceeded to a standardisation depending upon the

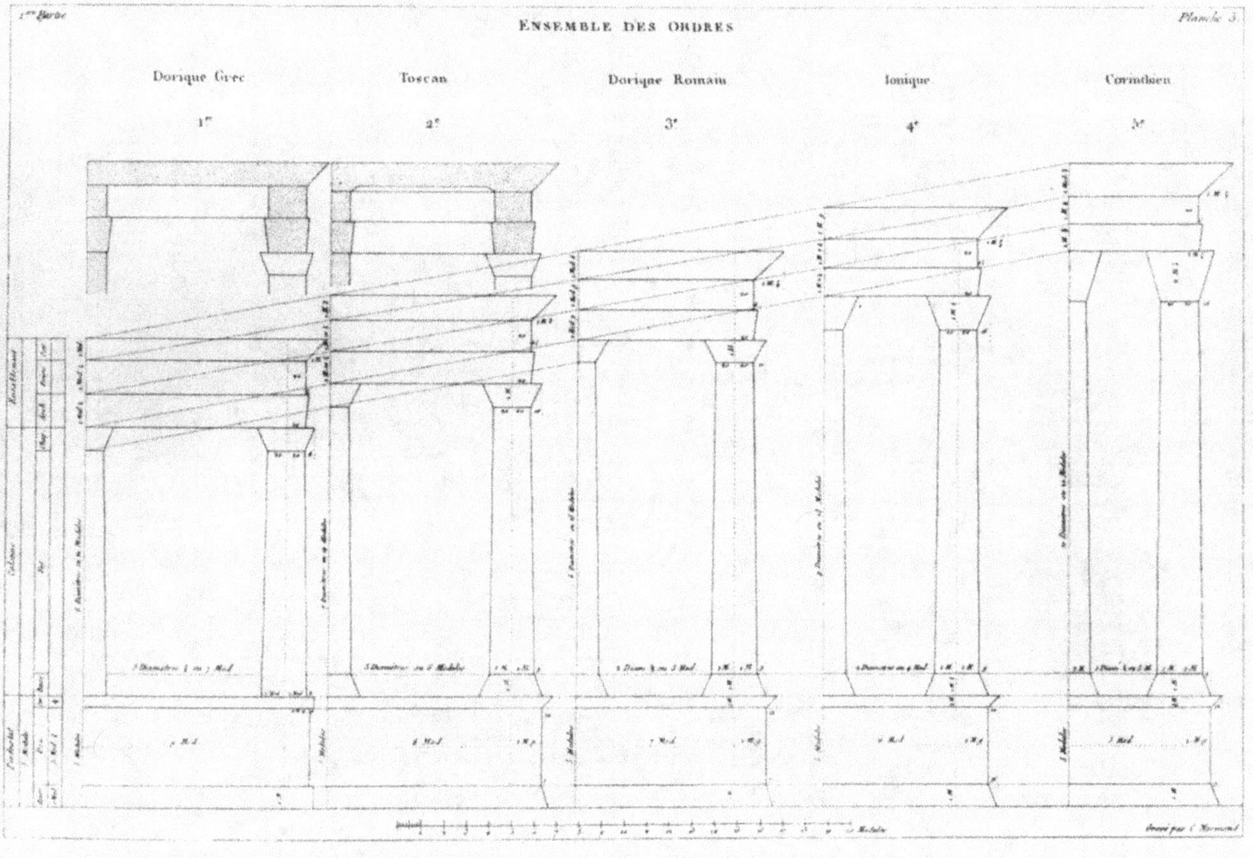

FIG. 118 J.-N.-L. Durand, *Précis* (1802–5, republished 1817–19), 'The whole series of the Orders'. Durand shows an even more pronounced concern with standardisation than Perrault had done.

arithmetical progression of proportions (see Fig. 118).[200] But the elements of edifices, 'which are to architecture what words are to discourse',[201] comprised above all the parts of the building: walls, floors, roofs, porches, courtyards, vestibules and rooms, which Durand passed in review, their classification serving as a prelude to the study of the problems of composition (see Fig. 119).

Composition played, as we have said above, a central role. It was a matter of standardising its procedures (see Figs. 120, 121). Durand rejected the poetics of revolutionary architecture. Nevertheless, he employed elementary

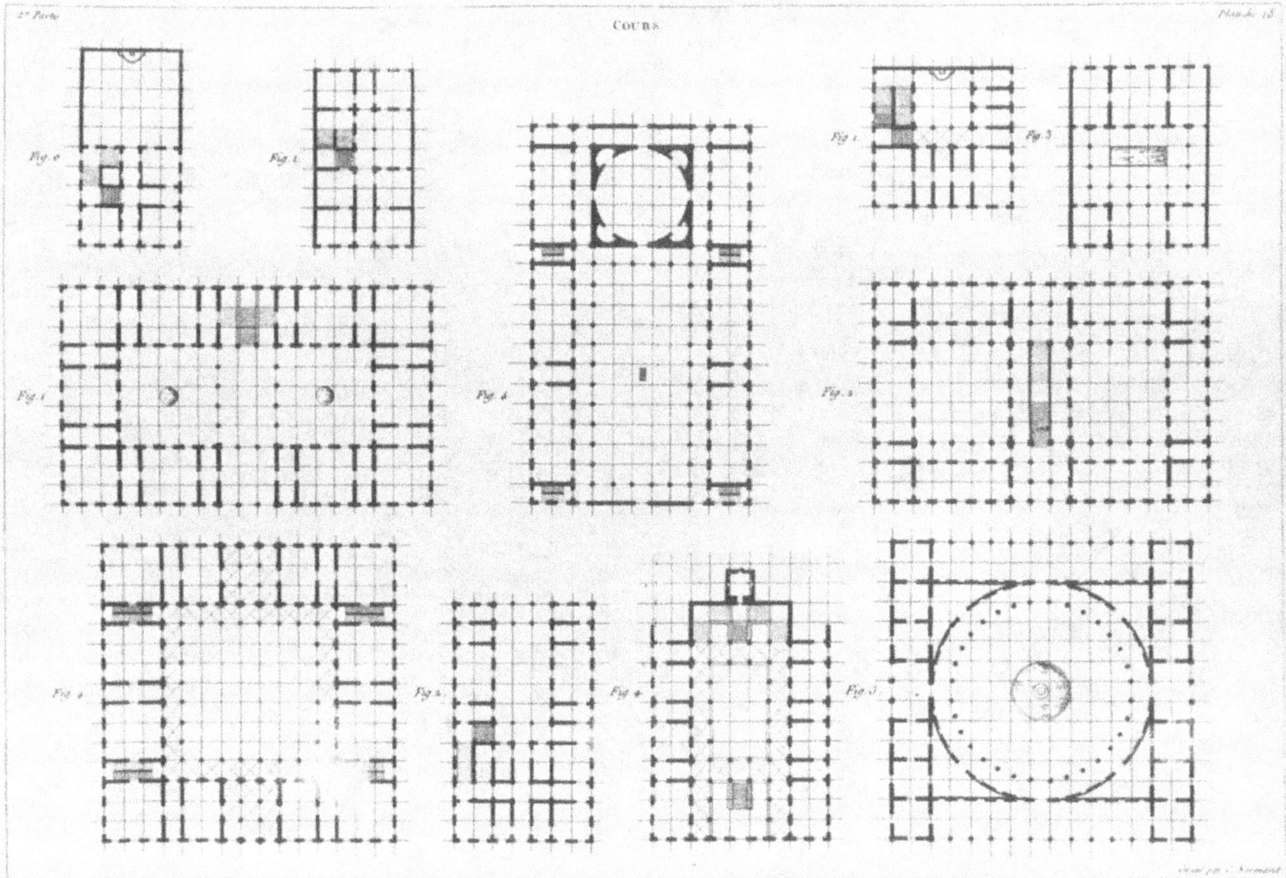

FIG. 119 J.-N.-L. Durand, different courtyards.

geometrical forms, circles or squares but the justification of this choice rested upon a concern for geometrical economy. A circular or square form, for example, demanded 'less contour' than a parallelogram; more generally, 'an edifice will be the less costly, the more symmetrical, the more regular and the simpler it is'.[202]

As a pupil of Boullée, Durand always put the emphasis upon problems of conception, so as to imitate the principles of nature rather than its creations. His highly rationalised neo-classicism nevertheless involved a rejection of the theory of simple forms in favour of an economical and calculated geometry.

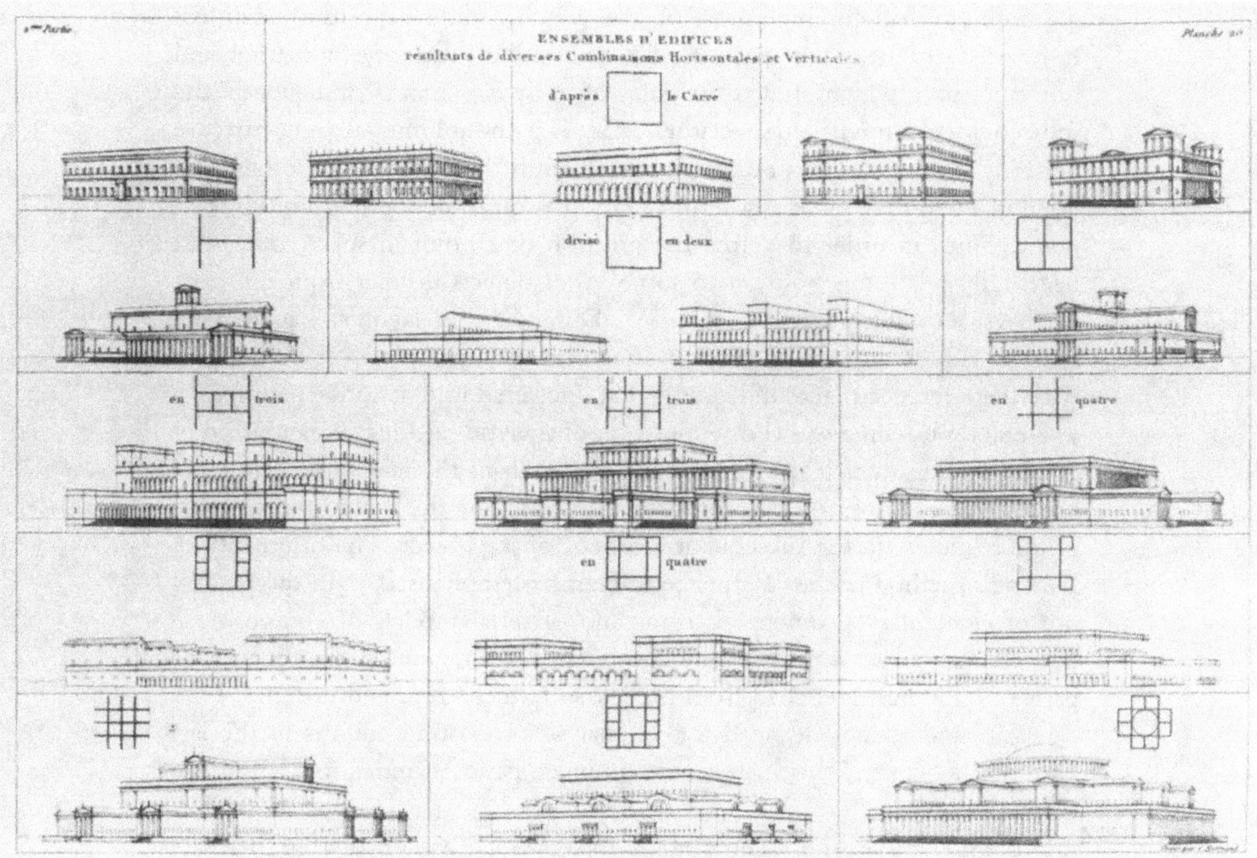

FIG. 120 J.-N.-L. Durand, 'Series of buildings resulting from various horizontal and vertical combinations'.

In the *Précis*, designs were also engraved as grid layouts,[203] which seemed to be inspired by the procedures of descriptive geometry, taught very successfully around the same time by Monge.

As Monge declared at the start of his course at the Ecole Normale, which he later repeated at the Polytechnique, 'descriptive geometry has two aims: first, to provide the methods which will represent on a drawing sheet in only two dimensions . . . all the masses in nature, which have three dimensions . . . provided that these masses are rigorously defined'.[204] Inspired by the same deductive spirit as was employed in the exact sciences, architecture ultimately did no more than combine those objects which could be rigorously defined. It referred to the same space as did descriptive geometry, a neutral and isotropic space which eluded the rules of perspective and consequently the subjectivity of a privileged spectator: 'Space is without limits; all of its parts are perfectly similar, there is nothing to distinguish them from each other, and none of them may serve as a term of comparison to define the position of a point. Thus, in order to define the position of a point in space, one must necessarily relate this position to some other objects distinct from the parts of the space which encloses them.'[205] Durand's grid layouts, which were symptomatic of a depersonalisation of architecture, present a certain analogy with the reference marks and connecting lines used in descriptive geometry.

Durand's teaching was also reminiscent of analytic calculus, a calculus rid of the ambiguities which still vitiated the approach of the *idéologues*. The chief problem of composition had consisted of reconciling the analysis of the parts of the edifice with the subsequent synthesis of the object. The originality of Durand's method lay, as Werner Szambien has emphasised,[206] in the parallel use of elementary geometrical forms and of serial models of composition. Squares and circles would thus be combined with fragments of ordinance or parts of buildings wrenched from history so as to be repeated in series. One can thus understand the need for an analysis of existing edifices in the last part of the course, an analysis providing a basic material for systematic recompositions.

Alongside the second volume of the *Précis*, which was devoted to 'particular *convenances*', that is to say, the characteristics of the different types of building, Durand could also refer his reader to his *Recueil et parallèle des édifices de tous genres, anciens et modernes, remarquables par leur grandeur ou par leur singularité, et dessinés sur une même échelle*, which was published in 1799–1801. Written in collaboration with Legrand, this book further extended the notion of the comparative history of architecture which Leroy had inaugurated with his *Histoire de la disposition et des formes différentes que les chrétiens ont données à leurs temples, depuis le règne de Constantin le Grand jusqu'à nous*. But whereas the latter had been concerned with a history which would integrate the analysis of sensations,[207] Durand restricted himself to geometry, and to a catalogue of forms whose common scale would facilitate comparison. Nevertheless, the *Recueil* may be seen as the expression of a typological ambition rounding off the cycle of reflections inaugurated by Boullée and Ledoux; temples, forums, basilicas, market-places, colleges and libraries were thus all passed in review. Although the theory of the art upon which Durand and Legrand had sought

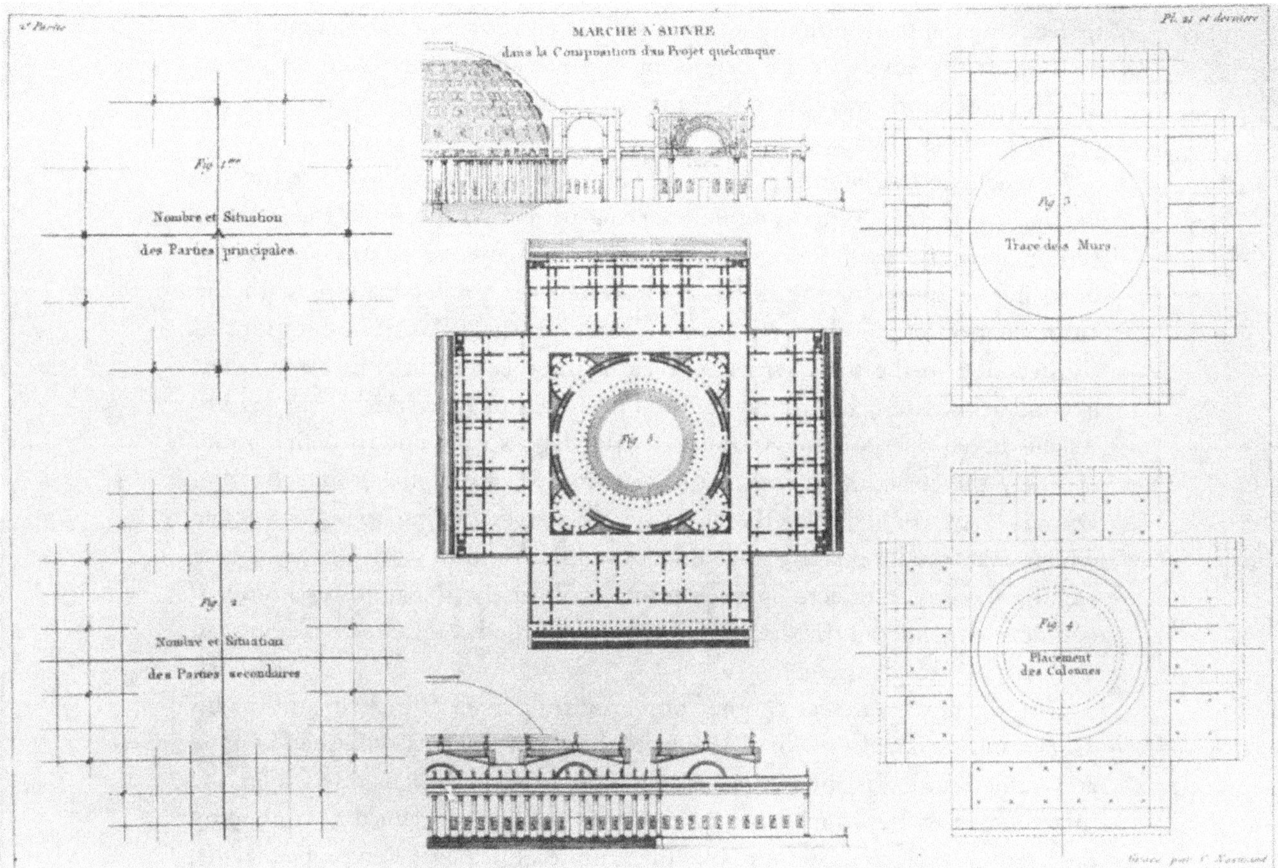

FIG. 121 J.-N.-L. Durand, 'Course to follow in the design of any project whatsoever'. Durand's method explained by means of examples. Analysis of the main parts of the building enabled one to define the major axes of the composition, to which the secondary parts related. The placing of walls and columns was virtually automatic. Nevertheless, in the corners of the central square the hemicycles of which Boullée had been so fond survive, and they were a kind of residue of the design within the standardisation of forms aimed at.

to base their historical approach was still a tentative one, the comparative method of the *Recueil* had a degree of success. Under the name of 'Grand Durand', the work was also to become one of the classical points of reference for the Ecole des Beaux-Arts.

As taught to the engineers, Durand's architecture was subordinated in the last analysis to their system, while adopting to its own purposes the spirit of certain scientific methods.[208] This subordination was to take a paradoxical turn, if one thinks of the distance separating the preoccupations with the rationalisation of architectural form expressed in the *Précis* from questions of construction in the strict sense. As regards these latter, Durand barely went beyond generalities, and simply referred his pupils to Rondelet's *L'art de bâtir.* At the Ecole des Ponts et Chaussées, Mandar also set out to define more precisely the structural scope of the teaching offered at the Polytechnique. Durand's approach should therefore be seen as belonging to a perspective which was predominantly spatial. But while Boullée and Ledoux were seeking to turn architectural space into a tool for the dynamic resolution of conflicts engendered by society, Durand resigned himself to playing an auxiliary role to planning.

History, however, was another powerful influence. Winckelmann, in his *Histoire de l'art chez les anciens* (see Fig. 122), which was first published in 1764 at Dresden and was quickly translated into French, had ironised over the 'magnificent works which have appeared up until the present day, containing descriptions of ancient statues', like the Comte de Caylus' *Recueil d'antiquités*, works in which one would look in vain for a 'reasoned history of art'. According to Winkelmann, this history should constitute 'the précis of a system of art'; it was a question of going back to its origins, of tracing its progress and its variations, of observing its decline and its fall, so as to isolate more effectively its governing principles, principles which were also operative in society.[209] Leroy, in his *Ruines des plus beaux monuments de la Grèce* (1758), had advanced a similar line of argument, his view being that one's appreciation of Greek monuments should be subordinated to one's knowledge of their philosophy, to that 'of the order and of the division which they introduce into the sciences, and of the rules which they observe in the arts', or to that of 'the general system formed by the Greeks as regards the sciences and the arts'.[210] In the closing years of the eighteenth century, art history in the strict sense was born, in relation to the general history which Volney had conceived, in a perspective as much physical as moral, in his lectures of the Year III at the Ecole Normale.[211] Likewise, art history was meant to retrace not only the evolution

FIG. 122 J.-J. Winckelmann, *Histoire de l'art chez les anciens* (1765, translated into French 1790–2), colossus of Memnon.

of forms but also the labour of the creative spirit. This history seized hold of architecture and sought to reorder it in terms of its own categories. At a time when Winkelmann was being republished, men such as Cassas, Dufourny, Alexandre Lenoir, Legrand and Molinos[212] made the first attempts to found a museum of architecture. Quatremère de Quincy was already controlling for architecture by regulations.[213] It was between the two poles of a 'system' of technology and a 'system' of art, expressing themselves through a history, that architecture was from then on to be elaborated, an architecture that was subject to contrary influences.

If we talk here of an architecture under the yoke, it is because the architect now suffered from an 'unhappy consciousness', being afflicted both by a nostalgia for the luminous origins of Greek art and by an apprehension at the boundless future of technology, so that he was haunted both by the unrivalled purity of the Doric and by the ineluctable dulling of forms in a society which reduced them to the status of commodities. An attempt would later be made to resolve this dilemma, by reconciling the beautiful with its history, and the quality of the object with its standard production.

Descriptive geometry meanwhile eliminated any reference to the gaze of a spectator ordering the world by means of a central vanishing point. Neo-classicism was to prove impersonal in much the same way. Relinquishing, for example, the fluctuating rules of classical proportion and contour in favour of a freedom that was in theory limitless, it thereby eliminated the possibility of an always singular 'expression' of the Orders and sections of architecture. Between the purity of origins and the precision of machines, everything became equivalent. The architect withdrew in some sense from his work. As François Beguin has put it, 'the visible order has become an invisible order . . . The post-classical period inaugurated an era in which . . . human power would no longer manifest itself except in the celebration of scenes from which the architect would always appear to be absent.'[214]

REDEPLOYMENTS

The French Revolution marked, as we observed above, the triumph of the engineers. Their sphere of competence was extended, in theory at least, to every branch of human activity. From Condorcet to Destutt de Tracy, *idéologie* traced a programme of instruction that was in line with their ambitions.[215] The Ecoles Centrales and, after them, the Ecole Polytechnique, provided a training closely resembling the kind which Prony had advocated. Following the recommendations of the author of the *Réflexions*, the engineers were everywhere, from the Commission du Système Metrique to the various cogs of an administration coping with a war which rendered technicians indispensable.

At the same time, the architects suffered a degree of neglect, their role being simply to illustrate the lofty deeds of the Republic, as the programmes for the famous competitions of the Year II would suggest: the festival for Bara and Viala imagined by David, the triumphal arches and the temples to Equality.[216] The Academy of Architecture was closed down in 1793; David Leroy may have gone on teaching in an official capacity at the Louvre, but it was not until the early years of the nineteenth century that a school of architecture in the full sense was reconstituted; it was only in 1819 that the Ecole des Beaux-Arts was definitively established.[217] The Polytechnique meanwhile trained the engineer-architects who were replacing the artists of the *Ancien Régime*. The latter were affected by the revolutionary events. De Wailly held a secure post on the Commission du Plan des Artistes, which considered ways and means of transforming Paris,[218] and whereas Boullée taught at the Ecole Centrale du Panthéon, Ledoux was imprisoned and Mique was executed during the Terror. The Institute, which was set up in 1795, had only one class for Fine Arts, of which architecture constituted merely a sub-section. The *Plan d'instruction des élèves de l'Ecole nationale des Ponts et Chaussées*, which was drafted by Prony in Year VII, recommended a form of teaching of construction subsuming architecture and reducing it to the art and manner of raising edifices.[219]

With the consolidation of the Napoleonic Empire, the situation was to some extent reversed. The early years of the nineteenth century saw a retreat from the scientific and technological Utopia of the Revolution. In the lycées, which replaced the Ecoles Centrales in 1802, the emphasis was upon the humanities, and the sciences were therefore neglected.[220] Nevertheless, Napoleon consolidated the corps of the Ponts et Chaussées, which was called upon to plan his Empire. The edicts of 27 Messidor and of 7 Fructidor Year XII (16 July and 25 August 1804) reorganised the Ecole and the corps of the Ponts et Chaussées, and likewise founded the Conseil Général des Ponts et Chaussées, which was supposed to proffer its advice on 'all questions touching upon art and accountancy which shall be submitted to it'.[221] But opposite the Conseil Général des Ponts there stood the Conseil des Bâtiments Civils, which largely consisted of architects, and which had been founded in 1795 and confirmed subsequently in its task of controlling public building.[222] Behind this institution, which featured the names of Brongniart, Chalgrin, Poyet and Rondelet, one can discern the intention to resuscitate the corps of the architectes du Roi.

The first half of the nineteenth century thus saw the coexistence of two complementary institutions, with the Conseil Général des Ponts being especially concerned with the planning of the territory and the construction of buildings, and the Conseil des Bâtiments Civils being responsible for the inspection of edifices of public utility. Both institutions may be seen as the expression of the same concern with rationalisation and state planning.

Nevertheless, the gulf between architects and engineers widened appreciably. The architects were soon to have the same ambiguous relation with the state as they had had with the former Academy. Compromise between artist and civil servant was as difficult as ever. Once formed into brigades by Napoleon, the engineers of the Ponts et Chaussées on the other hand would have no problem in adjusting to the new requirements of the administration. But the crucial factor was not so much such differences in institutional strategy as the distance between the cultural universes and professional practices of the persons involved. While the eighteenth-century engineers had still been fairly close to the architects, their successors only maintained strictly pragmatic relations with architecture. At the Institute, Prony, Girard and Navier belonged to the first class, that of the physical and mathematical sciences. The engineers were now primarily concerned to win the collaboration of the scientists.[223]

More important still, preoccupations regarding development had changed. Whereas the Enlightenment had tended to refer to the civic works of Antiquity, the engineers admired the efficiency of the English. This tendency was to become more pronounced after the Restoration. Thus Dutens, in his *Mémoires sur les travaux publics de l'Angleterre* (1821), emphasised the stability of Perronet's great stone bridges, but he immediately went on to observe that the solutions adopted on the other side of the Channel were generally less cumbersome.[224] Alongside the imperishable monuments of technology, it was therefore a question of inventing structures which were more ordinary and repetitive, and which represented a compromise between durability and large-scale development. The perfecting of roads and bridges, the introduction of tarmacadamed roadways and of Navier's suspension bridges[225] are evidence for this concern with economy. From now on, the engineers' slogans would be calculation and economy.[226]

It was no longer space in the strict sense which mattered, but its relationship with time, with the time of exchange, transportation, the circulation of commodities, flows of every kind, the time of the mortgaging of services, of the calculation of their profitability, and the limitless timescale of progress. Through the constant interplay of derivatives and combinations, analysis made it possible to produce models of time. The use of differential equations lent support to Laplace's determinism and enabled him to publish in 1814 his famous *Essai philosophique sur les probabilités*, in which the future itself had become the object of an approximate quantification, from election results to the average lifespan of citizens:

> We should . . . envisage the present state of the universe as the effect of its previous state, and as the cause of the one which is to follow. If there were an intelligence which at a given instant could know all the forces by which nature is animated, and the various situations of the beings of which it is composed, if moreover it was sufficiently immense to subject these data to analysis, it would be able to embrace in the same formula the movements of the largest bodies in the universe and that of the tiniest atom. Nothing would be uncertain for such an intelligence, and the future and the past alike would be present before its eyes.[227]

The contributions of engineers active in the first half of the nineteenth century should be understood in this context of a reversible time. The meaning of such contributions no longer lay in architecture, nor even in the conquest of space alone; it was a matter of mastering all the movement and flow of the universe, and thereby moving ever closer to the truth. Scientific calculation, economy, so many partial tools, the adaptation of the unique formula by which the world was regulated, a world whose fulfilment was anticipated in theory, as Laplace's superior intelligence had already discerned it.

With industrialisation, whose effects began to be felt under the July Monarchy, this race to complete the world accelerated. The civil engineers made their appearance alongside the state engineers. The Ecole Centrale, which was explicitly designed to rival the Polytechnique in the training of cadres more abreast of technical and economic realities, was founded in 1829; the Ecole des Ponts et Chaussées itself admitted pupils from outside the administration from 1851 on.[228] Since they were employed by banks and private businesses, such pupils were more concerned with questions of profitability.

But architecture and the future, architecture and progress, still had to be reconciled, a task to which the architects would soon apply themselves. The 1800s were a watershed. On the theoretical plane, while the traditional rules governing the correspondence between architecture and social hierarchy were gradually being eroded, forms of rationality having more to do with processes than with objects were emerging. Since this was essentially the rationality of a burgeoning capitalist society, science itself became a process involving the interrogation of nature. In disconnecting itself from this type of interrogation, architecture went into a long-term crisis. If the rise of neo-classicism cannot be readily related to that of architectural modernity, its related uncertainties prefigure it.

CONCLUSION

In spite of a context that was still by and large classical, the end of the Enlightenment saw a significant breach in the domain of architecture and planning. As a society of orders and of compromise, founded both upon partitioning and upon negotiation, broke down, the architectural theory of the seventeenth and eighteenth centuries gradually lost its meaning. Since this theory reflected in its own way the idea of absolute sovereignty, the hierarchy of conditions of the *Ancien Régime*, and the dream of an ordered and pyramid-shaped world, it proved to be ill-adapted to the issues of the day. Blondel's *Cours* ended with an admission of impotence with respect to the great conflicts of the modern age, such as the clashes between state and citizen, the confrontation between utility and beauty, and the from now on uncertain relationship between planning and design.

The technicians, by contrast, were confirmed in their vocation. The Revolution had consolidated the institutional, doctrinal and practical system of the engineers, who were increasingly regarded as the preferred instruments for the overall progress of man and society.

Using treatises, cartography and 'Utopian' drawings, I have sought in the present work to identify symptoms of the weakening of classical architectural theory and of the increasing domination of technology. It has thus been possible to contrast Blondel's desire for a reconciliation between architecture and society with the violence of the 'general will' championed by the engineers.

More generally, where classical thought tended to argue for a continuous hierarchy, for variations upon ideal models, and for negotiation between these models and routine obstacles, the engineers employed abrupt oppositions and similarities. This difference in attitude may be discerned in a whole number of different areas, from teaching and construction to the town and the country.

For the author of the *Discours sur la nécessité de l'étude de l'architecture*, architectural knowledge was a language accessible to all, albeit at various different levels, whereas for Prony it became a set of knowledges which were at once universal and narrowly specialised, being the expression of the ultimate regularities of nature, and, paradoxically, that of the profession of an engineer jealous of his prerogatives. The relationship between theory and practice was also modified. The *Cours d'architecture* allowed some distance between theory and practice, so as to preserve the possibility of an architectural critique preliminary to the design. In Prony's *Réflexions*, on the other hand, theory and practice were seen as a single continuous process running from the conception through to the realisation of particular works.

It was a process of this same type which served to lend substance to the structural model, being applied to Perronet's bridges in particular. While an approach to problems of construction in terms of structure was clarifying, calculation was also beginning to make an appearance, as the instrument of a rationalisation still to come. The visual, psychological and symbolic stability of the architects would soon be matched by the construction of the engineers, which employed 'indirect and powerful forces'.

Cartography then made it possible to tackle the relationship between planning and design. The territory of the engineers was alternately divided up and recombined. Within the contradictions of a nature which was by turns unchained and pacific, there emerged an aesthetic of conflict and of process, which transposed categories borrowed from painting and ethics to design. The sublime and the picturesque led, moreover, to an excess of representation, which itself favoured a radical reorganisation in favour of calculation.

If one were to compare the engineers' system with the rationalism – tempered by tradition – of Blondel, it would seem to possess a remarkable efficacy, albeit one that was paradoxical, since it rested upon a series of contradictions which generated as many processes destined in the long run to overcome them. The 'general will', through which state and citizen might be reconciled, the assimilation of theory and practice within a dynamic of conception and realisation of works, the aesthetics of the sublime were in reality processes. I want here to emphasise the new type of resemblance which was introduced between these different processes, resemblances which no longer concerned objects or configurations, as they had done before, but potentialities for future development, alternatives or choices to be made. These dynamic resemblances, which were replacing the static equivalences of the Classical Age, plainly revealed the existence of a new 'spirit', which would ultimately show that these different transformations were convergent.

This convergence was founded upon notions of development and progress. Bélidor or Patte's catalogue of techniques was reordered in terms of the possibility of a progress in which space and time were closely associated. As we have already emphasised, it was this interrelation between space and time which from then on mattered, with technology being primarily concerned with quantifiable flows and movements. This perspective also allows us to understand the failure of classical theory. For Blondel, built space was still the ultimate point of reference, with theory serving to express the stability of the principles involved. This was no longer the case in the future, since the object produced was to be increasingly subordinated to the production process.

In the Age of Enlightenment, only the actual possibility of progress had been established. In spite of the triumphant tone adopted by the engineers, their writings give the impression of a progress that was essentially limited, as if the world would never become altogether reasonable, because planning was forever encountering the same obstacles, whether it was the natural elements, human prejudice or the complexities of production. The engineers' optimism seemed therefore to indicate that the scope of their action consisted in this perpetual incompletion, in their pursuit of a final state which was forever eluding them. It is tempting to suppose that this surprising situation already prefigured the uncertainties of the following century.

Technological progress in fact depends upon a paradox that is even more fundamental than the contradictions of the engineers' discourse. For, to master nature by systematically consuming its resources, as such a concept of progress presupposed, would culminate in a sort of final exhaustion of the world. With the discovery, in the latter half of the nineteenth century, of the irreversibility of certain physical phenomena, such as exchanges of heat, this conception was to be rendered more precisely. Thus, in 1865, Clausius appended a famous cosmological statement to the first two principles of thermodynamics: 'Die Energie der Welt ist konstant. Die Entropie der Welt strebt einem Maximum zu' ('The energy of the world is constant. The entropy of the world tends towards a maximum'). Should one think in terms of entropy or should one characterise the situation as one of disorder in a system? Given the irreversibility which was then regarded as crucial in physics, the universe was seen as tending to a final state, which would also be its death. In this context, technology served as a measurement of the state of the world, a measurement of a kind of internal combustion.

Faced with this embryonic threat from the end of eighteenth century, with the presentiment of an irremediable violence exercised over nature, there was a great temptation to take refuge in a perpetual present, to conjure away the future at the same time as one was devouring space by covering it with services. The suspended time of the Enlightenment engineers, their progress which was at once incomplete and completed, already evokes the vision of the world of the superior intelligence featured in the *Essai sur les probabilités*, an intelligence in whose eyes past and future were condensed in the same mathematical formula. For the engineers, calculation was rightly identified with progress, the progress of the human mind, economic and social progress, with time and space interlocking and therefore neutralising each other.

The notion of progress thus helped to negate the specific dimension of space and time. What then seemed to emerge was an attempt at total control, an attempt in which one may already discern a degree of anxiety, as if the ultimate aim of progress were in question. If, to use the phrase formulated by the Frankfurt School, 'technology is a rational activity with a certain end in mind', this end remains curiously vague, unless one admits that the conquest of nature represents the very objective to be attained.

Perhaps one has to be content in the meantime with this constantly reiterated objective, as one explores its links with the social order. From *idéologie* to Saint-Simonianism and Positivism, these links invariably tend to be clothed in Utopian colours. Yet it remains the case that the engineers' system was above all else that of the enlightened bourgeoisie. In the course of the eighteenth century, the engineers of the Ponts et Chaussées had been subordinated to the administration of finance, while their doctrine was modelled upon that of the Physiocrats, and subsequently upon that of the *idéologues*. Later on, the bonds between capitalism and development would be as close as can be. Technological choices are always political, economic and social also. The anxiety which they generate might be simply the reverse of a doomed desire for domination, and the immobile progress of the Enlightenment engineers an expression of impatience towards the social and material complexities impeding their seizure of power.

This seizure of power was to be effected through a rationalisation of production, achieved by means of calculation and planning. The collapse of classical theory derived from its inability to embody this rationalisation.

In this perspective, neo-classicism undeniably offered the basis for a solution, by arriving at an accommodation with technological determinations, to such an extent, indeed, that it transformed itself into an effective auxiliary to development. But this attitude remained an ambiguous one, inasmuch as architecture claimed at the same time an artistic character, reinterpreting the productive rules in such a way as to further an original and always unusual form of expression. The tension which was thereby created between the certainties of the science of building and a history of the art, which was structured by other imperatives altogether, undoubtedly furnished one of the founding experiences of architectural modernity, a widespread malaise at the start which was gradually to spread.

This malaise particularly concerned the relationship between architecture and technology, which was to be increasingly fraught with conflict. Between form and process, between the completion of the built object and the boundless progress of science and technology, splits were gradually introduced, whose ultimate course we are acquainted with today. It is the origin of these splits that I have sought to trace in the present study. In the future, we shall do our best to overcome them, but at what price? With the collapse of classicism, there actually began the cycle of hope and disillusionment, truth and lies, that of the avant-gardes and the reactions to them, and that of an architectural 'debate' marked with the seal of a kind of impotence. We have yet to emerge from this cycle, for is not architecture still, when all is said and done, the art which Diderot, in his *Prospectus de l'Encyclopédie*, had already ranged under the heading 'Imagination'?

APPENDIX. BIOGRAPHICAL NOTES

JACQUES-FRANÇOIS BLONDEL (1705–74)

Jacques-François Blondel was born at Rouen in 1705. In spite of the homonym, his family was not related to François Blondel, the first director of the Academy of Architecture.

We do not know precisely when Jacques-François Blondel left his native town for Paris, but by the 1730s he was working alongside his uncle, François Blondel, a member of the Academy of Architecture, and together they decorated the Hôtel Rouillé. At the same time, the young architect embarked upon a career as an engraver, an activity which he was to pursue throughout his life.

Blondel's first book, *Distribution des maisons de plaisance*, appeared in two volumes in 1737–8, and this work still showed traces of rococo influence, as did his engravings for Mariette's edition of d'Aviler's *Cours d'architecture.*

Blondel seemed already destined to be a theoretician, although he built a number of buildings, among them an orangery in Florence, the terraces of a château on the road to Brittany, and a private hôtel at Besançon.

Around 1740, Blondel opened the Ecole des Arts, in rue de la Harpe, an institution which was designed to compensate for the shortcomings of academic teaching by increasing to a considerable extent the range of subjects studied by the 'young artists'. In spite of a bankruptcy, possibly fictitious, which occurred around 1754, the Ecole des Arts was to keep its good reputation up to Blondel's death. In 1743, Blondel was authorised to deliver public lectures and, in 1750, the King bestowed two scholarships upon those pupils at the Ponts et Chaussées who were training as architects. Links between the Ecole des Arts and Perronet's Ecole were, moreover, so close that many of Blondel's pupils ended up in the corps of the Ponts et Chaussées.

In the 1750s, Blondel thus became an official figure, whose teaching was known to be of the highest quality. As well as being in constant contact with the administration, he also frequented the *philosophes*, and through this connection he came to write the majority of the articles in the *Encyclopédie* on building and architecture.

The first four volumes of *Architecture françoise* appeared between 1752 and 1756. Blondel followed Mariette in presenting a description of the most beautiful buildings of Paris. The engravings were accompanied by commentaries and essays reflecting Blondel's predilection for a classicism adapted to contemporary taste, especially through the integration into its system of the art of distribution, which favoured those 'conveniences' seen by the eighteenth century as one of the most essential means of perfecting a habitation.

In spite of the *succès d'estime* which it enjoyed, the *Architecture françoise* was never to be finished, the great cost of publication serving to restrict the number of copies sold. This venture brought financial ruin upon Blondel, for his talents as a businessman did not match those as an author.

In 1755, however, Blondel's works earned him the honour of admission to the Royal Academy of Architecture and, through the publication of his *Discours sur la nécessité de l'étude de l'architecture* (1754), which summarised his pedagogic conceptions, he became one of the best-known teachers in the whole of Paris. His fame helped to emphasise the inadequacy of the teaching offered by men such as Jossenay and Loriot, who had succeeded to the post of Professor at the Academy. In order to introduce new blood into the course at the Academy, Loriot was ultimately forced to resign, being replaced by Blondel in 1762. Blondel, who was soon to be assisted by David Leroy, held this post until his death, while running the Ecole des Arts at the same time. It is extremely hard to distinguish between his role as professor at these academies. Be this as it may, Blondel was to train most French architects of the latter part of the eighteenth century, from Boullée and Ledoux to Peyre and de Wailly.

Blondel's purpose in bringing out his *Cours d'Architecture* (1771), a synthesis of his basic concepts, was to provide his pupils with a working tool for the study of their subject. Only the first four volumes appeared in his own lifetime, together with the first two volumes of plates, devoted to the decoration and distribution of buildings. The work, which was intended to include a third part devoted to problems of construction, was finished after Blondel's death by Patte. In his *Cours*, Blondel proved an energetic champion of French classicism, subjecting both the excesses of Italian baroque and the bolder flights of certain of his pupils, for example Ledoux, to criticism. Yet he presented himself as above all a rationalist, and one who sought to promote 'reasoning in art' even in the smallest details of ornamentation.

Although Blondel's chief preoccupation was with theory, he pursued a career as a designer, albeit to a lesser extent, so as to illustrate his concepts in practice. Thus, he undertook the transformation of the Hôtel de Choiseul in Paris, and made architectural drawings for the Archbishop of Cambrai, for the residence of a German prince, and for an Academy of Fine Arts in Moscow.

Blondel was even more involved in urban architecture. Invited to Metz in 1761 to draw up plans for the abbey of Saint-Louis, the building of which had just been decided upon by the King, Blondel went on to draft a plan for the whole of the town, which was partly realised. Although the Abbey of Saint-Louis, which had been his original reason for coming, was never to be built, he designed the town-hall and a Doric portal for Metz cathedral.

Blondel was summoned to Strasburg in 1763, and again commissioned to produce an overall plan for the town, which was begun by him. At both Metz and Strasburg, Blondel's approach to city planning was large scale. He broke with a tradition which involved making no more than limited improvements, and attempted to rationalise the town. Squares and streets were intended to form coherent sequences, especially in Strasburg, where his project was organised around the route d'Allemagne, which traversed the whole town. Blondel, like Patte, exemplifies both the strengths and weaknesses of late eighteenth-century planning, invoking both morphological coherence and the analysis of urban functions, a town-planning which, as Leroy Ladurie put it, merely 'grazed' the surface of things, since it was not able to impinge on the actual fabric of the city.

Towards the end of his life, Blondel suffered a series of setbacks. His last years were shadowed by ill-fortune and by declining health. Moreover, the generation of revolutionary architects seemed intent upon overturning his patient efforts to bring up to date and to synthesise the classical tradition. 'When I was delivered of that pedant Blondel', wrote Mathurin Crucy, 'I was able to give my work a broader scope', whereas Ledoux lost no opportunity to deride a master who was 'circumscribed by the five Orders'. Posterity has often remembered only the criticisms which have been levelled at Blondel, forgetting the innovatory character of the teaching of this director of the Ecole des Arts and professor of the Academy of Architecture, and of a doctrine which is not readily confined within the boundaries of the classical tradition as such.

PIERRE PATTE (1723–1814)

Pierre Patte's life is only known to us in broad outline. No portrait survives, and the only biographical details we have refer to the various stages of his professional career, which was, moreover, a fairly curious one. The odd nature of his career may be exemplified by the fact that he was never a member of the Academy of Architecture, that disputes over the *Encyclopédie* and the Panthéon occupy a disproportionately large place in it, and that his thought was strikingly original.

Pierre Patte was born in Paris in 1723. He was initially a pupil of Boffrand, and after his master's death he attended courses at the Academy, without ever winning any prizes or being accorded any honourable mention. Around 1750, he travelled in Italy, and worked for a time on the Cassinis' map of France. His talent as an engraver soon brought him to public notice. In 1752, he contributed several plates to Blondel's *Architecture françoise*; he became editor for the engravings for the *Encyclopédie* in 1757.

In 1759, Patte, who had just announced his resignation from the *Encyclopédie*, published a letter of denunciation in Fréron's *Année littéraire*, which marked the beginning of the controversy between himself and those responsible for the *Encyclopédie*. In his letter, he accused the latter of having plagiarised the engravings which Réaumur had assembled for the Academy of Sciences a few years before. The affair very quickly turned sour. The Academy of Sciences confiscated the relevant file and appointed a commission of enquiry, which issued a report in the *Encyclopédistes*' favour. Nevertheless, Patte reiterated his accusations in 1760, casting aspersions upon the reliability of the Academy's checks. Voltaire compared Fréron and Patte to Cartouche and Mandrin, while Diderot waxed ironical over the 'excessively clever and honest man', whose scruples he saw as stemming from the desire for easy fame.

Patte nevertheless persevered with his career as an engraver. Having supplied the plates for Leroy's *Ruines des plus beaux monuments de la Grèce*, and those for Goguet's *Origine des lois*, he devoted his time to the illustration of his own *Monumens érigés à la gloire de Louis XV*.

The year 1770 saw the opening skirmishes in the dispute over the Panthéon, which was once again instigated by Patte, who, in his *Mémoire sur la construction du dôme projeté pour couronner l'église de Sainte-Geneviève*, questioned whether it was really possible to build the dome in the way specified in Soufflot's project. His chief opponents were Soufflot, Perronet and Gauthey. Patte's position had been universally repudiated by 1771, and the construction of the church went ahead as planned. In 1778, the first cracks in the dome's piers were noticed, and Patte took up the dispute, but without enjoying much success. Nevertheless, he was a member of the commission whose task it was to examine the Panthéon in 1798, upon which occasion he published an *Analyse raisonnée de l'état alarmant du dôme du Panthéon français*, reiterating the same points as he had made in his *Mémoire*.

If one were to take into consideration only Patte's attitude towards celebrities such as the *encyclopédistes* or Soufflot, one would be tempted to endorse Diderot's negative judgement of him. The relative insignificance of his completed buildings serves to reinforce one's doubts about this 'excessively clever and honest man', who had been rejected by the Academy of Architecture and who was very probably embittered by his successive failures. Architect of the Elector of Zweibruggen, Patte had built little more than a wing of a château, a pavilion at Jeresbourg, together with the hôtel Charost in Paris and the church of Bolbec in Normandy.

The extreme positions which Patte took up cannot, however, be explained solely in terms of a simple hypothesis. One should not forget the complexity of his character, nor should one disregard the sophistication of his theoretical work, which itself disposed him to aggression.

The list of treatises and of pamphlets published by Patte is an impressive one. Out of this wealth of material, one should note the *Monuments érigés à la gloire de Louis XV* (1765) and the *Mémoires sur les objets les plus importants de l'architecture* (1769), in which Patte developed lines of argument about urban matters which were extraordinarily coherent for the period. Criticising, much as Laugier had done, the superficial nature of improvement, Patte adumbrated global solutions, integrating both urban morphology and questions concerning traffic and sanitation. In his *Mémoires sur les objets les plus importants de l'architecture*, he thus reviewed almost all of the problems that were then being debated. He dealt in turn with the creation of squares and avenues, with the waterworks and with street lighting. He fulminated against the filthiness of the streets and the inefficient operation of public hygiene, but he waxed lyrical in his proposals for an ideal city whose spatial economy seems to anticipate functionalist zoning. Le Corbusier was right to begin his course of reading at the Bibliothèque Nationale by unearthing Patte's works. By a curious coincidence, the Enlightenment theoretician was already defending the roof terrace, which was destined to supplant the all too inflammable wooden frame.

As well as urbanism, Patte was also concerned with construction. His *Mémoires sur les objets les plus importants de l'architecture* were also devoted to the description of the leading buildings of the period, from the completion of the colonnade of the Louvre to the construction of quays and bridges. It was Patte's pronounced taste for building which led to his being chosen to finish Blondel's *Cours*, for which he supplied the final two volumes of text and the final volume of plates, which were devoted to internal decoration and to construction. As far as Patte was concerned, construction, even more than decoration, represented the domain where an architect's thinking could most fruitfully be applied. His criticisms of Sainte-Geneviève, like those of the church of the Madeleine or those of Saint-Sulpice, were partly inspired by his desire for a public debate over problems of construction. Being a man of the eighteenth century, Patte endeavoured to reconcile the teachings of reason with public opinion.

Patte's views on decoration were much less clear-cut. From his *Discours sur l'architecture* (1754) to his final writings, his position did not alter much. Being a disciple of Perrault, from whom he had borrowed the idea of average proportions, he became the champion of a classicism that was adapted to the new needs of the period. These needs were in fact to inspire a remarkable *Essai sur l'architecture théâtrale*, published in 1782, in which he surveyed the most famous theatres in Europe, before going on to expound what, in his view, 'the genuine shape of a modern theatre', conceived in terms of the requirements of optics and acoustics, should be.

If Patte's career as an architect was none too illustrious, his theoretical work bears witness, as we have seen, to an ever alert curiosity and to a powerful originality. This originality was itself soon to condemn Patte to oblivion. He had retired to Mantes at the beginning of the Revolution, and he died, all but unknown, in 1814. His posthumous reputation therefore derives to a great extent from his rediscovery by the historians and architects of the present century, a rediscovery which is somewhat ambiguous, since it generally relates to the Modern Movement, and a desire to find precursors. It is worth emphasising that Patte is closer to Blondel than to Le Corbusier.

JEAN-RODOLPHE PERRONET (1708–94)

Jean-Rodolphe Perronet was born at Suresnes on 8 October 1708. His father, a former Swiss officer who had been in the service of France, left very little at his death. Although he had planned originally to enter the corps of the Génie, Perronet's lack of funds obliged him to turn to architecture instead. In 1725, at the age of seventeen, he entered the offices of Debeausire, first architect of the town of Paris, who employed him to check quantities, to settle workers' accounts and to supervise whatever structures were entrusted to him. Among these structures, one should note the design and the execution of the great sewer, a part of the quay of the Tuileries, as well as the cantilevered pavement of the Pelletier quay. At the same time, Perronet was in charge of the construction of the suburban roads, funded by the town of Paris. From this time on, the future director of the Ecole des Ponts et Chaussées focussed his attention upon civil engineering, while continuing to educate himself in mathematics and in mechanics, attending in particular Privat de Molière's lectures at the Collège de France. Perronet also frequented the milieux of the municipality of Paris which were influenced by the ideas of the *Encyclopédistes*.

With this training behind him, Perronet entered the corps of the Ponts et Chaussées in 1735. The following year he was appointed sub-engineer at Alençon, before becoming chief engineer for the district of Alençon, upon the recommendation of the Intendant Lallemant de Levignen, one of those 'enlightened' administrators whose importance the *Philosophes* all emphasised.

Between 1737 and 1747, Perronet was mainly concerned with the construction of roads, drawing his inspiration from the directives formulated by the Controller-General Orry, in his *Mémoire instructif sur la réparation des chemins* (1738). Orry's report advocated cancelling and starting over again with the road plans with a uniform scheme for all the regions, and this document inspired Perronet to call for a survey of all the roads and tracks in the region of Alençon, which would serve as a programme for whatever roadworks were necessary. At the same time, Perronet introduced some improvements in highway construction through a paving technique using cobblestones, which was to become commonplace in later years.

During this same period, Perronet took a strong interest in the work process, and two reports were written on the manufacture of pins at Laigle: *Description de la façon dont on fait les épingles à Laigle en Normandie* and *Explication de la façon dont on réduit le fil de laiton à différentes grosseurs dans la ville de Laigle en Normandie*. Although neither of these documents was ever published, they nevertheless had some influence upon Panckoucke's *Encyclopédie méthodique*. In his concern with a precise and factual description of the various stages involved in the work and of the tools required for the manufacture of pins, Perronet seems to prefigure the thought of Adam Smith. Nevertheless, his chief interest in the realm of work seemed to revolve around its technological aspects.

In 1747, Perronet was summoned by Trudaine to Paris to take charge of the Drawing Office, which had been set up in 1744 with the aim of rationalising planning policies. Perronet's arrival coincided with a considerable expansion in the Office's work, since it was being transformed from a simple instrument of coordination and technological supervision into an institution where future Ponts et Chaussées engineers might receive a training. The decree of 14 February 1747 made this point quite clearly, for it committed Perronet,

> that, being under the instructions of sieur contrôleur général des finances, and under those of sieur intendant des finances, he would be responsible for the running of the Ponts et Chaussées, for the supervision and inspection of the geographers and draughtsmen drawing plans and maps, and for the training of the aforesaid draughtsmen in the sciences and practices which are necessary to anyone wishing to take up successfully any of the various posts in the aforesaid Ponts et Chaussées.

By 1750 the Drawing Office was already known officially as the Ecole des Ponts et Chaussées, and Turgot was subsequently to confirm this title. Perronet continued to run the Ecole des Ponts et Chaussées to the time of his death, and he played a key role in its development.

About the time of his appointment as director of the Ecole des Ponts et Chaussées, Perronet was raised to the rank of first engineer of the Ponts et Chaussées. 'It was also during this period', wrote Lesage in his *Notice pour servir à l'élogue de M. Perronet*, 'that the reputation of this famous man, who has rendered illustrious this corps (itself noteworthy for the remarkable talents who have served it), began to grow.'

Perronet's reputation was acquired chiefly through his work as a structural engineer. Having finished the bridges at Orléans and at Mantes, which had been designed by first-engineer Hupeau, his predecessor, Perronet went on to build Nogent bridge, Fontanes bridge (which is near Chantilly), and the angled bridge at Bicheret. From 1768 to 1774, he was engaged in building the Neuilly bridge, from 1774 to 1785 he was working on the Sainte-Maxence bridge and between 1787 and 1792 he was building the pont de la Concorde.

These were the most famous bridges for which Perronet was responsible, but to this list should also be added Château-Thierry, Brunoi, Rosoi and Nemours, together with numerous, unrealised projects for bridges at Lyons, Nantes and even St Petersburg.

What earned the enthusiastic endorsement of Perronet's contemporaries was not so much the audacity of his conceptions, which had finally culminated in his famous system of 'stretched arches', as the methodical nature of his approach. The concern for exactitude which had already been apparent in his studies of pin manufacture was also evident in his rationalisation of tasks on the building-site, which turned bridge building into a problem to be resolved rather than a venture whose outcome was frequently in doubt.

As first engineer of the corps of the Ponts et Chaussées, Perronet had also to undertake tours of inspection and frequent journeys. During these trips, he would visit building sites, enquire if all was going according to plan, and give his advice to the engineers. Being responsible for the roads in the region of Paris, Perronet also took on the responsibility for coordinating roads policies for the whole of the realm.

In addition, Perronet took part in the scientific debates of the period. he was admitted to the Academy of Sciences in 1765, and to the Academy of Architecture in 1768. He was also a member of numerous provincial academies, including those at Metz, Lyons, Dijon and Bordeaux. Perronet produced relatively little theoretical work, consisting in the main of reports on technical subjects, such as the centring and decentring of bridges (1773), stone foundations in water (1775), the reduction of the width of piers and the form of arches (1777), and piles and piling (1779). In 1782, he published his *Description des projets et de la construction des ponts de Neuilly, de Mantes, d'Orleans, etc.; du projet du canal de Bourgogne . . . et de celui de la Bièvre à Paris*. This work, which was later republished, was hugely successful, and became a classic of the art of engineering.

Although heaped with honours under the *Ancien Régime*, and granted letters patent of nobility in 1763, Perronet's position was further reinforced during the Revolution. The National Assembly voted him an official residence and an especially high salary. As an unchallenged figure within a new political order, which aspired also to be scientific and technological, the engineer died in 1794 when at the very height of his glory. Nevertheless, at this date his administrative work seemed to be threatened. Thus, during the debate preceding the vote on the law on Bridges and Highways, the abolition of the corps of the Ponts et Chaussées was for a time contemplated. As events turned out, the administration to which Perronet had wholly devoted himself was preserved.

The consolidation of the corps of the Ponts et Chaussées was in fact largely due to its chief engineer. Thus, at the Ecole des Ponts et Chaussées, Perronet concerned himself with every detail of the operation of the institution, from the admission of pupils to the course of study. As a kind of middleman between the engineers and the management of the Ponts et Chaussées, guaranteed by the Trudaines, Perronet exercised a form of guardianship which offered a wholly paternal form of benevolence. Indeed, Lesage even went so far as to describe the pupils at the Ecole des Ponts et Chaussées as Perronet's 'adoptive children'. In 1778, some sixteen years before his death, these pupils presented to him a bust of himself bearing, in letters of gold on the pedestal, the inscription 'Patri carissimo familia'. The 'familial' character of the former administration of the Ponts et Chaussées was further reinforced by the masonic connections of a fair number of the pupils.

What Perronet had bequeathed to his successors was not so much his technological innovations, which were soon superseded, but the twofold aspect of the state engineer, at once a man dedicated to method and a wholehearted member of a community bound together by *esprit de corps*. Even today the advantages and the disadvantages of technical management may be said to descend from this dualism.

GASPARD RICHE DE PRONY (1755–1839)

Gaspard Riche de Prony was born in 1755 at Chamelet, in Beaujolais. His father had been a royal notary, a fiscal attorney and a member of the Parlement of Dombes. These antecedents mark out the future engineer as a member of that provincial bourgeoisie which would subsequently supply the Revolution with a fair number of its administrators and technicians. Having completed his classical studies, Prony spent one or two years gaining a deeper knowledge of mathematics, before entering, at the age of twenty-one, the Ecole des Ponts et Chaussées, which was then run by Perronet.

Prony's student years at the Ecole des Ponts seemed to prefigure his subsequent career. In a school whose main emphasis was upon teaching design, Prony came to the notice of Perronet because of his remarkable aptitude for mathematics and because of the order and clarity of his ideas. As early as 1779, Perronet had declared to him: 'Devote yourself to deepening your understanding of the principles of your art, for you are destined one day to become head of the Ecole des Ponts et Chaussées.' Having become Perronet's protégé, Prony was appointed sub-engineer in 1780, and was assigned in turn to the regions of Berry, Orléans and Paris. From 1783, Prony began to assist Perronet, who was beginning to feel the effects of his age. This collaboration, which was to last to 1791, provided him with the opportunity to take the first steps in his scientific career, through the publication of his *Mémoire sur la statique et la poussée des voûtes* (1783), which was intended to refute criticism of the Neuilly bridge. This report, which was read before the Academy of Sciences in the course of the same year, earned Prony the congratulations of Monge. At the same time, he was assisting Perronet with the administrative responsibilities which devolved upon him in his role as chief engineer of the corps of the Ponts et Chaussées, an experience which helped to prepare him for his future duties as director.

In 1785, Prony accompanied Perronet to Dunkirk, and then travelled with him to England. This journey brought him into contact with a number of the most famous structures of the English engineers, such as Smeaton's Eddystone Lighthouse or the Duke of Bridgewater's canal, which had been conceived by Brindley. Upon his return to France, Prony devoted all his efforts to the construction of the pont Louis XVI (pont de la Concorde). In 1787, he was appointed inspector of Ponts et Chaussées. In 1790, the first volume of his *Nouvelle architecture hydraulique* was published. This book, after rehearsing the basic principles of general mechanics, gave the equations for the equilibrium and movement of fluids. It was a huge success and very soon became a standard work of reference for engineers.

In 1791, Prony was appointed chief engineer of the Ponts et Chaussées, and director of the cadastral survey. During the revolutionary period, he made some significant contributions to science, and his expertise was much in demand on special missions. As director of the cadastral survey, he sat on the commission whose task it was to introduce the metric system; he was consulted on the division of French territory into *départements*; and, finally, he was involved in the drawing up of the new logarithmic tables, an important venture for the purpose of which it was imperative 'not only to compose tables whose accuracy leaves nothing to be desired, but to make of them the most immense and impressive monument to calculation which has ever been executed or even conceived'. In 1796, the second volume of Prony's *Nouvelle architecture hydraulique* was published, which was mainly concerned with steam engines, and in fact gave one of the very first systematic descriptions of them.

From now on, Prony was in contact with all of the most prominent scientific personalities of the day. He corresponded with Watt, and was personally acquainted with Monge, Laplace and Cuvier. Prony never sought to play a political role, and he would seem to have taken to heart the note which Carnot had written to him in 1793, and which read 'You have just escaped death; live circumspectly and in seclusion; if the danger recurs, you will be alerted.' This episode aside, Prony spent the revolutionary years unhindered, and concerned more with the progress of the sciences than with the course of events. In 1795, he played some part in the reorganisation of the former Academy of Sciences, which became the first section of the Institut de France, and he was himself elected a member of this same institution the following year. He was also appointed professor at the Ecole Polytechnique, to a chair in mechanics and in analysis which he shared with Lagrange. In 1798, he became Inspector General of Ponts et Chaussées, then, on 4 October of the same year, director of the Ecole des Ponts et Chaussées, a position that he was to hold up until his death.

During the Consulate and the Empire, Prony's life was one of frenetic activity. In spite of his refusal to take part in the Egyptian Campaign, and in spite of Bonaparte's rancour towards him, he soon became an indispensable element in France's annexationist policies. Prony travelled from one end of the Empire to the other, from Italy to the Low Countries, giving his advice to engineers and drafting countless reports. Whenever someone spelled out the difficulties involved in carrying out some planning project or other to Napoleon, it was his custom to ask: 'What does Prony think of it?' This was how he came to be called upon to study the damming of the river Po, or pass judgement upon works undertaken at the ports of Genoa, La Spezia, Ancona and Venice, and how in 1810 he set up a project for the draining of the Pontine Marshes which, shortly after its publication in 1822, earned him the congratulations of Pope Leo XII.

Prony still managed, while fulfilling these various missions, to find the time to continue with his scientific work. For example, he published a work on the pressures of earths, presenting simple and graphic formulae which allowed one to calculate the form and dimensions of retaining walls. He also showed an interest, when concerned with the works on the Ourcq canal, in problems arising out of the measurement of water. At the Institute, Prony often addressed questions concerning the engineer's art. As he was also known, since the publication of the second volume of his *Nouvelle architecture hydraulique*, as a specialist on the topic of steam engines, he was consulted about them also.

During the Restoration, Prony resigned his chair at the Polytechnique, but he kept his post of director of the Ecole des Ponts et Chaussées. His professional activities were still of the greatest significance. He took an interest in some engineering works on the Rhône. Prompted by a court case between a manufacturer of steam engines and one of his clients, in 1821 Prony invented the dynamometric brake, which served to measure the power generated by machines. This apparatus, which was first described in his *Rapport sur la nouvelle et l'ancienne machine à vapeur établie à Paris, au Gros-Caillou*, published in 1826, was to enjoy a lasting success, since it was still in use in the early years of the twentieth century. Prony, now honoured with the title of Baron and since 1820 a member of the Royal Society, was at the height of his glory. The whole of Europe recognised his talents as engineer, which were confirmed, if confirmation were needed, in a dazzling way by the arms which the King had granted him. Prony's arms consisted of a three-arched bridge on a field of azure, a ruler and a compass and, so as to represent the benefits which the engineer's efforts had spread, a cornucopia dispensing its fruits.

Prony was named Peer of France in 1835, and ended his career having won universal esteem. He was widowed in 1832 and, since he had never had children, he devoted himself almost entirely to his work and to his friends. He was, for example, very attached to Cuvier. He took an interest in music, and drafted a *Traité des intervalles musicaux* in 1832; he studied the problems raised by the advent of the railways, as his unpublished *Recherches sur les chemins de fer* proves. Prony finally died at the age of eighty-four, having survived and served with the same zeal each and every regime. It fell to Arago to deliver his funeral encomium at the Academy of Sciences, and upon this occasion he observed 'that he had really become the personification in France of the art of engineering'.

Notwithstanding the sheer range of his theoretical work and the admiration still felt by Arago, Prony fell very rapidly into oblivion. His published works consisted for the most part of treatises, reports and pamphlets, of mission logs or of expert surveys, none of which were enhanced by significant theoretical discoveries. From the *Nouvelle architecture hydraulique* to the *Recherches sur les chemins de fer*, his work, being orientated chiefly towards practical application, lacked the visionary dimension evident in, say, Cuvier or Laplace. Nowadays Prony is barely known at all, save for the dynamometric brake which bears his name. He was nevertheless a key figure, placed between two worlds. In the early stages of his career he had been influenced by Enlightenment thought, but his writings gradually came to be marked by the genealogical and systematic spirit of the nineteenth century. In the figure of Prony, one can discern the real advent of the modern engineer, a person concerned with the concrete application of knowledge resting both upon science and upon experience. If, therefore, Prony himself has fallen out of favour, it is, paradoxically enough, because, through his 'practical formulae' or his 'methods of graphic resolution', his inheritance now constitutes the modern engineer's daily routine.

NOTES

1 SOME GENERAL OBSERVATIONS

1 I have borrowed this phrase from A. Koyré, *Etudes d'histoire de la pensée scientifique* (1966; Paris, 1973), p. 14.

2 See E. Kaufmann, 'Three revolutionary architects, Boullée, Ledoux, Lequeu', *Transactions of the American Philosophical Society*, n.s., 42, part 3, 1953, 431–564.

3 A. Braham, *The architecture of the French Enlightenment* (London, 1980), p. 7.

4 See B. Gille, *Histoire générale des techniques* (Paris, 1978).

5 *Histoire générale des techniques*, edited by M. Daumas (Paris 1962–79), vol. 3, p. 34.

6 There is plainly no need here to stress in this respect the gap opening up between England, then undergoing its first technological 'revolution', and France.

7 See B. Groethuysen, *Philosophie de la Révolution française* (1956; Paris 1982).

8 See P. Chaunu, *La civilisation de l'Europe classique* (Paris, 1966).

9 The Academy of Architecture was involved in discussions regarding the nature of taste as early as its second meeting, on 7 January 1672.

10 W. Szambien, *Reflet de la connaissance, études sur la théorie et la terminologie de l'architecture à l'âge classique*, research report, SRA Paris, IERAU, 1983, p. 4.

11 See R. Mousnier, *Les institutions de la France sous la monarchie absolue* (Paris, 1974–80).

12 See N. Elias, *The court society*, translated by E. Jephcott (Oxford, 1983).

13 L. Hautecœur, *Histoire de l'architecture classique en France* (Paris, 1943–57).

14 J.-P. Epron, *L'architecture de la règle* (Brussels, 1981), pp. 161–8.

15 See W. Hermann, 'The problem of chronology in C.-N. Ledoux's engraved work', *The Art Bulletin* (New York), 42, no. 3, 1960, 191-210.

16 See J.-F. Blondel, *Architecture française* (Paris, 1752–6), vol. 3, p. 10.

17 For information regarding the teaching of architecture in the eighteenth century, consult L. Pelpel, *La formation architecturale au XVIIIe siècle en France*, research report, CORDA (Paris, 1980).

18 J.-F. Blondel, *Cours d'architecture* (Paris, 1771–7), vol. 1, p. xvii.

19 J.-P. Epron, *L'Ecole de l'Académie (1671–1793) ou l'institution du goût en architecture*, research report, CEMPA (Nancy, 1984), pp. 98–103.

20 See A. Querrien, *Devenir fonctionnaire ou/et le travail de l'Etat, lectures hypothétiques sur l'histoire du corps des Ponts et Chaussées*, research report, DGRST (Paris, 1977), pp. 96–105.

21 On the life and work of Jacques-François Blondel, see the biographical note in the Appendix. One should also consult: A. N. Dézallier d'Argenville, 'Jacques-François Blondel', *Vie des fameux architectes depuis la Renaissance des arts* (Paris, 1787), pp. 467–73; R. Etlin, '"Les dedans": Jacques-François Blondel and the system of the home', *Gazette des Beaux-Arts* (Paris), April 1978, 137–47; F. Franque, 'Elogue de J.-F. Blondel par M. Franque, membre de l'Académie Royale d'Architecture', *Journal des Beaux-Arts et des Sciences* (Paris), 1, 1774, 559–70; M. Gallet, 'Blondel (Jacques-François) 1705–1774', *Encyclopaedia Universalis* (Paris, 1984–5), vol. 3, pp. 735–7; J. Garms, 'Le plan d'urbanisme de Strasbourg dressé par Jacques-François Blondel en 1764–1769', *Cahiers Alsaciens d'Archéologie, d'Art et d'Histoire* (Strasburg), 21, 1978, 103–41; K. Harrington, 'Blondel, Jacques-François', *Macmillan Encyclopedia of Architects* (New York, 1982), vol. 1, pp. 220–4; E. Kaufmann, 'The contribution of J.-F. Blondel to Mariette's Architecture françoise', *The Art Bulletin* (New York), 31, no. 1, 1949, 58–9; B. Leclerc, *Le projet d'embellissement de J.-F. Blondel, Strasbourg* (1764–1774), Mémoire de fin d'études, Strasburg School of Architecture, 1982; J. Lejeaux, *La place d'Armes de Metz: un ensemble architectural du XVIIIe siècle, œuvre de Jacques-François Blondel* (Strasburg, 1927); R. Middleton, 'Jacques-François Blondel and the Cours d'Architecture', in *Journal of the Society of Architectural Historians of Great Britain* (London), 18, no. 4, 1959, 140–8; P. Patte, *Cours d'architecture . . . commencé par feu J. F. Blondel . . . et continué par M. Patte*, vol. 5, 'Avertissement', pp. v–xii; A. Prost, *J.-F. Blondel et son œuvre* (Metz, 1860); D. Remyt 'Urban beautification in 18th century France, Blondel in Metz and Strasbourg', *Lotus International* (Milan), 39, 1983, 94–101.

22 L. Pelpel, *La formation architecturale*, pp. 26, 27.

23 A. Prost, *J.-F. Blondel et son œuvre*, pp. 11, 12.

24 On the conflict between Blondel and Ledoux, see M. Gallet, *Ledoux* (Paris, 1980), pp. 8, 9.

25 On the life and work of Pierre Patte, see the biographical note in the Appendix. Also consult: F. Fichet, *La théorie architecturale à l'âge classique* (Brussels, 1979), pp. 387–407; M. Mathieu, *Pierre Patte, sa vie, son œuvre* (Paris, 1940).

26 Thanks to his fame as an engraver, Patte became director of engravings for the

Encyclopédie in 1757. He retired two years later and immediately published a letter in Fréron's *L'Année Littéraire*, denouncing the plagiarism of which his former employers had been guilty at his expense. A lengthy polemic ensued, in the course of which both Diderot and Voltaire intervened. For the details of this somewhat confused affair, see M. Mathieu, *Pierre Patte*, pp. 21ff.

27 See chapter 7.

28 See chapter 8.

29 These repeated failures are very probably attributable to the limited significance of the buildings actually built by him. The entirety of his work consisted of very little more than the wing of a château and a trianon for a German prince at Heresbourg, the Hôtel Charost at Paris, and the church of Bolbec in Normandy. See M. Mathieu, *Pierre Patte*, pp. 285ff.

30 On the life and work of Jean-Rodolphe Perronet, see the biographical note in the Appendix. One should also consult H. Cheguillaume, *Perronet ingénieur de la généralité d'Alençon (1737–1747)* (Alençon, 1891); F. de Dartein, 'La vie et les travaux de Jean-Rodolphe Perronet, premier ingénieur des Ponts et Chaussées, créateur de l'Ecole des Ponts et Chaussées', and 'Le pont de la Concorde à Paris', in *Annales des Ponts et Chaussées* (Paris), 4th quarter 1906, 5–87, 88–148; F. Fichet, *Le corps des Ponts et Chaussées du génie civil à l'aménagement du territoire*, research report, CORDA (Paris, 1982), pp. 101–6; P.-C. Lesage, *Notice pour servir à l'éloge de M. Perronet* (Paris, 1805); H. Marcel, 'Les débuts de Jean-Rodolphe Perronet ingénieur des Ponts et Chaussées de la généralité d'Alençon', *Annales des Ponts et Chaussées* (Paris), July–August 1948, 419–40; F. Moureau, 'Diderot et le portrait de Perronet: trois lettres inédites', *Dix-Huitième Siècle* (Paris), 16, 1984, 243–52; L. Petit, 'Le décintrement du pont de Neuilly (22 septembre 1772)', *Annales des Ponts et Chaussées* (Paris), 1st quarter, 1932, 142–55; A. Picon, 'Die Vorstellungskraft eines Konstrukteurs: Jean-Rodolphe Perronet und das Modell einer Brücke', *Daidalos* (Berlin), 18, 1985, 74–87; G. Riche de Prony, *Notice historique sur Jean-Rodolphe Perronet* (Paris, 1829). One could also consult his file as engineer for the Ponts et Chaussées, which is kept in the National Archives under the classification F14 2296/2.

31 P.-C. Lesage, *Notice*, p. 2.

32 See chapter 5.

33 Perronet actually continued to be the model for the practical engineer of the Ponts et Chaussées well into the nineteenth century. Indeed, Firmin Didot republished his complete works in 1820, which shows that his ideas still had some influence.

34 See chapter 7.

35 On the life and work of Gaspard Riche de Prony, see the biographical note in the Appendix. One should also consult the following: F. Arago, 'Prony', *Œuvres de François Arago* (Paris, 1854–63), vol. 3, pp. 584–92; M. Bradley, 'Prony the

bridge-builder', doctoral thesis, Coventry Polytechnic, Coventry, 1984; Y. Chicoteau, A. Picon and C. Rochant, 'Gaspard Riche de Prony ou le génie "appliqué"', *Culture Technique* (Paris), 12, 1984, 171–83; C. Dupin, *Eloge de M. le Baron de Prony* (Paris, 1840); J. Fieux, 'Prony mécanicien et le frein dynamométrique', *Bulletin de la Société d'Encouragement pour l'Industrie Nationale* (Paris), 139, 1940, 88–98; A. Lorion, 'Une mésaventure de Prony à Venise sous l'Empire', *Revue de l'Institut Napoléon* (Paris), 93, 1964, 164–8; C. Monteil, 'Prony hydraulicien', *Bulletin de la Société d'Encouragement pour l'Industrie Nationale* (Paris), 139, 1940, 83–7; *Notes sur les services et les travaux de M. de Prony, directeur de l'école impériale des Ponts et Chaussées* (Paris), Year XIII; M. Suquet, 'Prony, Ingénieur des Ponts et Chaussées', *Bulletin de la Société d'Encouragement pour l'Industrie Nationale* (Paris), 139, 1940, 79–83; C.-A. Vielh de Boisjolin, 'Prony', *Biographie universelle et portative des contemporains* (Paris, 1836), pp. 1,024–6; M. Walckenaer, 'La vie de Prony', *Bulletin de la Société d'Encouragement pour l'Industrie Nationale* (Paris), 139, 1940, 68–78. One could also consult his engineer's file, held at the National Archives under the classification F14 2304/2.

36 See Y. Chicoteau, A. Picon and C. Rochant, 'Gaspard Riche de Prony ou le génie "appliqué"'.

37 See M. Bradley, 'Prony the bridge-builder'. Also see chapter 6 and chapter 10, n. 181.

38 *Notes sur les services et les travaux de M. de Prony, directeur de l'école impériale des Ponts et Chaussées.*

39 As his *Fragments sur la théorie des beaux-arts*, drafted in 1783, testifies. G. Riche de Prony, *Fragments sur la théorie des beaux-arts, leur influence etc.*, manuscript, ENPC, MS 1892.

40 In 1777, Patte published vols. 5 and 6 of Blondel's *Cours*, which were devoted to interior decoration and to construction.

41 To make this situation official, the administration granted Blondel an annual stipend of 24,000 *livres*, to cover the costs involved in instructing pupils from the Ponts et Chaussées.

42 E. Kaufmann, *Three revolutionary architects.*

43 In particular, J.-K. Pérouse de Montclos, *Etienne-Louis Boullée (1728–1799). De l'architecture classique à l'architecture révolutionnaire* (Paris, 1969); M. Gallet, *Ledoux et Paris*, Cahiers de la Rotonde 3, Paris; Commission du Vieux Paris, 1979; it is also worth consulting A. Vidler, *Ledoux* (Paris, 1987), which expands upon themes in his article, 'The theatre of production: Claude-Nicolas Ledoux and the architecture of social reform', in *AA Files* (London), 1, no. 1, 1981–2, 54–63.

44 M. Gallet, *Ledoux*, pp. 39, 40.

45 M. Gallet, *Ledoux*, p. 12.

46 See J.-F. Chenais, Y. Chicoteau, P. Madec, J. Michel, B. Patarin, A. Picon and

C. Vie, *Raison–projet–représentation. La formation architecturale des ingénieurs à l'Ecole des Ponts et Chaussées du XVIIIe siècle*, SRA Research Report (Paris, 1983).

47 E.-L. Boullée, C.-A. Coulomb, J.-C. Périer and G. Riche de Prony, *Rapport sur les moyens de sauver les personnes renfermées dans les maisons incendiées, lorsque les issues ordinaires sont interceptées ou insuffisantes* (Paris, n.d.).

48 J.-M. Pérouse de Montclos, *Etienne-Louis Boullée*, p. 37.

49 P. Chaunu, Introduction to J.-P. Bardet, P. Chaunu, G, Desert, P. Gouhier and H. Neveux, *Le bâtiment, enquête d'histoire économique, 14e–19e siècles* (Paris, 1971), p. 13.

2 THE CLASSICAL TRADITION AND ITS AMBIGUITIES

1 See chapter 4.

2 J. Guillerme, 'Classicisme et répetition', *Poïetique III* (Paris, 1981), pp. 65–88.

3 J.-M. Pérouse de Montclos, *L'architecture à la française* (Paris, 1982), pp. 27–31.

4 See chapter 4.

5 Cf. M. Foucault, *The order of things* (London, 1970), translated by A. Sheridan, pp. 86–91.

6 Cf. Y. Belaval, *Leibniz, initiation à la philosophie* (Paris, 1975), pp. 124, 125.

7 On the doctrine of proportions, one should consult W. Szambien, *Reflet de la connaissance*, pp. 35–50, and E. Forssman, *Dorisch, Jonisch, Korinthisch, Studien über den Gebrauch der Säulenordnungen in der Architektur des 16.–18. Jahrhunderts* (Braunschweig-Wiesbaden, 1984).

8 F. Fichet, *La théorie architecturale à l'âge classique*, pp. 21-3.

9 Cf. A. Borie, P. Micheloni and P. Pinon, *Forme et déformation des objets architecturaux urbains* (Paris, 1984).

10 C. Mignot, Introduction to the facsimile edition of P. le Muet, *Manière de bien bastir pour toutes sortes de personnes* (Aix-en-Provence, 1981), p. xvi.

11 See P. Gresset, *L'écart du système, critique des relations entre les figurations et discours instaurateurs de bâtir* (Paris, 1977).

12 See chapter 8.

13 By detail we understand here what Jacques-François Blondel would call 'petites parties'. A detail in classical architecture may be either ornamental or structural, the order itself being described as a succession of details in relation to each other, or as a specific structural arrangement.

14 I have borrowed this distinction from Jacques-François Blondel also.

15 C.-A. d'Aviler, *Cours d'architecture* (1691; Paris, 1720).

16 Ibid.

17 See J.-F. Blondel, *Cours d'architecture*, vol. 4, pp. 200–8.

18 Ibid., pp. 206, 207.

19 E. Kaufmann, 'Die Architekturtheorie der französischen Klassik und des

Klassizismus', in *Repertorium für Kunstwissenschaft* (Berlin, Leipzig), 64, 1924, 197–237.

20 Basing his case on custom, the importance of which he emphasises, Perrault argues, against Blondel, that one should avoid increasing the high parts of a building; the eye, being accustomed to correct details in terms of distance, cannot help but be disagreeably struck by an immediately perceptible disproportion, of a kind that would be rendered still more shocking in so far as it leads one to place the weaker parts above the stronger ones. C. Perrault, *Les dix livres d'architecture de Vitruve corrigez et traduits nouvellement en François, avec des Notes et des Figures* (1673; Paris, 1684), pp. 204–6.

21 J.-M. Savignat, *Dessin et architecture du moyen-âge au XVIIIe siècle* (Paris, 1980), pp. 122-4.

22 Cf. T. Taton, *L'œuvre mathématique de Desargues* (Paris, 1951). In spite of the enthusiasm of Abraham Bosse, who in 1643 published his *Pratique du trait à preuves de M. Desargues . . . pour la coupe des pierres en l'architecture*, followed by his *Manière universelle de M. Desargues*, this mathematician's methods were very little used by those practically involved in architecture. Nevertheless, they mark an important stage in the history of geometry.

23 C.-A. d'Aviler, *Cours d'architecture*, plate A.

24 In the same spirit, Félibien described the procedure he had adopted when compiling his dictionary: 'I was yet again obliged to have recourse to the workers: I had to return to their shops, visit their workshops, consider their machines, and their tools, and consult them in regard to their various customs, and often confer with them as to the different names which they gave to the same thing' (A. Félibien, *Des principes de l'architecture, de la sculpture, de la peinture et des autres arts qui en dépendent, avec un dictionnaire des termes propres à chacun de ces arts* (Paris, 1676), preface).

25 See A. Dupire, B. Hamburger, J.-C. Paul, J.-M. Savignat and A. Thiébaut, *Deux essais sur la construction* (Brussels, 1981).

26 Cf. chapter 10.

27 See J.-P. Epron, *L'Ecole de l'Académie*, pp. 43, 44. The basic work on the Académie de l'Architecture is still *Procès-Verbaux de l'Académie Royale de l'Architecture*, published by H. Lemonnier (Paris, 1911–29).

28 On the life and work of François Blondel, see H. Lemonnier, *Procès-Verbaux*, vol. 1, Introduction, pp. xxii–xxxi; C. Mauclaire and C. Vigoureux, *Nicolas-François de Blondel ingénieur et architecte du roi (1618–1686)* (Laon, n.d.).

29 A. Desgodets, *Les édifices antiques de Rome, dessinés et mesurés très-exactement* (1682; Paris, 1779), pp. ix, x.

30 W. Hermann, *La théorie de Claude Perrault* (1973; Brussels, 1980); one should also consult A. Pérez-Gomez, *Architecture and the crisis of modern science* (Cambridge, Mass., 1984), pp. 18ff.

31 C. Perrault, *Ordonnance des cinq espèces de colonnes selon la méthode des anciens* (Paris, 1683), p. ii.

32 Ibid., pp. vi, vii.

33 One cannot help but wonder whether Perrault would have developed so radical an argument if he had not become embroiled in a polemic with Blondel. This was what most of the architects of the Enlightenment, in particular, supposed. See chapter 3.

34 See W. Herrmann, *La théorie de Claude Perrault*, p. 42. On the general context within which the Dispute between Ancients and Moderns unfolded, see P. Hazard, *La crise de la conscience européenne 1680–1715* (Paris, 1961), pp. 26ff.

35 In a tone of voice somewhat reminiscent of that of his brother, the author of the *Ordonnance* declared that, as far as he was concerned: 'It is truly cause for wonder just how extreme the awe and reverence of the architects is for those works which are called antique, in which they admire everything, but more especially the mystery of proportions, which they are prepared to contemplate with an attitude of profound respect, without daring to try and discover the reasons why the dimensions of a given moulding are a little smaller, or a little larger: which is something one can safely presume was not known, even by those who made them' (C. Perrault, *Ordonnance*, pp. xvii, xviii).

36 On Boileau's conceptions, one should read T.-A. Litman, *Le sublime en France (1660–1774)* (Paris, 1971), pp. 63ff.

37 See chapter 3.

38 C. Perrault, *Ordonnance*, p. xvi.

39 Ibid., p. 5.

40 Ibid., p. 6.

41 Ibid., p. 9.

42 F. Blondel, *Cours d'architecture* (Paris, 1675–83), fourth, fifth and final parts, pp. 768–74.

43 Three numbers A, B, C are in harmonic proportion if B–A is to C–B what A is to C. If one gives as values to A, B and C heights of parts of the order, one may see that harmonic proportion helps to reinforce the internal coherence of the latter.

44 C. Perrault, *Ordonnance*, p. xxii.

45 See chapters 6 and 7.

46 See A. Pérez-Gomez, *Architecture and the crisis of modern science*, p. 26. Perrault's pronounced taste for machines should be set under this same heading, in particular his system for raising stones for the façade of the Louvre. Cf. W. Herrmann, *La théorie de Claude Perrault*, p. 159. Do not the 'positive beauties' of the actual execution of a project already call to mind the representation of the construction process in the engineers' treatises of the eighteenth century, in which machines also had a privileged place? See chapter 7.

47 Cf. Y. Belaval, *Leibniz critique de Descartes* (1960; Paris, 1976), p. 502, where the author observes that 'of the mathematical model, Descartes and Leibniz only kept the deductive rigour, and not the precision, which was to be conquered by measure, by cleaving more and more to inductive approximation'.

48 See chapter 6.

49 F. Blondel and L. Savot, *L'architecture françoise des bastimens particuliers. Composée par Me Louis Savot, augmentée dans cette seconde édition de plusieurs figures, et des notes de Monsieur Blondel* (Paris, 1685), p. 341.

50 Cf. chapter 9. The Dispute of the Ancients and Moderns, which took place at the end of the seventeenth century, is of crucial importance for our understanding of the eighteenth also. Where Claude Perrault announced a new normative attitude, Boileau seemed to establish the conditions for an equally necessary freedom of invention. Norm and inventiveness would later have to be reconciled.

51 Like his brother Charles, like François Blondel and Antoine Desgodets, Claude Perrault was a protégé of Colbert.

52 Thus, in 1686, when the notion of establishing quantity surveyors was first considered (see chapter 7, n. 5), the Academy appealed to the Contrôleur général des finances 'to beg him to do nothing which would harm the Academy's interests' (*Procès-Verbaux*, vol. 2, p. 131).

53 Unlike Serlio, Desgodets was not content to relate all the proportions of buildings to the diameter of the column, but he also gave all the dimensions in royal feet. Architectural measurement thus tended to become an abstract system.

54 This course, which consisted of a *Traité des ordres de l'architecture*, a *Traité de la commodité de l'architecture, concernant la distribution et les proportions des édifices* and a *Traité du toise*, was never to be published in its entirety. Nevertheless, the *Traité du toise* was to be extracted from it in the course of the eighteenth century, when it appeared under the title *Loix des bâtimens*. Its publication had been undertaken at the behest of the Academy, which had resolved in 1715 to 'begin, as regards architecture in general and all the parts which depend upon it, with some deliberations which may prove useful' (*Procès-Verbaux*, vol. 4, p. 83). On Desgodets' course, one should also read W. Herrmann, 'Antoine Desgodets and the Académie Royale d'Architecture', in *The Art Bulletin* (New York), 61, no. 1, 1963, 157–64; H. Lemonnier, *Procès-Verbaux*, vol. 4, pp. 351, 352.

55 B. Lepetit, *Chemins de terre et voies d'eau* (Paris, EHESS), 1984, p. 18.

56 *Procès-Verbaux*, vol. 1, pp. 168ff.

57 Ibid., pp. 49ff. Around 1680–90, the Academy was to inspect the repairs of the pont de Nantes, the pont de Moulins and the pont de Lyon, and to undertake the construction of the pont de La Charité, the pont de la Ferté-sous-Jouarre and the pont de Saint-Pourçain, serving thus in effect as an upper council of the Ponts et Chaussées (H. Lemonnier, *Procès-Verbaux*, vol. 2, p. xxxii). With the creation, in

1716, of the corps of the Ponts et Chaussées, this role became of less importance. Nevertheless, in 1721, the Academy still studied Gautier's *Traité des ponts* (*Procès-Verbaux*, vol. 4, pp. 227ff.). Finally, one should note the existence of a project by Perronet for a 'wooden bridge with just one diameter arch for the crossing of the Seine, with respect to the village of Sèvres, where it is proposed to construct it', C. Perrault, *Recueil de plusieurs machines de nouvelle invention* (1700), pp. 40–1 and plates x, xl.

58 See, for example, M. Hamon, 'Colbert et le "Colbertisme"', in *Colbert 1619–1683*, a catalogue of an exhibition at the Hôtel de la Monnaie, Paris, 4 October–30 November 1983 (Paris, 1983), pp. 85–98.

59 *Procès-Verbaux*, vol. 1, p. 109.

60 L. Pelpel, *La formation architecturale*, p. 43.

3 THE IMPACT OF THE ENLIGHTENMENT ON ARCHITECTURE

1 See W. Herrmann, *La théorie de Claude Perrault*, pp. 134-41.

2 On the influence of Newton upon the Enlightenment, see E. Cassirer, *The philosophy of the Enlightenment*, translated by F. C. A. Koelln and J. P. Pettegrove (Princeton, 1951), 52–87; G. Gusdorf, *Les principes de la pensée au Siècle des Lumières* (Paris, 1971), pp. 151–79; A. Koyré, *Newtonian studies* (London, 1965); P. Casini, 'Le "newtonianisme" au siècle des Lumières', in *Dix-Huitième Siècle*, 1, 1980, 139–59.

3 G. Gusdorf, *Les principes de la pensée*, p. 152.

4 Voltaire's friend was in fact assisted in her task by the physicist Clairaut. See the *Principes mathématiques de la philosophie naturelle* (Paris, 1756).

5 See J.-L. Heilbron, *Elements of early modern physics* (Berkeley, 1982).

6 For further information on the *Encyclopédie*, one should consult R. Darnton, *The business of enlightenment: a publishing history of the Encyclopédie, 1775–1800* (Cambridge, Mass., London, 1979).

7 This was the *Encyclopédie méthodique*, which was conceived as the summa of twenty-six sub-encyclopaedias, each covering a separate branch of knowledge. The most prestigious authors of the late eighteenth century were employed for this task, namely, Monge, Lalande, Fourcroy, Guyton de Morveau, Lamarck, Prony and so on. See R.Darnton, *The business of enlightenment*, pp. 395–459. One should also consult S. Tucoo-Chala, 'La diffusion des Lumières dans la seconde moitié du XVIIIe siècle: Ch.-J. Panckoucke, un libraire éclairé (1760–1799)', in *Dix-Huitième Siècle*, 6, 1974, 115–28.

8 See the special number of *Milieux* devoted to the *Encyclopédie*, 'L'Encyclopédie ou la méchanique dans le boudoir', No. 19/20, 1984.

9 E. Cassirer, *The philosophy of the Enlightenment*, pp. 17–18.

10 F. Fichet, *La théorie architecturale à l'âge classique*, pp. 35–43; W. Szambien, *Reflet de la connaissance*, pp. 177–93.

11 G. Boffrand, *Livre d'architecture contenant les principes généraux de cet art* (Paris, 1745), p. 3.

12 Like Diderot with respect to the beautiful, Montesquieu simply listed the various categories of taste: pleasures of order, of variety, of symmetry, of contrast, of surprise or of rules.

13 C.-E. Briseux, *Traité du beau essentiel dans les arts, appliqué particulièrement à l'architecture et démontré physiquement et par l'expérience* (Paris, 1752), p. 1.

14 L. de Saint-Simon, *Mémoires*, edited by P. Galleret (Paris, 1974), p. 416.

15 On the rococo, one should read L. Hautecœur, *Histoire de l'architecture classique en France*, vol. 3; F. Kimball, *The creation of the rococo* (Philadelphia, 1943).

16 M. Eleb-Vidal and A. Debarre-Blanchard, *Architecture domestique et mentalités. Les traités et les pratiques XVIe–XIXe siècle* (Paris, 1984), pp. 56–79.

17 F. Kimball, *The creation of the rococo*, pp. 211-36.

18 See chapter 4.

19 See chapter 4.

20 E.-L. Boullée, *Architecture. Essai sur l'art*, texts assembled and presented by J.-M. Pérouse de Montclos (Paris, 1968), p. 52. Soufflot himself wrote as follows: 'Two of the most skilful architects of the last century had had the liveliest of disputes regarding the proportions of architecture. Their writings on this art, together with the superb monuments erected on the basis of their designs and according to their instructions, will serve for posterity, as for us, as the most certain guarantee of their real merit, and it would be hard to understand how two architects with so opposed a point of view regarding so crucial an aspect of their art could have produced things of equal beauty, if one did not know that scientists have sometimes, and indeed only too often, the shortcoming of wishing never to withdraw what they once advanced, even when in their heart of hearts they are aware of having been mistaken, and therefore they act in accordance with the principles against which they are prone to declaim' (J.-G. Soufflot, *Mémoire sur les proportions de l'architecture*, Manuscript in the Académie de Lyon, 1739, quoted by M. Petzet, in *Soufflots Sainte-Geneviève und der französische Kirchenbau des 18. Jahrhunderts* (Berlin, 1961), p. 131).

21 P. Patte, *Mémoires sur les objets les plus importants de l'architecture* (Paris, 1769), p. 82.

22 In his loyalty to the spirit of the Enlightenment, Patte in fact suggested the composition of a kind of catalogue of the ordinances of architecture, in P. Patte, *Mémoires*, pp. 80, 81.

23 Leroy's *Les ruines des plus beaux monuments de la Grèce* (1758) in fact emphasised the variety of forms and proportions which the Greeks had given to the Doric column. The appearance in 1764 of Dumont's engravings of the temples at Paestum merely served to confirm this diversity, which was shown to be wholly

irreducible to the rules. Thus, a little less than a century after the work by Desgodets, Antiquity suddenly became a wholly separate field of enquiry. The eclecticism of the eighteenth century was also linked with this rediscovery of the inventiveness of the Ancients. On this topic, one should consult D. Wiebenson, *Sources of Greek revival architecture* (London, 1969).

24 N.-H. Potain, *Traité des ordres d'architecture* (Paris, 1767), 'Extraits des registres de l'Académie Royale d'Architecture du 10 mars 1766'.

25 See chapter 7.

26 See chapter 7.

27 J.-F. Blondel, *Cours d'architecture*, vol. 1, p. 376.

28 J. Summerson, *The classical language of architecture* (London, 1963), p. 3.

29 On the influence of travel upon Enlightenment thought, see N. Broc, *La géographie des philosophes: géographes et voyageurs français au XVIIIe siècle* (Gap, 1975).

30 Chambers sought to exploit this fad by bringing out his *Designs of Chinese buildings, furniture, dresses, machines and utensils* (1757), which was a runaway success. Le Rouge then published it in France. On this topic one should read *Jardins en France, 1760–1820*, catalogue of the Caisse Nationale des Monuments Historiques (Paris, 1978), pp. 12–16.

31 On the theme of the primitive hut, see J. Rykwert, *Adam's house in paradise* (Cambridge, Mass., 1972). See chapter 10 also.

32 See chapter 4.

33 N. Le Camus de Mézières, *Le génie de l'architecture, ou l'analogie de cet art avec nos sensations* (Paris, 1780), pp. 1-3.

34 See chapter 10.

35 See J. Derrida, *On grammatology*, translated by G. C. Spivak (London, 1974).

36 E. Cassirer, *The philosophy of the Enlightenment*, p. 197.

37 J.-F. Blondel, *Cours d'architecture*, vol. 1, 'Abrégé de l'histoire de l'architecture', pp. 1–75.

38 See chapter 6.

39 This plate, which seems to prefigure Durand's comparative method, has been analysed in some detail by W. Szambien, *Jean-Nicholas-Louis Durand 1760–1834: de l'imitation à la norme* (Paris, 1984), pp. 27–30.

40 D. Leroy, *Histoire de la disposition et des formes différentes que les chrétiens ont données à leurs temples, depuis la règne de Constantin le Grand jusqu'à nous* (Paris, 1764), p. 46.

4 'TOWARDS A CLASSICAL ARCHITECTURE': JACQUES-FRANÇOIS BLONDEL AND THE 'COURS D'ARCHITECTURE'.

1 See chapter 8.

2 J.-F. Blondel, *De la distribution des maisons de plaisance et de la décoration des édifices en général* (Paris, 1737–8), vol. 1, p. v.

3 J.-F. Blondel, *Cours d'architecture*, vol. 3, pp. xxiii, xxiv.

4 J.-F. Blondel, *Discours sur la nécessité de l'étude de l'architecture*, Paris, 1754, pp. 83–9.

5 Blondel described the pont de Neuilly, built by Perronet, as 'a famous construction, whose combined size, magnificence and economy will teach future generations what the genius, taste and experience of this *savant*, as good an architect as he was a citizen, was capable of' (*Cours d'architecture*, vol. 1, p. 107).

6 J.-F. Blondel, *Architecture françoise*, vol. 1, p. 14.

7 J.-F. Blondel, *Cours d'architecture*, vol. 1, p. 378.

8 Ibid., vol. 3, p. 370.

9 Ibid., vol. 2, p. 311.

10 Ibid.

11 Ibid., p. 317.

12 Ibid., pp. 338–42.

13 J.-F. Blondel, *L'homme du monde éclairé par les arts* (Amsterdam/Paris, 1774), vol. 1, pp. 255ff.

14 J.-F. Blondel, *De la distribution des maisons de plaisance*, vol. 1, p. 1.

15 Blondel thus mentions a château, an orangerie and a pavilion on the outskirts of Florence, another château for a private person living at Besançon, a large house with its outbuildings on the route de Bretagne (*De la distribution des maisons de plaisance*, vol. 1, pp. 11ff, pp. 95ff, pp. 117ff).

16 J.-F. Blondel, *De la distribution*, vol. 1, plate 34.

17 Ibid., plate 12.

18 J.-F. Blondel, *Architecture françoise*, vol. 1, n. 21.

19 J.-F. Blondel, *Cours d'architecture*, vol. 1, pp. 3, 4.

20 This was especially true of Prony's *Réflexions sur l'organisation d'une académie qui auroit pour objet la perfection et l'enseignement de la construction*, which I will study in chapter 6.

21 J.-F. Blondel, *Cours d'architecture*, vol. 1, pp. 281–3.

22 Ibid., pp. 294-6.

23 Ibid., p. 376.

24 This at any rate was what Patte claimed in the notice which he placed at the beginning of the fifth volume of the *Cours*, for the editing and writing of which he was himself responsible. Listing the papers left at Blondel's death, Patte wrote: 'As for construction, all we found was the article on Masonry in the *Encyclopédie*, which was in his own hand, which formed part of the notebooks which he had dictated to his own pupils; an article which seemed to us to be mediocre; indeed, it was not in this particular area that M. Blondel's lectures shone: we know that he applied himself to training theoreticians more than practitioners, and it is probable that he would have rewritten this part of his *Cours*' (P. Patte, *Cours d'architecture . . . commencé par feu J.-F. Blondel . . . et continué par M. Patte*, vol. 5, p. x). To judge by the article in question, and by the chapters of *Architecture*

françoise which are devoted to the art of building, it does indeed seem as if Blondel had not gone very deeply into questions related to construction, or less deeply at any rate than Patte.

25 J. Starobinski, *Jean-Jacques Rousseau, la transparence et l'obstacle* (1957; Paris, 1971).

26 M. Gallet, *Ledoux et Paris*, p. 17.

27 In spite of the closeness of their two approaches, Blondel was fairly reserved with regard to Laugier. The commentary which he included in *L'homme du monde éclairé par les arts* is revealing in this respect: '(Father) Logier [Laugier], a Jesuit, gave to the world a few years ago an *Essai sur l'architecture* which was very much in vogue among fashionable people. One wishes that he had kept to what others have said, for where he recapitulates such works, albeit cursorily, he does very well. That part of this little book which is strictly his own does not have the same virtues, and serves to show that one has to be thoroughly trained in the precepts of this art, if one is to risk writing on these matters. When Laugier left the College de Louis-le-Grand to take holy orders, he gave us a number of other works, which have already been forgotten. He died a few years ago' (J.-F. Blondel, *L'homme du monde éclairé par les arts*, vol. 2, p. 314). What separates Blondel from Laugier was undoubtedly the clear-cut positions adopted by the latter, which led him to condemn, for example, pediments which did not correspond to a gable wall. Blondel was offended by Laugier's primitivism, and by his disregard for the classical tradition. But, aside from these doctrinal matters, it was above all the unrealistic character of Laugier's proposals which irritated Blondel.

28 J.-F. Blondel, *Cours d'architecture*, vol. 2, pp. xxxviii, xxxix.

29 Ibid., vol. 1, p. 159.

30 J.-F. Blondel, *Discours sur la nécessité de l'étude de l'architecture*, pp. 28, 29.

31 J.-F. Blondel, *Cours d'architecture*, vol. 2, p. xxv.

32 Ibid., vol. 1, pp. 260–3.

33 See W. Sazmbien, *Reflet de la connaissance*, pp. 40, 41.

34 J.-F. Blondel, *Cours d'architecture*, vol. 3, p. 8.

35 J.-F. Blondel, *Discours*, p. 29.

36 The twinning of Doric columns raised insurmountable difficulties, especially because it threatened the alternation of triglyphs and metopes. In the case of the portail des Minimes and the château of Maisons, François Mansart had resolved this problem by having base and capital of the columns interpenetrate. Libéral Bruant, however, with the frontispiece of the bureau des marchands, had combined the sections with their bases. In the *Cours d'architecture*, Blondel advanced a compromise solution, obtained by coordinating the proportions of the different elements of the ordinance; he was also interested in coupling the columns within the re-entrant angles of the centrepiece. Blondel would later apply these ideas to

the portal he designed for Metz Cathedral, built in 1764 (*Cours*, vol. 2, pp. 17–24, plates IV, V).

37 Here too, these liberties concerned the coupling and the interpenetration of the elements of the order (ibid., vol. 3, pp. 24–30, plate I).

38 Ibid., p. lviii.

39 Ibid., p. lxiii.

40 Ibid., vol. 2, pp. x, xi.

41 C.-N. Ledoux, *L'architecture considerée sous le rapport de l'art, des mœurs et de la législation* (Paris, 1804), p. 135.

42 J.-F. Blondel, *Cours d'architecture*, vol. 1, p. 136.

43 Ibid., p. 453.

44 Ibid., p. 450.

45 Ibid., vol. 4, p. xix.

46 Ibid., vol. 3, pp. 56ff, plate IV.

47 Ibid., vol. 4, pp. vi, vii.

48 Blondel thus criticised the façade of the château at Blois for having 'a domestic air' unsuited to a Royal residence (ibid., vol. 3, p. 92).

49 To convince oneself of this, one merely has to turn to the embarrassed explanation given in the *Cours* of the superimposed orders on Mansart's façade. According to Blondel, these orders were so well combined that they seemed to belong to a single shaft. The weakness of the argument reveals very clearly the contradictions which always subsist in the 'raisonnement de l'art' (ibid., p. 86).

50 Ibid., vol. 1, p. 377.

51 Ibid., vol. 4, p. xxxvi.

52 Blondel's particular complaint was that they had deviated from the precepts of the art 'without any other specific object save that of creating chimeras at will' (*Cours d'architecture*, vol. 3, p. 422).

53 See below.

54 J.-F. Blondel, *Cours d'architecture*, vol. 1, pp. 396, 397.

55 Ibid., vol. 1, p. 408.

56 Ibid., vol. 1, p. 398.

57 Ibid., vol. 3, p. 430.

58 'Should we not, in our decorations, choose to depict the national character rather than servilely imitate the productions of peoples who have preceded us' (*Cours d'architecture*, vol. 1, p. 346); 'Never shall we tire of being citizens and Frenchmen' (Ibid., vol. 4, p. 87).

59 Ibid., vol. 2, pp. xxx, xxxi.

60 *Histoire générale des sciences*, edited by R. Taton (Paris, 1964–6), p. 521. On the development of optical theories, one should also read A.-T. Sabra, *Theories of light from Descartes to Newton* (Cambridge, 1981).

61 On Father Castel's strange invention, which involved linking sounds and colours

in a perspective of physiological experimentation, see A.-K. Chouillet-Roche, 'Le clavecin oculaire de père Castel', in *Dix-Huitième Siècle*, 8, 1976, 141–66.

62 The decomposition of colours by means of the prism served in fact to contradict Descartes' hypotheses on the nature of light. As far as the latter was concerned, light was not a substance but a form of agitation to be attributed to variations of pressure within an incompressible atmosphere. It could not therefore possess any intrinsic qualities, and the colours were linked to the various ways in which bodies transmitted the luminous agitation. Newton's demonstration of the dispersion of light wrecked this system by eliciting different colours from one and the same transparent body.

63 There is an account of these projects in J.-M. Pérouse de Montclos, *Etienne-Louis Boullée*, pp. 198ff, illustrations 81–3, and pp. 155ff, illustrations 62–8.

64 E.-L. Boullée, *Architecture. Essai sur l'art*, pp. 90, 91.

65 J.-F. Blondel, *Cours d'architecture*, vol. 3, pp. 375–82.

66 Ibid., pp. 382–93, plate LIV.

67 Ibid., pp. 393–407, plates LV–LIX.

68 'Whatever the kind of building, but especially in our temples, nothing is more shocking than to overwhelm, so to speak, the grandeur of the place by an immoderate architecture or by affecting weightiness in the pillars' (*Cours d'architecture*, vol. 3, pp. 331–2).

69 We know that Laugier used to specify the use of an entablature above the columns instead of the 'evil' practice of using arcades (M.-A. Laugier, *Essai sur l'architecture* (1753; Paris, 1855), p. 29).

70 Blondel also emphasised the kinship between his projects and the churches of Flanders. The use of brick vaults, which enabled one to avoid carpentry and the attendant fire risks, was also derived from Flemish architecture (*Cours d'architecture*, vol. 3, pp. 393, 394, 405, 406).

71 Blondel spoke of a 'tempered light', which, he claimed, enhanced the simple and regular character of the ordinance (*Cours d'architecture*, vol. 3, p. 397).

72 The Royal Abbey of Saint-Louis replaced two abbeys of Benedictine nuns situated at the centre of Metz, whose amalgamation had been decided upon by the King. Blondel was entrusted with the project in 1761. Although the laying of the foundation stone took place in 1764, the new buildings were never completed (*Cours d'architecture*, vol. 3, pp. 412–20, plates LX, LXI). One should also consult A. Prost, *J.-F. Blondel et son œuvre*, p. 58.

73 J.-F. Blondel, *Cours d'architecture*, vol. 4, plates XLI–XLIV.

74 E. Kaufmann, *Von Ledoux bis Le Corbusier: Ursprung und Entwicklung der Autonomen Architektur* (Vienna, 1933).

75 This brings us to the other side of neo-classical art: 'that which is aware of the distance separating it from the forms it represents, which knows it is aiming at a

kind of absence' (J. Starobinski, *1789: les emblèmes de la raison* (Paris, 1979), p. 112).

76 For Boullée, the multiplication of columns in fact offered 'the picture of immensity', thanks to the laws of optics and to the effects of perspective (E.-L. Boullée, *Architecture. Essai sur l'art*, p. 83).

77 Quoted by M. Eleb-Vidal and A. Debarre-Blanchard, *Architecture domestique et mentalités*, p. 56.

78 J.-F. Blondel, *Cours d'architecture*, vol. 4, p. 106.

79 See M. Eleb-Vidal and A. Debarre-Blanchard, *Architecture domestique et mentalités*, p. 56.

80 See P. Ariès, *Centuries of childhood*, translated by R. Baldrick (London, 1962), pp. 398–9.

81 W. Szambien, *Reflet de la connaissance*, pp. 27, 28.

82 J.-F. Blondel, *Cours d'architecture*, vol. 3, pp. 69ff, plates VII, VIII.

83 J.-F. Blondel, *Discours sur la necessité de l'étude de l'architecture*, p. 43.

84 J.-F. Blondel, *Cours d'architecture*, vol. 4, p. lix.

85 Ibid., vol. 1, pp. xxii, xxiii.

86 Ibid., vol. 4, pp. 101, 102.

87 See J. Starobinski, *Jean-Jacques Rousseau, la transparence et l'obstacle*.

88 J.-F. Blondel, *Cours d'architecture*, vol. 4, pp. 103, 104.

89 Ibid., pp. 272, 273.

90 Ibid., p. 104.

91 Ibid., p. 5.

92 Ibid.

93 Ibid., pp. 83ff, plate XXIV.

94 Ibid., pp. 88ff, plates XXV, XXVI.

95 Ibid., p. 87.

96 See T. Mariage, *L'univers de Le Nôtre et les origines de l'aménagement du territoire*, research report, CORDA (Paris, 1984).

97 See *Jardins en France, 1760–1820*.

98 See chapter 9.

99 J.-F. Blondel, *Cours d'architecture*, vol. 4, p. 214.

100 Ibid., p. 199.

101 Ibid.

102 Ibid., pp. 199, 200.

103 Curved lines hardly feature in the *Cours d'architecture* save in these 'conveniences', which might be angled, sometimes creating the first hint of a diagonal in plans which were in other respects extremely rigid. Architecture thus proceeded by means of superimposed grids, in which could be discerned a systematic intention prefiguring the emergence of the notion of composition from Ledoux to Durand.

104 J.-F. Blondel, *Cours d'architecture*, plate XXXV.

105 Ibid., plate XXXVI.

106 Blondel's judgement may strike one as surprising, inasmuch as the H-shaped plan is certainly the most 'convenient' in the modern sense of the term. Yet, with this plan, one who approaches between the wings of the building finds himself contained between the right and the left wings, whereas the U-shaped plan offers him a single *enfilade*, which is more satisfying in terms of the laws of *bienséance*. In reality, the *Cours* was still referring here to perceptual codes inherited from the seventeenth century, and we certainly do not share its notion as to what constitutes 'the most advantageous arrangements'.

107 J.-F. Blondel, *Cours d'architecture*, vol. 4, p. 208.

108 State apartments were meant to 'announce the opulence of the owner', whereas social apartments corresponded to the more everyday activities, with the private apartments being reserved for domestic intimacy.

109 These were 'English' water-closets, whose absence in the case of the majority of urban dwelllings of the period was deplored by Patte. See R.-H. Guerrand, *Les lieux, histoire des commodités* (Paris, 1985), pp. 57, 58.

110 A member, like Blondel, of the Académie Royale d'Architecture, Franque delivered the funeral oration in praise of his friend before the members of the company, see chapter 1, note 21.

111 The abbot's house in question was that of Prémontrés à Villers-Cotteret: the design dates from 1765 (*Cours d'architecture*, vol. 4, pp. 357ff).

112 Ibid., pp. 360ff, plate XLVII.

113 This geometry was nevertheless confined to small spaces, in order to take advantage of them. One should again emphasise in this respect the importance of 'conveniences' in the development of the conception of the dwelling in the eighteenth century.

114 J.-F. Blondel, *Cours d'architecture*, vol. 4, p. 112.

115 C.-N. Ledoux, *L'architecture considérée sous le rapport de l'art, des mœurs et de la législation*, plate 33.

116 See chapter 5.

117 On *convenance* and *bienséance*, see W. Szambien, *Reflet de la connaissance*, pp. 83–92, 154–61.

118 J.-F. Blondel, *De la distribution des maisons de plaisance*, p. 3.

119 *Discours sur la nécessité de l'étude de l'architecture*, p. 42.

120 *Bienséance* was in fact related to the nobility, which served to limit its sphere of application. See W. Szambien, *Reflet de la connaissance*, p. 87.

121 '*Convenance* should be regarded as preliminary to every operation in the art of building. It indicates the *bienséance* which one should observe in every kind of building, sizes, forms, their ordinances, their richness and their simplicity; it

assigns the allegories *or* attributes *convenable* to every kind of building; it decides how lavish or economical one is to be, it determines the choice of materials, the use to which they are to be put, and their quality; finally, it is the spirit of *convenance* which, through the application of constant principles, enables one to give different forms to buildings raised with the same end in view, depending upon the rank, dignity or affluence of the owners' (J.-F. Blondel, *Discours sur la nécessité de l'étude de l'architecture*, p. 45).

122 J.-F. Blondel, *Cours d'architecture*, vol. 1, p. 128.

123 Ibid., vol. 2, p. 229.

124 There is a more exhaustive study of the notion of character in W. Szambien, *Reflet de la connaissance*, pp. 96–129.

125 Boffrand, in his *Livre d'architecture* (pp. 16ff), even went so far as to apply parts of Horace's *Ars Poetica* directly to architecture.

126 J.-F. Blondel, *Cours d'architecture*, vol. 2, pp. 230, 231.

127 Ibid., vol. 1, pp. 373ff.

128 Ibid., vol. 2, p. xxvii.

129 Boullée's *Recueil d'architecture privée* was to have been divided into notebooks, each containing 'a sequence in a particular genre', it being left to one's imagination and 'sometimes to one's character' to grasp it as a whole. Boullée's development of the notion of character was thus contemporary with the beginnings of a typological mode of thought.

130 J.-F. Blondel, *Cours d'architecture*, vol. 4, p. xl.

131 Ibid., pp. xli–xliii.

132 Ibid., pp. xliv–xlv.

133 Ibid., p. xlvi.

134 Ibid., p. xliii.

135 Ibid., p. 1.

136 Ibid., vol. 3, pp. xiv, xv.

137 See Jacques Derrida: 'One was not supposed to imitate nature, for, in assigning its rules to genius, it bent, folded back upon itself, and was reflected through art' (J. Derrida, 'Economimesis', in *Mimesis des articulations* (Paris, 1975), pp. 55–93, and p. 59 in particular).

138 J.-F. Blondel, *Cours d'architecture*, vol. 3, pp. lviii, lix.

139 J.-F. Blondel, *Discours sur la necessité de l'étude de l'architecture*, pp. 24, 25.

140 C.-N. Ledoux, *L'architecture considérée sous le rapport de l'art, des mœurs et de la législation*, p. 22.

141 J.-F. Blondel, *Cours d'architecture*, vol. 3, pp. liv–lvi.

142 See chapter 6.

143 C.-N. Ledoux, *L'architecture considérée sous le rapport de l'art, des mœurs et de la législation*, p. 135.

144 J.-F. Blondel, *Cours d'architecture*, vol. 4, plates XXIVff.

5 THE ENGINEERS' 'SYSTEM'

1 See B. Gille, *Les ingénieurs de la Renaissance* (Paris, 1964).

2 In the 1755 edition of d'Aviler's *Dictionnaire*, architecture was still divided into civil architecture and hydraulic architecture, with the latter being taken to comprise the construction of canals, roads and bridges. C.-A. d'Aviler, *Dictionnaire d'architecture civile et hydraulique* (Paris, 1755). On the notion of hydraulic architecture, see chapter 10 also.

3 On the officers of the Génie, see for example A.-M. Augoyat, *Aperçu historique sur les fortifications, les ingénieurs et sur le corps du Génie en France* (Paris, 1860–4); A. Blanchard, *Les ingénieurs du roy, de Louis XIV à Louis XVI* (Montpellier, 1979). See also chapter 6, n. 11.

4 See R. Taton (ed.), *Enseignement et diffusion des sciences en France au XVIIIe siècle* (1964; Paris, 1986).

5 The boundary between public and private domains was still, however, a permeable one, as is borne out by the career of Jeanson, an engineer with the Ponts et Chaussées who was placed at the disposal of the mines and foundries of Montcenis-Le Creusot, before going on to build some spinning-mills for the Prince de Ligne. P. Saddy, 'Un elève de Soufflot: Barthélémy Jeanson, architecte-ingénieur, 1760–1820', in *Cahiers de la Recherche Architecturale*, Proceedings of the Conference 'Soufflot et l'architecture des Lumières', supplement to 6–7, 1980, pp. 192–203. Another engineer, Charles Le Roy, a former pupil at the Ecole des Arts and at the Ecole des Ponts et Chaussées, was likewise released by the administration so that he might work on the construction of the château d'Aiguillon. See ANF14 2268/1.

6 On the ties between science and technology, and between scientists and technicians, one might consult C.-C. Gillispie, *Science and polity in France at the end of the Old Regime* (Princeton, 1980).

7 See *Les routes de France depuis les origines jusqu'à nos jours* (Association pour la Diffusion de la Pensée française, Paris, 1959, pp. 102, 103).

8 H. Cavaillès, *La route française, son histoire, sa fonction* (Paris, 1946), pp. 54, 55.

9 On 'the' map of France, see B.-H. Vayssière, '"La" carte de France', in *Cartes et figures de la terre* (exhibition catalogue published by the Pompidou Centre, Paris, 1980), pp. 252–65.

10 See P. Alliès, *L'invention du territoire* (Grenoble, 1980), pp. 156, 157.

11 On the history of the administration of bridges and highways, see A. Brunot and R. Coquand, *Le corps des Ponts et Chaussées* (Paris, 1982); A. Debauve, *Les travaux publics et les ingénieurs des Ponts et Chaussées depuis le XVIIe siècle* (Paris, 1893); F. Fichet, *Le corps des Ponts et Chaussées. Du génie civil à l'aménagement du territoire*; J. Petot, *Histoire de l'administration des Ponts et Chaussées 1599–1815* (Paris, 1958);

E.-J.-M. Vignon, *Etudes historiques sur l'administration des voies publiques en France au XVIIe et XVIIIe siècles* (Paris, 1862–4).

12 A decree of 1668 in fact stipulated that an architect or an engineer should be appointed by the King for the task of overseeing important sites, J. Petot, *Histoire de l'administration*, p. 74.

13 Daniel Trudaine (1703–69), Intendant at Riom, took charge of the Ponts et Chaussées in 1734. Together with Perronet, he was the person who did most to promote the corps of the Ponts et Chaussées. His son, Trudaine de Montigny, succeeded him in this post in 1769. The latter, who was a friend of both Lavoisier and Turgot, retired in 1777, after having played a part in the diffusion of Physiocratic thought among the engineers (see below). On these two figures, consult E. Choullier, *Les Trudaine* (Archis-sur-Aube, 1884).

14 One should bear in mind that the competence of the Royal corps of the Ponts et Chaussées did not extend into the pays d'Etats, such as Burgundy, Brittany or Languedoc. Corps of local engineers were responsible for the building of roads and bridges in these districts. The 1791 law on bridges and highways would, however, affiliate them to the corps of the Ponts et Chaussées. See J. Petot, *Histoire de l'administration*, pp. 259ff.

15 The rank of sub-engineer was equivalent to that of someone beginning their career with the corps of the Ponts et Chaussées. Pupils at the Ecole des Ponts et Chaussées were thus appointed sub-engineers at the end of their period of study.

16 On the Ecole des Ponts et Chaussées in the eighteenth century, read F. de Dartein, 'Notice sur le régime de l'ancienne Ecole des Ponts et Chaussées et sur sa transformation à partir de la Révolution', in *Annales des Ponts et Chaussées*, 1–2, 1906, 5–143; B. Fortier, 'La nascita dell'Ecole des Ponts et Chaussées; 1. Lo spazio', *Casabella* (Milan), 495, 1983, 40–7; 'La nascita dell' Ecole des Ponts et Chaussées; 2. Il progetto', *Casabella*, 496, 1983, 36–45; O. Selvafolta, 'Storia documentaria dell' Ecole des Ponts et Chaussées di Parigi' (architecture thesis at Milan Politecnico, 1974); G. Serbos, 'L'Ecole royale des Ponts et Chaussées', *Enseignement et diffusion des sciences au XVIIIe siècle*, pp. 345–63; *Regards sur la France* (Paris), L'Ecole Nationale des Ponts et Chaussées, 14, October 1961. Most manuscripts concerning the School are kept at the Centre Pédagogique de Documentation et de Communication de l'ENPC. See in particular MS 2154, 2629 bis, 2630, 2630 bis.

17 The decree of 14 February spelled out the substance of these two missions, committing Perronet 'to have, at the behest of his honour the Controller-General of Finances, and of his honour the Intendant of Finances, charge of the running of the Ponts et Chaussées, the conduct and inspection of geographers and designers of plans and maps, of the sciences and practices necessary for successfully and skilfully answering the requirements of the different jobs involved in the aforesaid

Ponts et Chaussées', (*Arrêt du conseil d'état qui nomme le sieur Perronet directeur du bureau des dessinateurs, etc., suivi de lettres patentes du 10 mars 1747, registrées en la cour des comptes le 27 mars 1747*, reproduced in M. Vignon, *Etudes historiques*, vol. 2, pp. 150–2.

18 *Instruction du contrôleur général des finances concernant les employés subalternes des ponts et chaussées et le bureau des dessinateurs*, reproduced in M. Vignon, *Etudes historiques*, vol. 2, pp. 170–3.

19 F. de Dartein, 'Notice sur le régime de l'ancienne Ecole des Ponts et Chaussées', p. 24.

20 Pierre-Charles Lesage (1740–1810) was a geographer-engineer attached to the French army, who transferred to the Ponts et Chaussées in 1766. He worked as a draughtsman at Bordeaux, and entered the Ecole des Ponts et Chaussées in 1774; he was an inspector at the School from 1778 up until his death in 1810. It was his responsibility to train pupils in drawing and in design, and he left numerous writings on the operation of the institution which Trudaine and Perronet had created. See his engineer's file, ANF14 2268/1.

21 P.-C. Lesage, *Mémoire sur l'établissement primitif de l'Ecole Nationale des Ponts et Chaussées, son état actuel, son organisation et l'emplacement d'un local plus avantageux qu'il conviendrait de lui assigner*, a report presented by Lesage in the name of the Commission of public works, 20 Nivôse Year III, Manuscript ENPC, MS 2629 bis.

22 This system of self-instruction did, however, have its limitations. Surveying and the drawing of maps were taught in the School by professional geographers, whereas pupils went outside to attend the architecture courses of Blondel and, after 1774, of Daubenton. This arrangement also involved instruction at a number of other faculties, from mineralogy and the natural sciences to physics. Self-instruction in the strict sense of the term concerned above all else mathematics, geometry, mechanics and hydraulics. See F. de Dartein, 'Notice sur le régime de l'ancienne Ecole des Ponts et Chaussées', pp. 25, 26.

23 The list at this date comprised the following: three competitions in mathematics, one in stone-cutting, one in surveying for a plan, one in surveying, one in estimating quantities for a building, one style competition, three drawing competitions, two writing competitions. Turgot, *Instruction concernant la direction des élèves, des sous-ingénieurs et des inspecteurs des ponts et chaussées*, reproduced in F. de Dartein, 'Notice sur le régime de l'ancienne Ecole des Ponts et Chaussées au XVIIIe siècle', pp. 107–23.

24 See chapter 6.

25 On the practice of design at the Ecole des Ponts et Chaussées, see J.-F. Chenais, Y. Chicoteau, P. Madec, J. Michel, B. Patarin, A. Picon and C. Vie, *Raison–projet–représentation. La formation architecturale des ingénieurs à l'Ecole des Ponts et Chaussées au XVIIIe siècle*; Y. Chicoteau, A. Picon and C. Rochant, *Les concours*

techniques du XVIIIe siècle à l'Ecole des Ponts et Chaussées (CNRS research report, Paris, 1984).

26 See A. Dupire, B. Hamburger, J.-C. Paul, J.-M. Savignat and A. Thiébaut, *Deux essais sur la construction*, pp. 29ff.

27 S.-P. Timoshenko, *History of the strength of materials* (1953; New York, 1983), pp. 41ff.

28 I shall return below to this discrepancy between scientific results and their concrete application, a discrepancy which would disappear only in the course of the nineteenth century.

29 It is worth noting that decisions taken regarding the corps of the Ponts et Chaussées tended generally to coincide with measures concerning the guilds. 1716 therefore saw both the foundation of the corps of the Ponts et Chaussées and the appointment of inspectors whose function was to liquidate the corporation's debts. Likewise, Turgot's *Instruction* of 1776 coincided with a bid to abolish the guilds. Finally, the Le Chapelier law and the law regarding bridges and highways were passed in the course of the same year. The stages through which the profession of engineer passed would seem to have corresponded to the convulsions experienced by the guild system in its struggle with the state. On the subject of the guilds, read E. Martin Saint-Léon, *Histoire des corporations de métiers* (Paris, 1897).

30 See chapter 6.

31 See chapter 7.

32 See chapter 9.

33 On the attitude of state engineers, A. Querrien, *Devenir fonctionnaire ou/et le travail de l'Etat* is essential reading.

34 The stylistic competitions are in manuscript form in the Centre Pédagogique de Documentation et de Communication de l'ENPC, in boxes labelled 'Pupils' literary competitions'; these have not been the object of a systematic inventory. I have reproduced here in abbreviated form the conclusions of my article, A. Picon, 'Les écritures de la technique, une étude des Concours de style de l'Ecole des Ponts et Chaussées à la fin du XVIIIe siècle', in *Raison–projet–représentation*, pp. 243–75.

35 Few of the pupils showed any sign of first-hand philosophical knowledge in their compositions, having acquired it more by osmosis than anything else. Most of the references were to the writings of Voltaire and to the empiricism of Condillac; a few pupils had read Rousseau, while one or two quoted Plato or Aristotle, a sign of a wider classical culture. Prony, however, was exceptional in having more definite theoretical ambitions than his fellows. His chief inspiration while studying at the Ecole des Ponts et Chaussées had been Condillac and Rousseau, but his subsequent intellectual evolution brought him nearer to the *idéologues*. Thus, his essay 'De l'intelligence moyenne et de ses progrès' (ENPC manuscript, MS

1771) invoked the ideal, dear to Condorcet, of the infinite perfectibility of the human species. The influence of Destutt de Tracy was still clearer in the *Réflexions sur l'organisation d'une académie qui auroit pour objet la perfection et l'enseignement de l'art de la construction*, a work which was drafted by Prony during the Revolution. See chapter 6.

36 Millard, pupil's dissertation submitted for the style competition of 1778, ENPC manuscript. The reference to the effects of famine is related, of course, to the debate over the trade in grain. Should one do as the physiocrats advised and free this trade, or should one rather adopt the more prudent policy recommended by the Abbé Galiani, in his *Dialogues sur le commerce des blés* (1770)? Whatever solution was adopted, the engineers were wholly persuaded of the utility of communications, since those provinces which were threatened with dearth had to be helped. On the question of grain in the eighteenth century, one should read L. Kaplan, *Bread, politics and political economy in the reign of Louis XV* (The Hague, 1976, 2 vols).

37 See, for example, L. de Lavergne, *Les économistes français du dix-huitième siècle* (Paris, 1870); R. L. Meek, *The economics of physiocracy* (Cambridge, Mass., 1963).

38 P. Planier, pupil's dissertation submitted to the style competition of 1779, ENPC manuscript.

39 'We ask: 1. What should be the physical constitution and the early education of young people aspiring to join the corps of the Ponts et Chaussées; 2. The moral and physical qualities that one should desire of them; 3. What are the knowledges which it would be important for them to cultivate and what is the order which they should observe in their studies, setting aside any notions they may have derived in this respect from their reading of the School's regulations' (topic for the style competition of 1788 at the Ecole des Ponts et Chaussées)?

40 De Bourges, pupil's dissertation, style competition of 1779, ENPC manuscript.

41 G. Riche de Prony, pupil's dissertation, style competition of 1779, ENPC manuscript.

42 J.-L.-H. L'Escaille, pupil's dissertation, style competition of 1782, ENPC manuscript.

43 We should note in passing that this authority chiefly concerned the allocation of the *corvée*, which was one of the engineers' great problems. See chapter 7.

44 While Perronet was in charge, this *esprit de corps* continued to be very familial in nature. The director of the Ecole des Ponts et Chaussées saw himself as a kind of father to his pupils; in 1778 the latter presented him with a marble bust of himself, which bore the inscription 'Patri carissimo familia' (P.-C. Lesage, *Notice pour servir à l'éloge de M. Perronet*, pp. 14, 15). The *esprit de corps* of the administration

of the Ponts et Chaussées was further reinforced by the Masonic affiliations of a number of pupils and engineers. At Paris, the Urania Lodge, which was run by Lesage, met under the noble protection of Perronet in the years 1787–90 (see manuscript BN FN2 126).

45 G. Riche de Prony, pupil's dissertation for competition of 1779.

46 The engineers' utilitarianism of course derived from that of the philosophers and the economists. On the concept of utility in the eighteenth century, see G. Gusdorf, *Les principes de la pensée au Siècle des Lumières*, pp. 428–43.

47 Engineers frequently employed legal metaphors in their writings. Thus, the style competition of 1785 asked 'which [was] more necessary to society, a great captain or a great magistrate, assuming that each were endowed with all the virtue of which humanity were capable'. The pupils without exception opted for the magistrate. An engineer's actions were in fact based upon primary needs and upon a natural right to happiness and to progress. Prony himself compared the corps of the Ponts et Chaussées to a 'tribunal of reason', which served to judge whatever conflicts might arise between its members (G. Riche de Prony, pupil's dissertation for the style competition of 1779). This latter attitude represented the transposition on to the professional plane of the legal aspect which permeated Enlightenment thought, from Montesquieu to Beccaria. One should bear in mind that the French translation of the latter's *Dei delitti e delle pene*, by the Abbé Morellet, was published in 1765, and was an immediate success.

48 J.-R. Perronet, *Etude d'un projet d'arche en fer de 200 pieds d'ouverture. Rapport fait au nom d'une commission composée de MM. Trudaine de Montigny, Perronet et Vaucanson, membres de l'Académie des Sciences, sur un projet de M. de Montpetit* (1779, ENPC manuscript, MS 2609); *Rapport à l'Assemblée des Ponts et Chaussées sur un projet de pont en fer proposé par le sieur Racle, pour être construit sur la rivière Reysouse à Pont-de-Vaux en Bresse* (1789, ENPC manuscript, MS 1424). One should also read B. Lemoine, 'L'origine des ponts metalliques en France', in *Annales des Ponts et Chaussées*, special 150th anniversary number, 19, 1981, 44-52.

It was by no means just the rules of *convenance* which caused Perronet to distrust iron. Like Patte and the majority of his contemporaries, the engineer contrasted the durability of stone with the fragility of metal, which was subject to corrosion. Nevertheless, a competition for iron bridges was held at the Ecole des Ponts et Chaussées in 1788.

49 See chapter 9.

50 In *La collection des prix que la ci-devant Académie d'architecture proposait et couronnait tous les ans*, which was published at the beginning of the Revolution, there also featured a number of projects by pupils of the Ponts et Chaussées. J.-M. Pérouse de Montclos, *'Les prix de Rome', concours de l'Académie royale d'architecture au XVIII siècle* (Paris, 1985), pp. 12, 13.

6 THE PRINCIPLES OF TRAINING AND THE ECONOMY OF KNOWLEDGE

1 L. Pelpel, *La formation architecturale au XVIIIe siècle en France.*

2 After Blondel's death, the ties between the Ecole des Arts and the Ecole des Ponts et Chaussées loosened somewhat. Daubenton, who succeeded Blondel in 1774, did not enjoy the administration's confidence to anything like the same degree.

3 'Although the loftiest mathematical speculation is not absolutely necessary to the architect, he should nevertheless be familiar with all the applications of nature' (J.-F. Blondel, *Discours sur la manière d'étudier l'architecture et les arts qui sont relatifs à celui de bastir* (Paris, 1747), p. 8). Around 1750, alongside architecture and drawing, the Ecole also taught the ways and customs of the building trade, the surveying of plans, quantity surveying, stone- and wood-cutting, mechanics and hydraulics (J.-F. Blondel, *Discours*, pp. 11–14).

4 De Courtonne, who was a mediocre theoretician, had tried without success to draft an architecture course; he proved no more able than his successor Jossenay in giving to the Academy a new lease of life. As Monique Mosser and Daniel Rabreau have observed, one of the major failures of the academic institution consisted in its inability to translate its ambitions into book form. M. Mosser and D. Rabreau, 'L'Académie Royale et l'enseignement de l'architecture au XVIIIe siècle', in *Archives d'Architecture Moderne* (Brussels), 25, 1983, 47–67, especially p. 55.

5 L. Pelpel, *La formation architecturale au XVIIIe siècle*, p. 26.

6 There were in fact three different kinds of pupil, namely, those who were studying with Blondel so as to gain admission to the Ecole des Ponts et Chaussées, the pupils of Perronet, who came at their own expense to perfect their knowledge of architecture, and finally, those who, though belonging to the Ecole des Ponts et Chaussées, lodged at the Ecole des Arts. See Y. Chicoteau, *Introduction du calcul à l'école des Ponts et Chaussées: enseignement et pratique de l'ingénieur au XVIIIe siècle* (ENPC, Paris, 1981).

7 J.-F. Blondel, *Discours sur la nécessité de l'étude de l'architecture*, p. 8.

8 Blondel succeeded Loriot, who had resigned a short time before. A new phenomenon was the simultaneous appointment of an assistant professor, David Leroy, who went on to succeed Blondel in 1774.

9 It is hard to decide just where the dividing line lay between Blondel's teaching at the Academy and at his Ecole des Arts. Numerous pupils tended anyway to attend his courses in both institutions.

10 The public architecture course at the Academy was generally held once a week at the Louvre. The staggering of this course across a three year period meant that new pupils could only be taken on at the end of this time.

11 It is worth noting in this respect that Mézières' Ecole du Génie was perhaps the

most 'modern' school of the eighteenth century, and that it was superior from a strictly scientific point of view to the Ecole des Ponts et Chaussées, for Abbé Bossut held the post of professor from 1752, while Abbé Nollet came in 1761, and Monge was appointed deputy to the latter shortly afterwards. Upon its foundation in 1794, the Ecole Centrale des Travaux Publics, the future Ecole Polytechnique, was to inherit some of the traditions of Mézières' school. Furthermore, many of Mézières' pupils played a part in the development of the sciences, following in the footsteps of Dubuat and, more particularly, Coulomb. On Mézières' School of Engineering, one should read Dardonville, *L'Ecole Royale du Génie de Mézières 1748–1948* (Nancy, 1948); R. Taton, 'L'Ecole du Génie de Mézières', in *Enseignement et diffusion des sciences en France au XVIIIe siècle*, pp. 559–615. See in particular the files kept at the Archive du Génie under the heading Art. 18, Sect. 1, §2.

In the eighteenth century, the corps of the Génie and that of the Ponts et Chaussées were often in competition with each other, especially on account of the desire of the engineers belonging to the latter to work on canals and on the planning of docks, which had traditionally been the fiefs of those in the Génie. Given this context, Perronet's school and that of Mézières gave the impression of being to some degree rival institutions. Mézières may have had the upper hand in scientific matters, but the reputation of the Ecole des Ponts et Chaussées rested upon the realism of a training which was orientated towards building and was undoubtedly more effective on the ground. For further information on the rivalry between the engineers of the Génie and those of the Ponts et Chaussées, and on the arguments which each used, consult J.-C.-E. Michaud D'Arcon, *De la force militaire considérée dans ses rapports conservateurs* (Strasburg, 1789), and *Réponse de la force civile à la force militaure ou observations sur le Mémoire du Colonel d'Arcon* (1790, ENPC manuscript, MS 3056).

12 The Etats des élèves, which are kept in manuscript form at the Centre Pédagogique de Documentation et de Communication of the ENPC, provide a fair amount of information about the operation of the former Ecole des Ponts et Chaussées, giving the names of the pupils present at the School, the nature of their activity, the competitions which were held during this period, together with a succinct description of the projects which each of them submitted. In the meantime, one should read F. de Dartein, 'Notice sur le régime de l'ancienne Ecole des Ponts et Chaussées', pp. 16–17, 126–43.

13 Ibid., p. 35. This form of academic training dealt with real issues. Through the assiduous practice of drawing and of architecture, the future engineers were also preparing themselves for the exercise of a career as constructor, supplementing their meagre pay with the fees they earned from works performed outside their own service. The engineers of the Ponts et Chaussées thus conceived public equipments, barracks or presbyteries, competing with architects by adopting a

cut-price policy. Whereas the latter ordinarily earned from one to two *sous* a pound from the amount paid for projects situated outside their own area, engineers were generally satisfied with half this rate. On this subject, one should read *Mémoire sur les honoraires qui sont dûs aux ingénieurs des Ponts et Chaussées pour les travaux étrangers à leurs services* (1780, ENPC manuscript, MS 2636).

14 *Etat des grades et des talents des élèves de l'Ecole des Ponts et Chaussées et de l'emploi de leur temps pendant le 2e quartier de l'année 1780*, reproduced in F. de Dartein, 'Notice sur le régime de l'ancienne Ecole des Ponts et Chaussées', pp. 130–1.

15 G. Riche de Prony, *Réflexions sur l'organisation d'une académie qui auroit pour objet la perfection et l'enseignement de la construction* (ENPC manuscript, MS 1056). This manuscript is not dated, although Prony refers to it in his *Nouvelle architecture hydraulique*, which appeared in 1790. The founding of the Polytechnique in 1794 constitutes another terminal date.

16 See below.

17 The *Discours sur la nécessité de l'architecture* dates from 1754. Blondel's conceptions did not alter much in later years, and the *Cours d'architecture* was to adopt pedagogic positions very close to those set out in the *Discours*. One should also note that the criticisms which Blondel's pupils levelled at him did not strictly concern his teaching methods, but rather his attachment to architectural forms which they saw as being out-of-date.

18 J.-F. Blondel, *Cours d'architecture*, vol. 3, pp. xiiiff.

19 J.-F. Blondel, *Discours*, Preface.

20 Ibid., pp. 10, 11

21 Ibid., pp. 13–15.

22 Ibid., pp. 14, 15.

23 G. Riche de Prony, pupil's dissertation, the style competition for 1779.

24 J.-F. Blondel, *Discours*, p. 17.

25 Let us note for the moment that we are faced with one of the most fundamental characteristics of the engineers' system. The direct but always limited effects of architectural knowledge have their counterpart in the indirect but universal effects of mathematics. The content of technical knowledge falls behind the secrets of the trade, whereas its consequences were widely broadcast. A complete interplay of analogies could now be established between the indirect effects of this knowledge and the 'indirect forces' which served to guarantee the stability of the large-scale engineering works, such as Perronet's bridges. In the case of construction also, mind was held to work matter in occult ways, only the results of which were perceptible to the uninitiated. See chapter 7.

26 G. Riche de Prony, *Réflexions*.

27 On the *idéologues*, the basic work is still F. Picavet, *Les idéologues* (Paris, 1891).

28 Prony's *Réflexions*, in particular, would seem to have been inspired by the thought of Destutt de Tracy, who was drafting his *Elémens d'idéologie* during this

same period, a work which was published for the first time in 1801, and was then republished many times.

29 'All of our knowledge', Condillac had written in his *Traité des sensations* (1754), 'derives from our senses', and Destutt de Tracy had followed him by asserting that 'thinking is nothing else but feeling'. The latter had acknowledged the virtues of Condillac's method, but had however rebuked him for 'all these parasitical operations' which he admits into his thought. A. L. C. Destutt de Tracy, *Elémens d'idéologie. Première partie. Idéologie proprement dite* (1801; Paris, 1817), pp. 226–7.

30 G. Riche de Prony, *Réflexions*.

31 Diderot, 'Système figuré des connaissances humaines', *Prospectus de l'Encyclopédie* (1750), republished in *Œuvres complètes de Diderot* (Paris, 1875–7), vol. 13, pp. 145ff.

32 G. Riche de Prony, *Réflexions*.

33 G. Riche de Prony, *Réflexions*.

34 'Two kinds of art may be distinguished, one of which is known as liberal, the other as mechanical. The first requires the theory of the sciences which are related to it; the other merely seems to demand experiment and practice. So it is that a man who exercises a liberal art is called an artist, whereas one who professes the mechanical arts is known as an artisan. All of the liberal arts are known under the name of the fine arts, among which are numbered architecture, sculpture, painting, engraving, etc. Masonry, carpentry, joinery and locksmithery, etc. are mechanical arts' (*Discours*, p. 17).

35 Ibid., p. 44.

36 Ibid., p. 51.

37 Ibid., p. 55.

38 Blondel in fact taught theory at the Ecole des Arts; he arranged visits to famous buildings, together with 'specific lectures on practice, in the various workshops of the buildings which are being constructed' (Ibid., p. 11).

39 If it seems hard to be certain about this latter point, the *Réflexions* do nevertheless allow some understanding of the context in which the founding of the Polytechnique was decided upon, together with the intentions behind its founding. As Prony saw it, construction was a universal discipline which legitimised the intervention of the engineer in virtually every area of social life, since it answered to the whole of man's needs.

40 According to Prony, the teaching of architecture 'ought not to be without some practical use', for it enables one 'to combine pleasing forms' and to impart 'the imprint of good taste' to buildings. Architecture thus represented an art of embellishment rather than a necessity (*Réflexions*). In his *Fragments sur la théorie des beaux-arts* (1783), Prony had in fact criticised the classical theory of imitation: 'If one were to seek for models for the Orders of architecture in nature, one would simply produce bizarre assemblies', since they derive their beauty solely 'from the

immediate perception of the proportions of which they are composed'. Prony went on to distinguish between those arts which please immediately, those which please through habit, those which derive their power from the imagination, and, finally, those which proceed from the spirit of comparison. Architecture, the engineer reckoned, belonged to the first category, much like rhythm in music or colour in painting. Thus, architecture was already characterised by its gratuitousness (G. Riche de Prony, *Fragments sur la théorie des beaux-arts, leur influence*).

41 G. Riche de Prony, *Réflexions.*

42 One should note the affinity between this way of breaking construction down into juxtaposed sciences, which was followed by a reorganisation of an operational nature, and Destutt de Tracy's *idéologie*. *Idéologie* in fact proposed to discover within the complexity of thought the simple elements of which it was composed, then to proceed in the opposite direction, 'from the simplest perception, from sensation that was pure and devoid of all judgement, to the most abstract ideas'. The engineer and the philosopher were agreed that there is no real knowledge except in a process which alternated analysis and synthesis, rational decompositions and recompositions. A.-L.-C. Destutt de Tracy, *Elémens d'idéologie. Seconde partie. Grammaire* (1803; Paris, 1817), pp. 17, 18.

43 J.-F. Blondel, *Discours*, p. 11.

44 Ibid., pp. 58ff.

45 Ibid., pp. 83ff.

46 G. Riche de Prony, *Réflexions.*

47 Ibid.

48 There should be no need to emphasise here the continuity between Enlightenment methods of academic teaching and those of the Ecole des Beaux-Arts, a continuity for which the enduring nature of the prix de Rome constitutes the most tangible emblem. See R. Chafee, 'The teaching of architecture at the Ecole des Beaux-Arts', in A. Drexler (ed.), *The architecture of the Ecole des Beaux-Arts* (London, 1977), pp. 61–109.

49 G. Riche de Prony, *Réflexions.*

50 Ibid.

51 J.-F. Blondel, *Discours*, pp. 71, 72.

52 Ibid., p. 72.

53 B. Forest de Bélidor, *La science des ingénieurs* (Paris, 1729), Book 1, pp. 2, 3.

54 G. Riche de Prony, pupil's dissertation, style competition for 1779.

55 Aubry, *Mémoire sur différentes questions de la science des constructions publiques et économiques* (Lyons, 1790), p. 1.

56 See chapter 7.

57 Lagrange's *Mécanique analytique* was first published in 1788. Lagrange achieved a synthesis between Newton's works and eighteenth-century ideas regarding analysis. R. Dugas, *A history of mechanics* (1950; Neuchatel, 1955), pp. 332ff.

58 Aubry, *Mémoire* , p. ii.

59 J.-F. Blondel, *Cours d'architecture*, vol. 3, pp. xxxii, xxxiii.

60 C.-J. Panckoucke, *Trois lettres à de Prony relatives à l'impression du dictionnaire des ponts et chaussées* (1788, ENPC manuscript, MS 1013); G. Riche de Prony, *Correspondance relative au dictionnaire des ponts et chaussées* (1788, ENPC manuscript, MS 1015).

61 G. Riche de Prony, *Discours préliminaire de la partie de l'encyclopédie par ordres de matières, intitulée: Science de l'ingénieur des Ponts et Chaussées, turcies et levées, canaux et ports maritimes*, p. 7.

62 Le Corbusier, *Towards a new architecture*, trans. F. Etchells (London and New York, 1959).

63 The law of 30 Vendémiaire Year IV (22 October 1795) in fact decreed the reconstitution of the Ecole des Ponts et Chaussées, which had for a time been subsumed within the Ecole Polytechnique, as a school of instruction for the latter (F. de Dartein, 'Notice sur le régime de l'ancienne Ecole des Ponts et Chaussées', pp. 84ff.

64 Between Perronet, who died in 1794, and Prony, Lamblardie and Chézy held the post of director. But the revolutionary troubles prevented them completing a reform whose need they had been only too aware of, Lamblardie even going so far as to draft a plan for a set of regulations prefiguring that of Prony. J.-E. Lamblardie, *Mémoire ou note historique sur l'Ecole des Ponts et Chaussées, projet de règlement*, 2 Brumaire Year VI (ENPC manuscript, MS 2629 bis).

65 See chapter 10.

66 The chair in mechanics was entrusted to Eisenmann, and that in stereotomy to Bruyère, while Mandar was responsible for the course in construction. Within this educational framework, architecture was subsumed by construction, although it represented little more than a third of it. With the promulgation of the decree of 7 Fructidor Year XII (25 August 1804), regarding the organisation of the corps of the Ecole des Ponts et Chaussées, a number of modifications were introduced. The stereotomy course, in particular, was grouped with the teaching of the construction of roads, bridges, canals and locks, while Mandar no longer taught architecture, since it had again become a discipline on its own.

67 These tables serve to illustrate the *plan d'instruction* which Prony had drafted and which had been approved by the Ministry of the Interior, upon which the Ecole des Ponts et Chaussées depended at this period. G. Riche de Prony, *Plan d'instruction des élèves de l'école nationale des Ponts et Chaussées pour l'an 7* (Paris, Ventose, Year VII).

68 G. Riche de Prony, 'Tableau synoptique du cours de la mécanique appliquée', in *Plan d'instruction*.

69 Ibid.

70 On the scope and implications of this analytic method, see chapter 10.

71 Practice of design and the periods of probation continued, nevertheless, to be fundamental, but the existence of masters' courses brought about a fundamental change in the operation of the School. The spirit of emulation, which had prevailed when Perronet had been director, disappeared. The pupils, who were now recruited the moment they had left the Polytechnique, were guaranteed a place in the corps of the Ponts et Chaussées as soon as their student days were over. Prony thus estimated that, from 1750 to 1790, 704 pupils had passed through the Ecole des Ponts et Chaussées, while only 320, or slightly less than half, had subsequently joined the corps of the Ponts et Chaussées. By contrast, in the first half of the nineteenth century, virtually all the pupils entered the service. In such a context, selection through competition, and therefore by the practice of design, lost all of its importance. G. Riche de Prony, *Note sur l'Ecole Royale des Ponts et Chaussées* (1828, ENPC manuscript, MS 2629 bis).

72 The teaching of architecture, in particular, returned to a more classical formulation when, from 1804 on, Mandar was given special responsibility for it. His 'Tableau synoptique d'un cours d'architecture civile', which was drafted for the Council of the Ecole des Ponts et Chaussées, revived the framework first developed by Prony, but it placed most emphasis upon programmes of types, upon the parts of the building and upon the corps of the trades. *Registre des Procès-Verbaux des séances du Conseil de l'Ecole des Ponts et Chaussées commencé en vendémiaire an 13* (ENPC manuscript, session of 9 Brumaire XIII).

73 One should also bear in mind that it was at the end of the eighteenth century that a technology plan, that is a 'plan for a scientific treatment, having as its object technological operations', was introduced. Prony's preoccupations should be seen in the context of such developments. See J. Guillerme and J. Sebestik, 'Les commencements de la technologie', in *Thales* (Paris), 12, 1968, 1–72.

7 SOLIDITY AND CONSTRUCTION

1 P. Patte, *Mémoires sur les objets les plus importants de l'architecture*, p. iii.

2 On the Abbé Laugier's theories, one should consult W. Herrmann, *Laugier and eighteenth-century French theory* (London, 1962).

3 See H. Klimrath, *Etudes sur les coutumes* (Paris, 1837).

4 J. Glissen, *Introduction historique au droit* (Brussels, 1977–8).

5 Introduced in 1660, in order to counterbalance the power of the guilds, the *architectes experts bourgeois*, who were twenty-five in number, were responsible for presenting expert opinion in the law-courts. A. Desgodets, *Les loix des bâtimens, suivant la coutume de Paris*, with notes by M. Goupy (1748; Paris, 1766), pp. 28, 29.

6 N. Le Camus de Mézières, *Le guide de ceux qui veulent bâtir* (1781; paris, 1786), p. 23.

7 See E. Martin Saint-Léon, *Histoire des corporations de métiers*.

8 Ibid.,pp. 295ff.

9 Enlightenment architectural treatises contained far more upon the ways and customs of the building trade than upon proposals for reorganising production. Patte, in his *Mémoires sur les objets les plus importants de l'architecture*, even recommended that a young architect should listen to skilled workmen, who, having only ever studied one object, 'will sometimes notice things which had escaped you' (*Mémoires*, pp. 106, 107).

10 P. Patte, *Mémoires*, pp. 269ff. The analogy stems from the arrangement of discontinued supports which transmit the horizontal thrusts to the extremities of the building, as was also the case with bridges, where the piers played approximately the same role as columns. On the crucial problem of lintels, one should also read P. Saddy, 'Ein "Machwerk" der Moderne: der armierte Sturz', in *Daidalos* (Berlin), 8, 1983, 54–65.

11 When Soufflot resumed work on the colonnade around 1755, the rust in the tie-beams placed in the entablature had actually caused certain voussoirs to shatter.

12 P. Patte, *Mémoires*, pp. 342ff.

13 Ibid., pp. 192ff.

14 Gabriel's two buildings also feature colonnades that were buttressed by lateral pavilions and reinforced by iron tie-beams, as at the Louvre. P. Patte, *Mémoires*, pp. 278ff. One should also read Y. Bottineau and M. Gallet, *Les Gabriel* (Paris, 1982), p. 276.

15 See M.-K. Deming, *La halle au blé de Paris 1762–1813* (Brussels, 1984), pp. 83–90.

16 On Perronet's proposals, one should read M. Petzet, 'Un projet de Perrault pour l'église Sainte-Geneviève à Paris', in *Bulletin Monumental* (Paris), 115, 1958, 81–96.

17 Numerous publications have been devoted to this problem, more particularly F. Hamon, 'Les églises parisiennes du XVIIIe siècle: théorie et pratique de l'architecture culturelle', in *Revue de l'Art* (Paris), 32, 1976, 7–14; W. Herrmann, *Laugier and eighteenth-century French theory*; R. Middleton, 'The abbé de Cordemoy and the graeco-gothic ideal', in *Journal of the Warburg and Courtauld Institutes* (London), 25, 1962, 278–320 and 26, 1963, 90–123; M. Petzet, *Soufflots Sainte-Geneviève und der französische Kirchenbau des 18. Jahrhunderts*; 'Soufflot et l'architecture de Sainte-Geneviève', in *Cahiers de la recherche architecturale*, proceedings of the conference 'Soufflot et l'architecture des Lumières', pp. 12–15.

18 As we shall see below, eighteenth-century bridges had much narrower piers than structures of the Renaissance or of the classical age had done. This lighter support required more stable foundations than had previously been the case. One had, in particular, to link more effectively the piers upon which the masonry supports rested.

19 J.-M. Pérouse de Montclos, *L'architecture à la française*, p. 178.

20 See P. Saddy, 'Ein "Machwerk" der Moderne'; B. Lemoine, *Construire en fer, éléments d'une connaissance historique et technique des bâtiments en fer et fonte du XIXe siècle* (SRA research report, Paris, 1982), p. 9.

21 This fineness of the jointings was also one of the causes of the difficulties at the Panthéon, where almost contiguous stones broke upon contact with each other. On this topic, one should consult J. Heyman, 'The crossing piers of the French Panthéon', in *The Structural Engineer* (London), 64A, no. 8, August 1985, 230–4.

22 Read A. Guillerme, *De la chaux au ciment. La révolution industrielle dans le génie civil fin XVIIIe–milieu XIXe siècle* (Paris, 1985), pp. 4ff.

23 M.-A. Laugier, *Essai sur l'architecture*, p. 127.

24 P. Patte, *Mémoires*, p. 99.

25 See chapter 3.

26 M. Guillerme, 'Des influences subis et exercées par Condillac en matière de théorie de la connaissance', in *Sciences et Techniques en Perspective* (Nantes), 2, 1983, 6–21, especially p. 10.

27 Chapter 6, n. 42.

28 S.-P. Timoshenko, *History of the strength of materials*, pp. 41ff. For a balance sheet of the experiments in the resistance of materials during the Enlightenment, one might also consult the great treatises on construction from the beginning of the nineteenth century, such as Rondelet's *L'art de bâtir*, Gauthey's *Œuvres*, which were published by Navier, and Sganzin's *Cours de construction*, etc.

29 E.-M. Gauthey, *Œuvres de M. Gauthey* (Paris, 1806–9), vol. 2, p. 149.

30 Ibid., vol. 1, pp. 267–77.

31 It was not until the publication of Lagrange's *Mécanique analytique*, in 1788, that this framework really began to become apparent among engineers. Numerous problems would remain unresolved, for example the effects of friction, not to speak of the phenomena of elasticity.

32 See J. Kérisel, 'Histoire de la mécanique des sols en France jusqu'au 20e siècle', in *Annales des Ponts et Chaussées* (Paris), July–August 1958, 503–31 and of course C. Stewart Gillmore, *Coulomb and the evolution of physics and engineering in eighteenth-century France* (Princeton, 1971), pp. 100–7.

33 See G. Mouret, 'Antoine Chézy: histoire d'une formule d'hydraulique', in *Annales des Ponts et Chaussées* (Paris), March–April 1921, 165–269.

34 One should read Y. Chicoteau, A. Picon and C. Rochant, 'Gaspard Riche de Prony ou le génie "appliqué"'. See also chapter 10.

35 One will gain a better sense of the scale of the problems of planinng if one compares the four million or so *livres* estimated for the construction of Perronet's pont Louis XVI with the 70,000 *livres* which, according to the estimate given by Le Camus de Mézières in his *Guide de ceux qui veulent bâtir*, a town house would cost.

36 B. Lepetit, *Chemins de terre et voies d'eau*, pp. 18ff.

37 André Brunot and Roger Coquand have estimated that a network of around

30,000 km of road was actually built under the *Ancien Régime*. A. Brunot and R. Coquand, *Le corps des Ponts et Chaussées*, p. 39.

38 P.-M.-J. Trésaguet, *Mémoire sur la construction et l'entretien des chemins faits en rachat des corvées dans la généralité de Limoges depuis 1764* (1775, ENPC manuscript, MS 1886). This report was reproduced in the *Annales des Ponts et Chaussées* (Paris), second term, 1831, pp. 243–8. On Trésaguet's work, one should also read M. Yvon, 'Pierre-Marie-Jérôme Trésaguet, ingénieur des Ponts et Chaussées (1716–1796); créateur d'une méthode rationelle pour construire et entretenir les chaussées empierrées au XVIIIe siècle', in *Les routes du sud de la France de l'antiquité à l'époque contemporaine* (Paris, 1985), pp. 295–318.

39 C.-L. Navier, *Considérations sur les principes de la police du roulage et sur les travaux d'entretien des routes* (Paris, 1835); J.-B. Schwilgué, *Mémoire sur les routes et le roulage* (Paris, 1832); G.-G. de Coriolis, *Des circonstances qui influent sur le tirage des chevaux et sur la conservation des voitures* (Paris, n.d.). On advances in the science of road building at the beginning of the nineteenth century, see A. Picon, 'L'apprentissage de la route à l'Ecole des Ponts et Chaussées dans la première moitié du XIXe siècle', in *Les routes du sud de la France de l'Antiquité à l'époque contemporaine*, pp. 319–29.

40 P. Orry, *Mémoire instructif sur la réparation des chemins du 13 juin 1738*, reproduced in Vignon, *Etudes historiques*, vol. 3, pp. 3–7. This report, which was inspired by a concern for rationalisation which was none too realistic for the period, also proposed a standardised bridge model which remained a dead letter. The standardisation of the road network may have begun during the Enlightenment, but there was virtually no trace of a similar development in bridge building.

41 See chapter 9.

42 The *corvée*, which was introduced in 1738, was without a doubt the major social problem with which the administration of the Ponts et Chaussées had to deal. Trudaine was personally opposed to it, and wished to be able to give cash payment for all works carried out under his direction. See J.-R. Perronet, *Journal de l'Assemblée des Ponts et Chaussées depuis 1747* (ENPC manuscript, MS 266, session of 17 March 1754). In 1776, Turgot was to try to suppress it, but it was only finally abolished in 1787. Many texts by engineers refer to the problems caused by the *corvée*, and recommend taking steps to replace it with a system of taxation. See, for example, *Arrêts et lettres concernant la corvée* (ENPC manuscript, MS 1922); *Mémoire et observations sur différents moyens de suppléer à la corvée* (ENPC manuscript, MS 2156); Coluel, *Examen de mémoires sur l'amélioration des corvées pour soulager le peuple* (1778, ENPC manuscript, MS 2483).

43 Y. Chicoteau, A. Picon and C. Rochant, *Les concours techniques du XVIIIe siècle à l'Ecole des Ponts et Chaussées*, p. 52.

44 On technical drawing, the standard work of reference is still Y. Desforges, *Le graphisme technique: son histoire et son enseignement* (Seyssel, 1981).

45 B. Forest de Bélidor, *La science des ingénieurs*, book 6, p. 2.

46 A. Dupire, B. Hamburger, J.-C. Paul, J.-M. Savignat and A. Thiébaut, *Deux essais sur la construction*, p. 23.

47 Bélidor was quite explicit on this point. The estimate was the 'engineer's masterpiece', and was supposed to supply 'very precise details' of everything involved in the construction, even going so far as to anticipate any accidents which were liable to happen, so as to be able to supervise at every moment the contractors and the workmen. It was on the estimate that 'the good or bad execution of the design which one [had] in mind' would in the last analysis depend. B. Forest de Bélidor, *La science des ingénieurs*, book 6, p. 2.

48 Whether it was a matter of phased drawings or of estimates, it was always a question of representing 'the order and conduct of the world', of which the edifice was merely one consequence, the most important one, no doubt, but a consequence nevertheless.

49 On the imaginary technology of water in the Enlightenment, see A. Guillerme, *Les temps de l'eau. La cité, l'eau et les techniques* (Seyssel, 1983), pp. 188f.

50 G. Riche de Prony, *Mémoires*, pp. 102-3.

51 J.-R. Perronet, *Description des projets et de la construction des ponts de Neuilly, de Mantes, d'Orléans, etc.; du projet du canal de Bourgogne . . . et de celui de la conduite des eaux de l'Yvette et de Bièvre à Paris* (1782–3; Paris, 1788), p. 178.

52 The construction of Blois bridge was staggered between 1716 and 1724, that of the Moulins bridge between 1750 and 1763, and that of the Saumur bridge between 1736 and 1770. See C. Rochant, *Les ponts modèles du XVIII siècle* (Versailles, 1983).

53 See M. Champion, *Les inondations en France depuis le VIe siècle jusqu'à nos jours* (Paris, 1858–9).

54 Linking the foundation piers was therefore an absolute necessity for the engineers.

55 This development should itself be related to the gradual shift, in the works of Clairaut, d'Alembert, Euler, Bossut and Dubuat, from hydrostatics to hydrodynamics. Consult S. Ince and H. Rouse, *History of hydraulics* (Iowa Institute of Hydraulic Research, 1957), pp. 99-111.

56 J.-B. de Voglie, *Mémoire pour servir à la fondation d'une des piles du pont de Saumur, sur la rivière de Loire, sans batardeaux ni épuisements* (ENPC manuscript, 1757, MS 2242); L-.A. de Cessart, *Mémoire descriptif de la scie employée aux fondations du pont de Saumur* (ENPC manuscript, 1764, MS C 2534). There is a detailed description of these two projects in L.-A. de Cessart, *Description des travaux hydrauliques de Louis-Alexandre de Cessart* (Paris, 1806–8).

57 L. de Regemortes, *Description du nouveau pont de pierre construit sur la rivière d'Allier à Moulins* (Paris, 1771).

58 In the session of 1 March 1774 of the Ponts et Chaussées Assembly, Perronet is

supposed to have declared that Moulins, Orléans, Blois, Tour, Saumur and Westminster bridges would all collapse totally if one of their piers were overturned. Quoted by F. de Dartein, 'La vie et les travaux de Jean-Rodolphe Perronet', p. 19.

59 This affinity is apparent in the evident fascination which the Enlightenment engineers felt for the Chinese bridge of Loyang, in the province of Fokien, where upright monoliths, in the form of piers, supported other monoliths laid horizontally, composing the roadway. Perronet quoted this example in support of his own researches. As there were no stones in the environs of Paris that were either sufficiently long or sufficiently resistant to lend themselves to this kind of construction, it was thought appropriate to replace them with club-shaped arches instead. This then raised a problem similar to that posed by the great entablatures of the Louvre or of place Louis XV. For a description of Loyang bridge, see E.-M. Gauthey, *Œuvres*, vol. 1, pp. 39, 40 and plate III, figures 52, 53.

60 In his *Mémoire sur la reduction de l'épaisseur des piles, et sur la courbure qu'il convient de donner aux voûtes le tout pour que l'eau puisse passer plus librement sous les ponts*, which was read to the Royal Academy of Sciences on 12 November 1777, Perronet even went so far as to compare the working of modern bridges with the structure of Gothic churches which 'would necessarily collapse if just one of the small pillars, or one of the columns were to fall' (J.-R. Perronet, *Mémoire*, p. 623).

61 On Perronet's work, see F. de Dartein, 'Notice sur le régime de l'ancienne Ecole des Ponts et Chaussées', pp. 115–17.

62 G. Simondon, *Du mode d'existence des objets techniques* (Paris, 1969), pp. 19–49.

63 Read F. de Dartein, 'Le pont de la Concorde sur la Seine à Paris'.

64 The pier-heads of the old bridges had generally had elongated outlines shaped in the form of a ship's prow. Their form evolved in the course of the eighteenth century, and became more rounded. However, this arrangement too was designed to hinder the flow of water as little as possible.

65 On 6 March 1787, the Assembly of the Ponts et Chaussées was in fact to decide that he should modify his design, in particular by eliminating the spaces between the columns marked out for the piers. See F. de Dartein, 'Le pont de la Concorde', pp. 115–17.

66 It was to this same origin that Boullée and Ledoux had wished to return when they gave their bridge piers the form of boats, making explicit reference to the model of the pontoon bridge. See chapter 10.

67 In the case of Perronet, this notion of process was in fact derived from the study of manufacture. His interest in industrial production had been stirred very early, as we know from his studies of the making of pins at Laigle in Normandy, which he had carried out early on in his career. Reproduced in part in the *Encyclopédie méthodique*, these same studies must have influenced the opening chapter of Adam

Smith's *The wealth of nations*, which is concerned with the division of labour. The interest of manufacture as far as Perronet was concerned lay precisely in the standardisation of procedures, in the definition of tasks, and in a division of labour which was reconstituted through the process of manufacture. His own workshops were later to be modelled upon those of the factories. See J.-R. Perronet, *Description de la façon dont on fait les épingles à Laigle en Normandie* (ENPC manuscript, 1740, MS 2385); *Explication de la façon dont on réduit le fil de laiton à différentes grosseurs dans la ville de Laigle en Normandie* (ENPC manuscript, 1739, MS 2383).

During the Enlightenment, many in the Ponts et Chaussées were drawn to the kinds of thinking that were current among manufacturers. The Trudaines, for example, frequently commented upon questions having to do with the regulation and development of factories. Their concepts, together with the personal experience of Perronet, very probably influenced engineers. See *Trois mémoires relatifs à l'amélioration des manufactures de France sous l'administration des Trudaine*, which were published by P. Boissonade, having been selected from the *Revue d'Histoire Economique et Sociale* (Paris), 7th year, 1914, no. 1. The model of the factory was of especial significance for the late eighteenth-century designs for arsenals, which were in effect machine-towns, developed in order to rationalise the tasks involved in naval construction. See chapter 8, n. 64.

68 These elastic centres, which allowed the arches to begin to settle before their discentring, were supported by piers instead of resting upon intermediate supports in the river. One was supposed to remove all of them at once, so as not to unbalance the arches. There is a description of these centres in J.-R. Perronet, *Mémoire sur le cintrement et le décintrement des ponts, et sur les différents mouvements que prennent les voûtes pendant leur construction* (Paris, 1777). On the spectacular simultaneous striking of all the centring of the arches, one should read L. Petit, 'Le décintrement du pont de Neuilly'. This system necessitated very delicate handling, and was subsequently severely criticised, by Gauthey and Lamandé in particular. For the latter's criticisms, see F. Lamande, *Mémoire de M. Lamandé sur le système de centres à adopter pour le pont d'Iéna* (ENPC manuscript, 1809, MS 2799).

69 Fer de la Nouerre, who subsequently wrote *La science des canaux navigables*, had submitted a report to the Academy of Sciences in which he had cast doubt upon the solidity of the pont de Neuilly, basing his argument on the case of the pont de la Mulatière near to Lyons, which had just collapsed.

70 G. Riche de Prony, *Application de la mécanique des voûtes à la construction du pont de Neuilly* (ENPC manuscript, MS 2215); *Calcul sur la stabilité des voûtes du pont Louis XVI* (ENPC manuscript, MS 2124).

71 Prony was anyway not so much concerned with the problem of the stability of the arches as with the equilibrium of the horizontal thrusts. This was a clever

switch in the question, since La Hire's theory did not enable one to make a decisive contribution to the demonstration of the solidity of Perronet's bridges.

72 The only genuinely original contributions to the theory of arches were works by Couplet and by Coulomb, in particular the latter's essay 'sur l'application des règles de maximis et minimis à quelques problèmes de statique relatifs à l'architecture'. This essay, which was published in the *Mémoires* of the Academy of Sciences for 1773, made hardly any stir at all, and La Hire's wedge theory continued to be the standard reference throughout the eighteenth century. See J.-M. Delbecq,'Analyse de la stabilité des voûtes en maçonnerie de Charles-Augustin Coulomb à nos jours', in *Annales des Ponts et Chaussées* (Paris), special 150th anniversary issue, 1981, 36–43.

73 It was Couplet, in two reports published in 1729–30, who was the first to introduce 'the friction and the linking of the voussoirs, which stopped them from slipping against each other'. See J.-M. Delbecq, 'Analyse de la stabilité des voûtes', p. 37.

74 Perronet referred to this same theory when compiling his tables for the dimensioning of bridges.

75 This was one of the first applications of analysis to the calculation of bridges. Prony's report, which was read to the Academy of Sciences, earned him the congratulations of Monge, who had been impressed by the clarity of the exposition. Prony in fact pioneered this notion of an aesthetics of engineering rationality, which was subsequently to assume such importance.

76 J.-R. Perronet, *Mémoire sur la recherche des moyens que l'on pourroit employer pour construire des grandes arches de pierre de deux cents, trois cents, quatre cents & jusqu'à cinq cents pieds d'ouverture, qui seroient destinées à franchir de profondes vallées bordées de rochers escarpées* (Paris, 1793).

77 J.-M. Pérouse de Montclos, *L'architecture à la française*, pp. 115–16.

78 To convince oneself of this, one has merely to consider the limited success of the system invented by Perronet in England, a country which was however technologically more advanced at the time. On the English bridges of this period, see T. Ruddock, *Arch bridges and their builders 1735–1835* (Cambridge, 1979).

79 Frézier, who had been an acknowledged authority on stereotomy and construction since the time of the publication of his *La théorie et la pratique de la coupe des pierres et des bois* (1737–9), had in fact launched himself in July 1754 into a polemic with Laugier, as one may judge from his 'Remarques sur quelques livres nouveaux concernant la beauté et le bon goût de l'architecture', *Mercure de France* (Paris), July 1754, 7–59. The engineer took particular pains to emphasise the difficulty of building projections above columns at the angles of the entablature, as at the crossing of the transepts of a church (p. 31).

80 P. Patte, *Monumens érigés en France à la gloire de Louis XV* (Paris, 1765), p. 7.

81 It is one of the ironies of fate that it should have been Soufflot who recommended Patte to Chambers.

82 P. Patte, *Mémoire sur la construction de la coupole projetée pour couronner la nouvelle église de Sainte-Geneviève à Paris* (Amsterdam, Paris, 1770, pp. 19, 20).

83 One of Frézier's letters to Patte was thus published in the *Mercure de France* of October 1770. Soufflot then wrote to Frézier, who instantly retracted, admitting that his knowledge of the project was very limited.

84 M. Mathieu, *Pierre Patte, sa vie, son œuvre*, pp. 181ff.

85 Patte's attitude may probably be explained in part by his taste for scandal and notoriety, which had been evident at the time of his dispute with the authors of the *Encyclopédie* also.

86 P. Patte, *Mémoire*, p. 36.

87 This was the complete title of the work.

88 P. Patte, *Mémoire*, p. 5.

89 In his writings, Patte frequently emphasised this point. This was especially the case with the treatise on construction which was appended to Blondel's *Cours d'architecture* (vol. 6, p. 2).

90 Perronet would seem to have been as reluctant as Frézier to meddle in a question which did not directly concern him. This reluctance was shared by the majority of the engineers, with the exception of Gauthey, who saw it as an opportunity to enter the architectural and structural debate at the front door.

91 See M. Petzet, *Soufflots Sainte-Geneviève und der französische Kirchenbau des 18. Jahrhunderts.*

92 J.-R. Perronet, letter to Soufflot, 26 January 1770, reproduced in M. Mathieu, *Pierre Patte*, pp. 400–1.

93 G. Riche de Prony, pupil's dissertation for the style competition for 1779. Prony placed particular emphasis upon the resemblance between Gothic architecture and 'the animal economy'.

94 P. Patte, *Mémoires*, pp. 221ff.

95 J.-F. Blondel, *Cours d'architecture*, vol. 6, pp. 186–90.

96 J.-R. Perronet, letter to Soufflot, 26 January 1770.

97 It is significant that Galileo was one of the first to use the comparison with the animal skeleton in order to illustrate the disproportion between a structure's resistance and its dimensions, in his *Discorsi e dimostrazioni matematiche intorno a due nuove scienze attenenti alla mecanica e i movimenti locali* (Leiden, 1638). See S.-P. Timoshenko, *History of the strength of materials*, pp. 13, 14.

A military engineer, on the other hand, explained the consequences of this disproportion between causes and effects, in a report with a particularly significant title, which could well have been a response to the arguments elaborated by Patte. See Pierron, *Réflexions sur la multiplicité des circonstances qui se rencontrent dans la construction des voûtes, qui démontrent l'impossibilité de donner par des règles constantes une proportion qui détermine dans tous les cas l'épaisseur des pieds-droits des voûtes relativement au poids et à la poussée. Je les communique en vue de désabuser ceux qui, moyennant d'augmenter un peu les épaisseurs que ces proportions établies donnent, ont confiance à ce que différents auteurs ont écrit et déterminé théoriquement sur cet objet* (Marseilles, 1777, a manuscript held in the Archives du Génie, Article 21, Section 7, § 6, 'Thrust and construction of arches'). This refusal 'to increase a little the widths' given by the proportions marks the point at which the classical rules of building really began to fall apart.

98 E.-M. Gauthey, *Mémoire sur l'application des principes de la méchanique à la construction des voûtes et dômes* (Dijon, Paris, 1771), pp. 4, 5.

99 E.-M. Gauthey, *Mémoire*, pp. 40ff.

100 Gauthey rebuked Patte in a somewhat similar fashion for having failed to take account of effects of scale in his reasoning, although this did not prevent him from later conceiving his church at Givry in Burgundy as a reduced model of Sainte-Geneviève, with the aim of demonstrating its solidity. See J. Guillerme, 'Brèves remarques à propos d'un monument expérimental', in *Revue de l'Art* (Paris), 58–9, 1982–3, 85–90.

101 See J. Guillerme, 'Soufflot, Sainte-Geneviève et les limites de l'invention technique', *Soufflot et son temps*, a catalogue published for the exhibition of the Caisse Nationale des Monuments Historiques et des Sites (Paris, 1980), pp. 158–66.

102 *Histoire générale des sciences*, vol. 2, pp. 450–60. See chapter 10 also.

103 I have of course borrowed the term 'paradigm', although transposed to a different context, from T.-S. Kuhn, *The structure of scientific revolutions* (Chicago, 1962).

104 See W. Szambien, *Reflet de la connaissance*, pp. 272–84.

105 See chapter 4.

106 See J.-M. Pérouse de Montclos, *L'architecture à la française*, pp. 219–24.

107 J.-F. Blondel, *Cours d'architecture*, vol. 4, p. 315.

108 See chapter 10.

109 E.-L. Boullée, *Essai sur l'art*, p. 49.

110 In fact, even if Boullée believed that 'The art of building was only a secondary art', his projects, even where they were immense, could always be built. See chapter 10.

111 See G. Teyssot, Introduction to E. Kaufmann, *Trois architectes révolutionnaires*, pp. 12–31.

112 *Séances des Ecoles Normales, recueillies par des sténographes et revues par les professeurs* (Paris, n.d.).

113 G. Gusdorf, *Dieu, la nature, l'homme au siècle des Lumières* (Paris, 1972), pp. 243–4.

114 I. Prigogine and I. Stengers, *La Nouvelle Alliance* (Paris, 1979), p. 94. On Diderot and his relation to the chemistry of his time, see also J.-C. Guédon, 'Chimie et matérialisme. La stratégie anti-newtonienne de Diderot', in *Dix-huitième siècle* (Paris), 11, 1978, 185–200.

115 See G. Pigeaud, 'La résistance des matériaux et l'elasticité au cours du dernier siècle', in *Annales des Ponts et Chaussées* (Paris), March–April 1931, 43–72. See chapter 10 also.

116 C.-F. Viel, *De l'impuissance des mathématiques pour assurer la solidité des bâtimens et recherches sur la construction des ponts* (Paris, 1805), p. 25.

117 B. Poyet, *Nouveau système de pont en bois et fer forgé . . . comparé avec les ponts ordinaires pour la solidité et l'économie* (Paris, n.d.). This was probably the first system of cable-stayed bridges invented in France.

118 *Nouveau système de construction de ponts en bois et fer forgé, par M. Poyet . . . Rapport du Conseil general des Ponts et Chaussées sur ce système; suivi de la réfutation de ce rapport par M. Poyet* (Paris, n.d.), p. 17. For more details regarding this curious affair, one should read B. Poyet, *Nouveau système de ponts en bois et fer forgé de son invention. Rapports, notes, etc.* (ENPC manuscript, MS 1414).

119 Read H. Lipstadt, *Architecte et ingénieur dans la presse, polémique, débat, conflit* (CORDA research report, Paris, 1980).

120 See chapter 9.

121 The advent of the engineers would seem to be linked to the state's preoccupation with large-scale investment, in war in the case of the seventeenth-century corps de Génie, and in planning in the case of the eighteenth-century engineers of the corps of the Ponts et Chaussées.

8 PIERRE PATTE AND THE CONCEPT OF THE RATIONAL TOWN

1 The *Cours d'architecture* marked a particular stage in this transition, for in it town houses featured alongside churches and palaces. Blondel's architectural theory thus sought to present a town that was unified under the aegis of *convenance* and through the agency of dimensional coordination brought about by the use of the orders and of their expression.

2 See, for example, F. Boudon, A. Chastel, H. Cousy and F. Hamon, *Système de l'architecture urbaine. Le quartier des Halles à Paris* (Paris, 1977).

3 See A. Desgodets, *Les loix des bâtimens.*

4 M.-A. Laugier, *Essai sur l'architecture*, p. 209.

5 'In examining carefully a large town, what strikes one first of all is the sight of refuse flowing down uncovered gutters before disappearing into sewers, giving off all kinds of unhealthy odours as it passes; then one notes the blood from the butchers streaming down the middle of the streets, and offering hideous and revolting spectacles at every step. On the one hand, one may see a quarter infested with the emptyings from the latrines; on the other hand, there may be a quantity of encrusted dung-carts plying the streets in order to collect refuse, which, aside from its filthy and disgusting appearance, occasions all kinds of obstacle. Further

on, one may observe, located in the midst of the most frequented places, the hospitals and the cemeteries, perpetuating epidemics and exhaling into houses the germ of illness and of death. Elsewhere you will notice that the rivers which flow through the towns, and which provide water for the inhabitants to drink, are continually used as a receptacle for all the cesspools and for all the refuse. Sometimes, owing to the narrowness of the streets or to the drawbacks in their positioning, citizens risk being trampled underfoot by horses or crushed by vehicles; finally, when it rains, you will see crowds of people drenched in unclean, salty water, issuing from the swill on the roofs which, because of their location, multiply the waters from the sky a hundredfold, or else spattered with mud from the horses' hooves, or from the turning of the vehicles' wheels in the gutters. In a word, the towns will strike you as being on every side the home of uncleanliness, infection and discomfort' (*Mémoires*, pp. 5, 6).

6 They were in fact linked to a literary tradition of denunciation of urban disorder, of which Juvenal and Boileau, with their famous descriptions of traffic jams, are the most notable examples. The classical town was anyway to a great extent supervised and regulated, as is proved by Nicolas de La Mare's impressive *Traité de la police*, which was published in instalments between 1705 and 1739. See, on this question, P. Saddy, 'Le cycle des immondices', in *Dix-Huitieme Siècle* (Paris), 9, 1977, 203–14, especially p. 212. In spite of their stereotyped nature, Mercier or Patte's descriptions betray a shift in sensibility, which was soon to result in a redefinition of what was clean and dirty, or healthy and unhealthy. On this development as regards smell and the perception of urban miasmas, see A. Corbin, *Le miasme et la jonquille. L'odorat et l'imaginaire social XVIIIe–XIXe siècles* (Paris, 1982).

7 Voltaire, *Le temple du goût* (Paris, 1733), and in addition, 'Des embellissements de Paris', *Œuvres complètes de Voltaire* (Kehl, 1784–9), vol. 29, pp. 163–74.

8 P. Monzani, 'Guides et plaisirs parisiens au XVIIIe siècle', in *VRBI* (Brussels), 9, 1984, lix-lxviii.

9 Read B. Fortier, *La politique de l'espace parisien (à la fin de l'Ancien Régime)* (CORDA reserch report, Paris, 1975).

10 J.-L. Harouel, 'L'alignement des rues', in *Dix-Huitième Siècle* (Paris), 9, 1977, 135–49.

11 On Mesmer and Mesmerism, read R.Darnton, *Mesmerism and the end of the Enlightenment in France* (Cambridge, Mass., 1968).

12 R. Etlin, 'L'air dans l'urbanisme des Lumières', in *Dix-Huitième Siècle* (Paris), 9, 1977, 123–34.

13 B. Fortier, *La politique de l'espace parisienne*, pp. 57, 58.

14 'It is a recognised truth that the most perfect water would soon alter were it not for the movement that sustained it, in its own purity, and which, putting it in a condition to impart freshness and resilience to the air, make it a health-giving

drink for both men and animals' (A.-A. Parmentier, *Dissertation sur la nature des eaux de la Seine* (Paris, 1787), p. 21).

15 Numerous details on the civic structures of the *Ancien Régime* may be found in A. Babeau, *La ville sous l'Ancien Régime* (Paris, 1880). Read also J.-C. Perrot, *Genèse d'une ville moderne, Caen au XVIIIe siècle* (Paris, 1975).

16 See P. Lavedan, *L'urbanisme à l'époque moderne XVIe–XVIIIe siècles* (Paris, 1982).

17 M. Mosser and D. Rabreau, 'Paris en 1778: l'architecture en question', in *Dix-Huitième Siècle* (Paris), 11, 1979, 141–64.

18 Since Lavedan's study, numerous works have been devoted to the problems raised by improvement. See, for example, J.-M. Dudot, B. Flouzat, M. Malcotti and D. Remy, *Le devoir d'embellir. Essai sur la politique d'embellissement à la fin de l'Ancien Régime* (CORDA report, Nancy, 1976).

19 See E. Leroy Ladurie, 'Un urbanisme frôleur', in *Histoire de la France urbaine* (Paris, 1980–5), vol. 3, pp. 439ff.

20 M.-A. Laugier, *Essai sur l'architecture*, p. 209.

21 Ibid., p. 210.

22 Ibid., p. 222.

23 'Anyone who knows how to design a park will have no trouble in tracing the plan according to which a town should be built, in conformity with its size and location. One needs squares, crossroads and streets. One needs regularity and oddity, relations and oppositions, accidents which lend variety to the picture, a high degree of order in the details, some confusion, din and tumult overall' (M.-A. Laugier, *Observations sur l'architecture* (The Hague, Paris, 1765), pp. 312–13).

24 The decoration of buildings was supposed to be narrowly subordinated to the urban effects which one set out to create. Laugier's proposed simplifications must be seen in relation to his desire to ordinance the town effectively.

25 M. Tafuri, *Projet et utopie* (1973; Paris, 1979).

26 J.-F. Blondel, *Cours d'architecture*, vol. 4, pp. 393ff.

27 Ibid., pp. 393, 394.

28 On Blondel as a town planner at Metz and Strasburg, read J. Garms, 'Le plan d'urbanisme de Strasbourg dressé par Jacques-François Blondel en 1764–1769'; B. Leclerc, *Le projet d'embellissement de J.-F. Blondel, Strasbourg (1764–1774)*; J. Lejeaux, *La place d'Armes de Metz*; A. Prost, *J.-F. Blondel, et son œuvre*; D. Remy, 'Urban beautification in 18th century France, Blondel in Metz and Strasburg'. I am especially indebted to Benedict Leclerc's exhaustive study.

29 J.-F. Blondel, *Cours d'architecture*, vol. 4, plate L.

30 On this topic, one should read L. Marin, *Le portrait du roi* (Paris, 1981).

31 See chapter 9.

32 The idea of creating a royal square, in which an equestrian statue of Louis XV would be erected, dated from 1748. Once the merchants' provost and the

municipal magistrates of the Ville de Paris had requested the king's authorisation, a competition was announced that same year, although the precise location of the square had yet to be fixed. Numerous architects presented their designs, but none was adopted, for fear of the costly expropriations involved. In 1753 another competition was held, which upon this occasion was restricted to Academicians. Their designs were rejected also, but Louis XV requested Gabriel to produce a synthesis of them. The latter, in its definitive version, was approved by the King on 9 December 1755. For the details of this operation, one should read Y. Bottineau and M. Gallet, *Les Gabriel*, pp. 254ff; B. de Montgolfier, 'La place Louis XV', in *Monuments Historiques* (Paris), 120, March–April 1982, 38–41; *De la place Louis XV à la place de la Concorde*, catalogue for the exhibition at the Musée Carnavalet (Paris, 1982).

33 P. Patte, *Monumens érigés en France à la gloire de Louis XV*, plate XXXIX.

34 Ibid., p. 221.

35 Ibid., p. 223. With his liking for formulae, Patte even went so far, in his attempts to convince his readers, as to expound the theory of compound interest.

36 Ibid., pp. 225, 226.

37 B. Fortier, *La politique de l'espace parisien*, p. 48.

38 On the problems created by the changes in scale of urban layouts, see the analysis of the goose foot at Versailles, an expansion of the tripartite arrangement at the Piazza del Popolo which undermined its original meaning, in J. Castex, P. Celeste and P. Panerai, *Lecture d'une ville: Versailles* (Paris, 1979), pp. 7, 8.

39 P. Patte, *Mémoires*, p. 8.

40 Ibid., p. 10.

41 Ibid., p. 11.

42 Ibid., p. 11.

43 Ibid., p. 13.

44 Ibid., p. 53.

45 Ibid., pp. 19ff.

46 A. Deparcieux, *Mémoire lu à l'assemblée publique de l'Académie Royale des Sciences, le samedi 13 novembre 1762 par M. Deparcieux . . . sur la possibilité d'amener à Paris, à la même hauteur à laquelle y arrivent les eaux d'Arcueil, mille à douze cents pouces d'eau, belle et de bonne qualité, par un chemin facile et par un seul canal ou acqueduc* (Paris, 1763). Deparcieux's design was revived by Perronet, and after him by Fer de la Nouerre, who began work on it on the eve of the Revolution. It was then abandoned, and the Ourcq canal was dug instead. For further information on this topic, consult J. Bouchary, *L'eau à Paris à la fin du XVIIIe siècle. La Compagnie des Eaux de Paris et l'enterprise de l'Yvette* (Paris, 1946).

47 See J. Bouchary, *L'eau à Paris*; B. Fortier, 'La maîtrise de l'eau', in *Dix-Huitième Siècle* (Paris), 9, 1977, 193-201.

48 P. Patte, *Mémoires*, p. 58.

49 Such demands were by no means unusual during this period. The Abbé de Lubersac, for example, who was more specific than Patte in this respect, called for the building of two main sewers on either side of the Seine, and for the installation of a huge reservoir at the tip of the Ile de la Cité. C.-F. de Lubersac de Livron, *Discours sur les monumens publics de tous les âges et de tous les peuples connus* (Paris, 1775), 'Observations regarding the chief modern monuments of the town of Paris, and several projects for the decoration or for the public utility of this capital city', pp. lxv–lxvi. On the introduction of drainage networks, see G. Dupuy and G. Knaebel, *Assainir la ville hier et aujourd'hui* (Paris, 1985).

50 P. Patte, *Mémoires*, plate II.

51 Ibid., p. 230.

52 Ibid., pp. 71ff.

53 Patte's approach to matters of ordinance was very similar to that of Blondel. See, for example, his *Discours sur l'architecture* (Paris, 1754).

54 G. Teyssot, Introduction to E. Kaufmann, *Trois architectes révolutionnaires*, p. 29.

55 As is well known, the enlightened elites were much preoccupied with this question. From the 1760s onwards, a whole series of measures inspired by hygienic aims were adopted. In 1763, for example, a decree issued by the Parlement of Paris prohibited burial in churches. Between 1780 and 1782, the Parisian cemeteries of the Innocents, of the Chaussée d'Antin, of Saint-Sulpice and of the Ile Saint Louis were closed. The cemeteries were then shifted to the outskirts of the towns, while the architects outdid each other in imagining new funerary domains. See P. Ariès, *The hour of our death*, translated by H. Weaver (London, 1981), pp. 475ff; R. Etlin, 'Landscapes of eternity: funerary architecture and the cemetery, 1793–1881', in *Oppositions* (New York), 8, 1977, 14–31; O. and C. Hannaway, 'La fermeture du cimetière des Innocents', in *Dix-Huitième Siècle* (Paris), 9, 1977, 181–91. Patte was to return to the problem of cemeteries in his *Mémoires qui intéressent particulièrement Paris*, Year IX, with an essay on the 'transfer of cemeteries' outside of the town.

56 P. Patte, *Mémoires*, pp. 41ff.

57 Ibid., p. 43.

58 Ibid.

59 Ibid., p. 15.

60 Ibid., p. 19.

61 Patte had taken part in the competition inaugurated by the Academy of Sciences on the topic of street lighting in Paris. Although Lavoisier won the competition, Patte went on in 1766 to publish his report, which was entitled *De la manière la plus avantageuse d'éclairer les rues d'une ville pendant la nuit, en combinant ensemble la clarté, l'économie et la facilité du service.*

62 See F. Choay, *La règle et le modèle* (Paris, 1980), p. 264.

63 'In imitation of those architects who divide the half-diameter of the column into

a given number of equal parts, in order to determine the members of an order, we have divided the width of the locks into twelve modules in order to decide upon theirs' (B. Forest de Bélidor, *Architecture hydraulique* (1737–9; Paris, 1810), vol. 3, p. v.).

64 The rationalisation of the arsenals was designed from the outset to lower the costs of boat building, through the manufacture and rational storage of the pieces involved in their construction. The French engineers took their inspiration from the model arsenal of Karlskrone in Sweden and, on the basis of that model, elaborated designs for Cherbourg, Calais, Saint-Malo, La Hougue and the Italian ports. These designs were marked by tendencies which would later characterise town planning, namely, the redefinition of the soil, so that it was conceived as a surface for the handling of flows, the fragmentation of spaces in terms of the various logics of production, and the dematerialisation of the framework. A. Demangeon and B. Fortier, *Les vaisseaux et les villes* (Brussels, 1978).

65 P. Patte, *Mémoires*, p. 7.

66 E. Hénard, *Etudes sur les transformations de Paris* (1903–9; Paris, 1982). The same author's *Les villes de l'avenir* contains a cross-section of a street reminiscent of the one in Patte's *Mémoires*, as Jean-Louis Cohen emphasises in his presentation of the new edition of the *Etudes*, p. viii.

67 I have borrowed this term from G. Teyssot, 'La ville-équipement. La production architecturale des bâtiments civils, 1795–1848', in *Architecture–Mouvement–Continuité* (Paris), 45, 1978, 86–94.

68 P. Patte, *Mémoires*, pp. 63ff.

69 See. B. Fortier, 'Logiques de l'équipement. Notes pour une histoire du projet', in *Architecture–Mouvement–Continuité* (Paris), 45, 1978, 80–5.

70 See B. Fortier and B. Vayssière, 'L'architecture des villes. Espaces, cartes et territoires', in *VRBI* (Brussels), 3, 1980, liii–lxii.

71 A. Vidler, 'The theatre of production: Claude-Nicolas Ledoux and the architecture of social reform'; also J. Perrault, 'Le concept de machine et de système chez Ledoux, Sade et Caucanson', in *Culture Technique* (Paris), 7, 1982, 112–23. Ledoux, who was Perronet's deputy at the saltworks of the eastern region, shared his interest in manufacture and in the notion of process serving to order design. See chapter 7, n. 67.

72 See chapter 10.

73 M.-A. Laugier, *Essai sur l'architecture*, pp. 237ff.

74 See P. Mesplé, *Joseph-Marie de Saget. Ingénieur des Travaux Publics de la Province de Languedoc* (Toulouse, 1946).

75 J.-G. Legendre, *Description de la place Louis XV que l'on construit à Reims, des ouvrages à continuer aux environs de cette place, & de ceux à faire dans la suite pour l'utilité & l'embellissement de cette ville* (Paris, 1765).

76 J.-R. Perronet, *Plan de Nantes*, ANF14 172/8.

77 The Intendant for the Ponts et Chaussées, Jules-François de Cotte, the successor to Trudaine de Montigny, reaffirmed the priority that should be given to the territory in a circular of 1778. After reproaching certain engineers for their negligence, he expressed the wish that they might reside at a distance from the large towns and at the centre, geographically speaking, of the road districts for which they were responsible. J.-F. de Cotte, *Circulaire aux ingénieurs des Ponts et Chaussées* (ENPC manuscript, 1778, MS 2636).

78 *Projet d'un canal de dérivation proposé pour faciliter la navigation dans les environs de la ville de . . .*, pupils' designs for the competition for 1756, ENPC manuscripts and drawings kept in the files labelled 'Concours des élèves Technique'.

79 I would note that a doctoral thesis is being prepared on this subject at the University of Paris XII, Y. Chicoteau, 'Les ponts urbains en France de 1750 à 1850'.

80 See, for example, R. Favre, 'Du "médico-topographique" à Lyon en 1783', in *Dix-Huitième Siècle* (Paris), 9, 1977, pp. 151–9.

81 See M. Foucault, B. Barret-Kriegel, F. Béguin, B. Fortier and A. Thalamy, *Les machines à guerir (aux origines de l'hôpital moderne)* (Brussels, 1979).

82 See M. Foucault, J.-M. Alliaume, B. Barret-Kriegel, F. Béguin, D. Ranciers and A. Thalamy, *Politiques de l'habitat (1800–1850)* (CORDA research report, Paris, 1977).

9 A PRODUCTIVE COUNTRYSIDE

1 J.-L.-H. L'Escaille, pupil's dissertation.

2 See, for example, J.-W. Konvitz, 'Remplir la carte', *Cartes et figures de la Terre*, pp. 304-15.

3 See N. Delavaud, 'L'ingénieur des Ponts et Chaussées: cartographe ou joueur des cartes', *L'ingénieur et le dessin*, the texts of lectures given to a seminar organised by the ENPC on 26 February 1982, pp. 36–48. The maps of the Ponts et Chaussées do not feature in the general catalogue of manuscripts, but are stored in a file of drawings labelled 'Concours des élèves' at the Centre Pédagogique de Documentation et de Communication de l'ENPC. In general, these maps are untitled. Their present labels are therefore the work of the curators at the Centre.

4 This is particularly the case with the 'Trompe-l'œil with Knave of Spades and King of Diamonds' of 1792, ENPC map, figure 73. In a similar spirit, maps designed by pupils at the Ecole des Ponts et Chaussées often featured invitation cards, see figures 73 and 74.

5 'Utopian map with blue knot', map competition for 1783, ENPC, map, figure 71.

6 'Trompe-l'œil with butterflies', map competition for 1793, ENPC map, figure 72.

7 See chapter 7.

8 See, for example, Buchotte, *Les règles du dessin et du lavis pour les plans particuliers des ouvrages et des bâtimens* (Paris, 1722); Dupain de Montesson, *L'art de lever les plans de tout ce qui a rapport à la guerre, & à l'architecture civile & champêtre* (Paris, 1763); *La science des ombres par rapport au dessin. Avec le dessinateur au cabinet et à l'armée* (Paris, 1750).

9 The 'theoretical' character of the maps of the Ponts et Chaussées was in fact reinforced by the existence of numerous 'study maps', drawn by Lesage or by one of the geographers temporarily assigned to the School for educational purposes. First of all, there are model sheets, as in the collection of 1793 (see note 13 below). Others contain much more elaborate compositions, which often served to inspire the pupils, sometimes indeed to such an extent that they would copy some detail or other. It is sometimes difficult to distinguish these documents from pieces actually submitted in the competitions. I have therefore decided to place them alongside the drawings of the pupils of the Ponts et Chaussées.

10 On 'picturesque' gardens in France in the eighteenth century, read *Jardins en France 1760–1820*; M. Mosser, *Monsieur de Marivgny et les jardins*, an extract from the *Bulletin de la Société de l'Histoire de l'Art Français* (Paris), 1973; D. Wiebenson, *The picturesque garden in France* (Princeton, 1978).

11 See, for example, Dupain de Montesson, *La science des ombres*, pp. 99ff.

12 The dividing line between convention and the imitation of nature was anyway a fluctuating one, even when it was a question of the most official of all maps, that of the Cassini. See G. Alinhac, *Historique de la cartographie* (Paris, 1973), p. 91.

13 Study map, 1793, ENPC manuscript, special collection of 353 manuscripts extracted from the 1886 Ponts et Chaussées catalogue, no. 103. It is a fictitious album of eighty-eight sheets, consisting of various maps with commentaries explaining how one should represent trees, rocks or coastlines, etc. There is a more precise description in *Dessin et sciences XVIIe–XVIIIe siècles*, exhibition catalogue for the Drawings Department of the Louvre, 22 June–24 September 1984, pp. 16, 17.

14 *Programme de pont de pierre proposé aux élèves des Ponts et Chaussées pour le concours de 1791*, ENPC, a manuscript which has not been entered in the general catalogue, and which is stored in the files labelled 'Concours des élèves Technique'.

15 Advenier, pupil's dissertation.

16 The plan to stabilise the dunes of Gascony by means of planting was first formulated by the engineer Bremontier at the end of the eighteenth century. It was carried out subsequently, and met with great success. See N. Bremontier, *Mémoire sur les dunes de Gascogne*, Year IV (ENPC manuscript, MS 1897 – a document which was reproduced in the *Annales des Ponts et Chaussées* (Paris), first quarter 1833, 145–225). French engineers in Italy were very much involved in the damming of the River Po. The library of the Ecole des Ponts et Chaussées thus contains numerous manuscripts referring to one of the great planning and

civil engineering ventures of the Empire period. One should, for example, consult G. Riche de Prony, *Rapport à Son Altesse le vice-roi d'Italie sur les causes des inondations du Pô, et sur les moyens de rendre les crues de ce fleuve moins dangereuses* (ENPC manuscript, MS 1432). For a panoramic survey of the works carried out by the engineers in the early nineteenth century, read A. Lorion, 'Quelques grands travaux publics du premier Empire et leurs maîtres d'œuvre', in *Annales des Ponts et Chaussées* (Paris), March–April 1959, 345–63.

17 T. Mariage, *L'univers de Le Nôtre*, pp. 78, 79.

18 It is from this angle that the countless references made by the pupils of the Ponts et Chaussées to the park at Versailles could also be interpreted. One should bear in mind the efforts made by Louis XIV to supply his water jets and fountains, even if it were at the price of works such as Marly's machine or the aqueduct of Maintenon. In the seventeenth century, Versailles was in effect a laboratory for hydraulic architecture. See P. Verlet, *Le château de Versailles* (1961; Paris, 1985), pp. 180ff.

19 T. Mariage, *L'univers de Le Nôtre*, p. 26.

20 I am referring here to M. Tafuri's well-known article, 'Giovanni-Battista Piranesi: l'utopie négative en architecture', in *Architecture d'Aujourd'hui* (Paris), 184, 1976, 93–108.

21 This crisis of representation was in fact general. Around this same period, the drawings for bridge designs ceased to be of such importance as explanatory devices, and written texts, which themselves assumed a more graphic quality, began to be of more significance instead. It was as if the engineers were using a sharpened aesthetic to investigate the limits of representation. See Y. Chicoteau, A. Picon and C. Rochant, *Les concours techniques du XVIIIe siècle*, pp. 62ff.

22 On the notion of the sublime in the eighteenth century, read A.-F.-B. Clark, *Boileau and the French classical critics in England* (Paris, 1925), and, above all, S.-F. Monk, *The sublime: a study of critical theories in eighteenth-century England* (New York, 1935).

23 The first French translation of the *Philosophical enquiry* dates from 1765, but the philosophical references of the engineers were almost exclusively drawn, as we have seen, from Condillac and Rousseau. Thus Prony, in his *Fragments sur la théorie des beaux-arts*, observes that the Abbé Dubos 'takes as the basis of his doctrine the secret anxiety which impels us to give nourishment to our mind and activity to our senses'. Nowhere, however, does he make any mention of Burke.

24 The landscape drawing competitions, which had been anticipated by Turgot in his *Instruction*, served to complete a pupil's apprenticeship to the landscape, albeit from a more anecdotal perspective. As a counterpoint to cartographic work, the 1789 competition, for example, allows us to appreciate a view of the Colosseum in Chinese ink, or else a small meadow scene in gouache.

25 The sublime had in fact first appeared at the end of the seventeenth century, in

the writings of French theorists, and it occasioned a controversy between Boileau and Huet over the nature of *fiat lux* in Genesis. Boileau considered this moment to be 'sublime', whereas Huet denied it, on the grounds that the Bible could not, in his view, have recourse to artistic procedures. The dispute rested upon a misunderstanding, in so far as the two protagonists were agreed as to the existence of an essential beauty which was not subject to the ordinary rules and which was terrible in its manifestation. See T.-A. Litman, *Le sublime en France (1660–1714)*, pp. 78ff.

26 The Abbé Dubos, in his *Réflexions sur la poésie et la peinture* (1719), had in fact stressed the paradoxical nature of aesthetic emotion, 'this pleasure which generally resembles an affliction, and whose symptoms are sometimes the same as those accompanying the most intense pain' (J.-B. Dubos, *Réflexions critiques* (1719; Paris, 1770), p. 1). This work was translated into English in 1748. Prony and, in all probability, the most cultivated of the engineers, were familiar with it.

27 On the doctrine of *ut pictura poesis*, which mainly concerned painting, see R. W. Lee, 'Ut pictura poesis, the humanistic theory on painting', in *The Art Bulletin* (New York), 12, 1940, 197–269. We have already seen how Boffrand and Blondel tended to apply this doctrine to architecture.

28 E. Burke, *A philosophical enquiry into the origin of our ideas of the sublime and the beautiful* (edited by J. T. Boulton, London, 1967), p. 39.

29 This concept bears comparison with Schopenhauer's famous text on architecture considered 'merely as a fine art', where its purpose was defined as being to express the most rudimentary representations of the will: gravity, cohesion, rigidity and hardness. 'Even at this low stage of the will's objectivity, we see its inner nature revealing itself in discord; for, properly speaking, the conflict between gravity and rigidity is the sole aesthetic material of architecture; its problem is to make this conflict appear with distinctness in many different ways. It solves this problem by depriving these indestructible forces of the shortest path to their satisfaction, and keeping them in suspense through a circuitous path; the conflict is thus prolonged, and the inexhaustible efforts of the two forces become visible in many different ways' (A. Schopenhauer, *The world as will and representation* (1819; translated by E. F. J. Payne, New York, 1969, 2 vols.), vol. 1, p. 214).

30 See chapter 7. Patte's positions, and his attacks upon Sainte-Geneviève, also arose out of his distrust of structural procedures which, so to speak, 'violated' the material.

31 Smeaton gave a description in *A narrative of the building and a description of the construction of the Eddystone Lighthouse* (London, 1791). Perronet and Prony may well have admired this construction at the time of their journey to England in 1785.

32 P.-C. Lesage, *Recueil de divers mémoires, extraits de la Bibliothèque des Ponts et Chaussées, à l'usage des élèves ingénieurs* (Paris, 1806–10), part 2, pp. 67ff.

33 Ibid., plate 1. This engraving is a literal plagiarism of the frontispiece to Smeaton's treatise.

34 E. Burke, *A philosophical enquiry*, p. 76.

35 See chapter 7.

36 W. Gilpin, *Three essays on picturesque beauty* (1792). Consult also C. Hussey, *The picturesque: studies in a point of view* (1927; London, 1983).

37 In contrast to the sublime, the picturesque did not really constitute a definite aesthetic category, but rather a register of effects and of sensations which were applied primarily to gardens, landscapes and voyages, and then to architecture. In his *Three Essays*, Gilpin listed some of the criteria of the picturesque, such as the combination of simplicity and variety, the contrast between the smooth and the rough, the luminous and the sombre, the clear and the vague. He thus borrowed a good number of its conflicting pairings from the sublime, as if he were opting for a discipline of the gaze which was increasingly taking the place of nature as a unificatory principle. See M. Conan, 'Le pittoresque: une culture poetique', Postface to the French edition of W. Gilpin, *Three Essays, Trois essays sur le beau pittoresque* (Paris, 1982), pp. 119–50.

38 It is worth emphasising that this was a dimension of the sublime that was peculiar to the engineers, and one whose existence Burke had never suspected. Planning which aspired to be 'natural' laid claim to the same effects as nature.

39 I. Kant, *The critique of judgement* (1790), translated by J. C. Meredith (Oxford, 1952), p. 168. On Kant's approach to nature and its planning, see B. Bourgeois, 'L'idéalisme allemand devant la beauté naturelle et l'embellissement de la nature', *Mort du paysage? Philosophie et esthétique du paysage*, proceedings of the conference held at Lyons, 3–4 December 1981 (Seyssel [1981], 1982), pp. 161–77; S. Trottein, 'Pour une aesthétique des jardins (sur une note de Kant)', in *VRBI* (Brussels), 9, 1984, xxxvi–xlv.

40 Deforestation assumed alarming proportions in the course of the eighteenth century. It reached its height at the beginning of the nineteenth century. L. Badre, *Histoire de la forêt française* (Paris, 1983), pp. 71–99. On canalside plantings, see A. Kriegel, 'Les plantations françaises', in *Un canal des canaux*, catalogue for an exhibition organised by the Caisse Nationale des Monuments Historiques et des Sites, 7 March–8 June 1986, pp. 245–50.

41 'A man who, with an infinite number of other forms of knowledge, has studied architecture as a philosopher, has observed that all the various details of a column, such as its entasis, its tori and its moulding, etc. are located in precisely those places where a column would be liable to break or to crumble' (G. Riche de Prony, pupil's dissertation).

42 See A. Delbecq, 'Analyse de la stabilité des voûtes en maçonnerie de Charles-Augustin Coulomb à nos jours'.

43 For this calculation, Bélidor began by using mechanics to eliminate the flat angle

and the right angle, then he arbitrarily selected the average of these two values, observing that if the flat angle enabled one to diminish at most the width of the gates, the right angle was the better from the point of view of the thrust of the waters (B.-F. de Bélidor, *Architecture hydraulique*, vol. 3, pp. 76ff).

44 The maintenance of French naval forces in the seventeenth and eighteenth century required by itself the entire yield of 100,000 hectares of oak forest, and the stripping of 500 to 600 hectares of plantations from about 150 to 160 years old. L. Badre, *Histoire de la forêt française*, p. 76.

45 Canals at summit level provide a good illustration of the energy aspect of the relief, since their layout was dictated by two major preoccupations, namely, that the summit level should be located at a sufficient height and power and that, at the same time, it should be able to collect waters available at a still higher altitude. The need to contain the water in order to economise it related to a concept of the potential. The term 'loss of pressure' was anyway used with respect to leakages at canal reaches before being applied to potential energy. On canal layouts, see B. Forest de Bélidor, *Architecture hydraulique*, vol. 4; E.-M. Gauthey, *Œuvres*, vol. 3; J. de Lalande, *Des canaux de navigation et spécialement du canal de Languedoc* (Paris, 1778). On the technology involved in the production of hydraulic energy in the classical age, see S. Reynolds, *Stronger than a hundred men. A history of the vertical water wheel* (Baltimore, 1983).

46 On Enlightenment agronomy, see A.-J. Bourde, *Agronomie et agronomes en France au XVIIIe siècle* (Paris, 1967).

47 A Poitrineau, 'Physiocrates", *Encyclopaedia Universalis*, vol. 14, pp. 608–10, and p. 608 in particular.

48 See chapter 7. It was in the eighteenth century also that urban canals were first set underground. A. Guillerme, *Les temps de l'eau*, pp. 188ff.

49 See R. Rappaport, 'The geological atlas of Guettard, Lavoisier and Monet: conflicting views of the nature of geology', in *Towards a history of geology* (Cambridge, Mass., 1969), pp. 272–87.

50 This was particularly true of a road competition for 1751. Y. Chicoteau, A. Picon and C. Rochant, *Les concours techniques du XVIIIe siècle*, pp. 47ff.

51 A.-O. Ospovat, 'Reflections on A. G. Werner's "Kurze Klassifikation"', in *Towards a history of geology*, pp. 242–56.

52 The Mines engineers had been trained until then at the Ecole des Ponts et Chaussées. For further information on the teaching of mineralogy and on the founding of the Ecole des Mines, see A. Birembaut, 'L'enseignement de la minéralogie et des techniques minières', in *Enseignement et diffusion des sciences en France au XVIIIe siècle*, pp. 365–418; *Regards sur la France* (Paris), Les Ecoles Nationales Supérieures des Mines, 15, November 1961.

53 J. Steuart, *An enquiry into the principles of political oeconomy: being an essay on the science of domestic policy in free nations* (London, 1767, 2 vols.), the French

translation of which was published in Paris in 1789–90. A detailed analysis of this phenomenon of parochialism may be found in P. Rosanvallon, *Le capitalisme utopique. Critique de l'idéologie économique* (Paris, 1979), pp. 95ff. It is also tempting to draw a parallel between the impetus towards recentring which was evident in the latter half of the eighteenth century and the attempts at rationalising the centres of towns which the town planners, beginning with Patte and his design for the improvement of the Ile de la Cité, were then making. See chapter 8 and, above all, A. Demangeon and B. Fortier, *Les vaisseaux et les villes*, pp. 11ff.

54 Brulé, study map of 1793, ENPC map, fig. 75. In this respect too, the Ponts et Chaussées maps may be compared with gardens. Do not the Egyptian, Gothic and Chinese buildings with which they teem remind one of an abbreviated journey in space and time? See *Jardins en France, 1760–1820*.

55 Gruber, *De l'utilité des cartes hydrographiques*, ENPC manuscript, MS 2454.

56 A topographical commission comprising military engineers, representatives of the Eaux et Forêts, of the Mines and of the Ponts et Chaussées, actually met in 1802 in order to propose the standardisation of cartographic techniques (G. Alinhac, *Historique de la cartographie*, pp. 101ff). Another source for the study of cartographic standardisation in the nineteenth century would be Berthaut, *La carte de France, 1750–1898, une étude historique* (Paris, 1898–9).

57 W. Szambien has analysed in much greater detail this 'Portfolio', which consisted of a set of drawings and plans realised by pupils in the course of their training, 'Architekturdarstellung an der Pariser Ecole Polytechnique zu Beginn der 19 Jahrhunderts', in *Daidalos* (Berlin), 11, 1984, pp. 55–64.

58 Under the Empire and the Restoration, Prony was therefore entrusted with numerous missions both in France and abroad, the draining of the Pontine Marshes being but one of a series. See A. Lorion, 'Une mésaventure de Prony à Venise sous l'Empire (juillet 1805)'.

59 In a report of 1775, Chézy had supplied the first formula for the uniform flow of water in open air or drains, a formula which linked the speed of the water to the pitch, to the area of the cross-section and to the wetted perimeter. Dubuat had returned to this same problem in his *Principes d'hydraulique* (1779), where he developed a theoretical formula and a practical formula which were both very close to those of Chézy. See G. Mouret, 'Antoine Chézy, histoire d'une formule hydraulique'.

60 S. Ince and H. Rouse, *History of hydraulics*, p. 142.

61 G. Riche de Prony, *Des Marais Pontins* (Paris, 1818), pp. 14ff.

62 G. Riche de Prony, *Atlas des Marais Pontins* (Paris, 1822).

63 This design, which may be said to have an exemplary status for the period, was drawn up after the passing of the 1807 law on drainage, which replaced the earlier legislation, which had been deemed somewhat ineffectual. See AN F14 6390. Upon the publication of the *Atlas des Marais Pontins* (1822), Prony was

congratulated by Pope Leo XII. There was much land reclamation in the course of the first half of the nineteenth century, as with the marshes of Rochefort, Arles, etc. Prony was often consulted over this draining. Thus, we hear of him again in relation to the Compagnie Générale de Dessèchement des Marais, Lacs, Etangs, Lais et Relais de Mer, founded in 1828 for the purposes of speculation.

64 One should once again note the influence of voyages, from which people were now expected to garner ever more detailed and systematic information. Travellers were increasingly belaboured with advice of this sort. Volney, for example, was to publish a model of the genre with his *Questions de statistique à l'usage des voyageurs* (1795), which provided a very detailed outline for an enquiry. Climate, the nature of the soil, industry and trade were tackled in turn. The state approached in much the same way the space that it sought to control. One could also consider Volney's *Questions* as a kind of 'programme' for the Egyptian campaign. See N. Broc, *La géographie des philosophes*, pp. 473, 474; J. Gaulmier, *Un grand témoin de la Révolution et de l'Empire, Volney* (Paris, 1959), pp. 176, 177.

65 J.-M. Roland, a circular for 18 October 1792, reproduced by the Commission de la Recherche et de Publication des Documents relatifs à la Vie Economique de la Révolution. *Mémoires et documents, IX. Fragments d'une enquête économique du ministre Roland auprès des ingénieurs des Ponts et Chaussées, 1792–1793*, published by G. Bourgin (Besançon, 1942), p. 14.

66 The Inspection des Manufactures was abolished in 1791. On this relatively unknown technical corps, see F. Bacquie, *Les Inspecteurs des Manufactures sous l'Ancien Régime* (Toulouse, 1927).

67 J. Guillerme, 'Roland de la Platière et les patiences de l'inspection méthodique. Contribution à l'archéologie de la condition d'ingénieur civil', *L'ingénieur dans la société française* (Paris, 1985), pp. 13–18.

68 G. Teysso, 'La ville-équipement', p. 87. On the 'technological' enquiries of the period, see also J. Guillerme and J. Sebestik, 'Les commencements de la technologie', pp. 55ff.

69 See J. and M. Dupaquier, *Histoire de la démographie* (Paris, 1985), pp. 251ff.

70 Ibid., pp. 129ff.

71 G. Ardant, *Histoire de l'impôt* (Paris, 1971), pp. 168ff.

72 See J. Petot, *Histoire de l'administration des Ponts et Chaussées*, pp. 338ff.

73 The idea of a geometrical division of the countryside would seem in fact to have come from Sieyès, a member with Thouret of the constitutional committee that was responsible for problems of administrative reorganisation. See E. Lebègue, *Thouret (1746–1794)* (Paris, 1910), p. 191.

74 Ibid., pp. 190, 191. For this debate, see F. Dagognet, *Pour une théorie générale des formes* (Paris, 1975), pp. 73ff.

75 The Land Survey Office had been instructed to assess the total area of France and its population, so as to provide an approximate basis for the allocation of the new

taxes. It was also involved in works preparatory to the drawing-up of the general trigonometric map for the *département* of the Seine, together with the composition of large-scale trigonometric tables capable of rivalling English publications (see ANF14 1393). This institution's mission thus reflected the territorial preoccupations of the revolutionary leaders. Prony had asked to head this office, and his request had been granted. This was because he did not wish to be too far removed from Paris, a city to which he was tied by his scientific interests, and by the writing of his *Nouvelle architecture hydraulique* in particular. On the basis of his own experience at the Land Survey Office, Prony was to publish the *Table des populations spécifiques des départements français* and his *Table des superficies des départements français évalués en kilomètres carrés*, which was a part of the widespread concern with enumeration evident in the first half of the nineteenth century.

76 G. Riche de Prony, *Rapport à l'Institut sur le projet d'une nouvelle division du territoire français par le citoyen Lhote* (1796; ENPC manuscript, MS 1350).

77 See F. Etner, 'Le calcul économique et le corps des Ponts et Chaussées entre les deux Napoléon', in *Annales des Ponts et Chaussées* (Paris), special issue, 1981, 104–9.

78 The quest for a new balance between the state and private interests was especially evident in the plan for the completion of the canals, drawn up by Becquey in 1821–2, which proposed mixed financial backing and responsibility shared between the administration of the Ponts et Chaussées and the finance companies involved. See A. Kriegel and P. Pinon, '"L'achèvement des canaux" sous la Restauration et la monarchie de Juillet', in *Annales des Ponts et Chaussées* (Paris), special number, 1981, 72–82, especially pp. 73, 74. There were later to be very similar debates regarding both roads and railways.

79 Particularly telling in this respect are Michel Chevalier's articles in *Le Globe* in 1832, together with his great work of synthesis, *Des intérêts matériels en France*, published in 1838. For further information on him, see J. Walch, *Michel Chevalier économiste saint-simonien, 1806–1879* (Paris, 1975). On Saint-Simon and Saint-Simonian doctrine, consult R.-P. Fehlbaum, *Saint-Simon und die Saint-Simonisten, vom 'Laissez-Faire' zu Wirtschaftplannung* (Tübingen, 1970); F.-E. Manuel, *The new world of Henri Saint-Simon* (Cambridge, 1956).

10 FROM THE REVOLUTIONARY 'GENIUS' TO NEO-CLASSICISM

1 M. Crucy, letter to the architect Baraguey, 1807, reproduced in *Mathurin Crucy 1749–1826, architecte nantais néo-classique*, exhibition catalogue from the Musée Dobrée, 15 May–17 August 1986, Nantes, 1986, pp. 146–7, especially p. 147. 'When I had escaped from the clutches of the pedant Blondel', added Crucy,

referring to his *Premier Grand Prix*, which he won the same year as Blondel died, 'the design for public baths which I had drafted at the Academy took wing to a greater extent than before.'

2 A. Braham, *The architecture of the French Enlightenment*, pp. 37–48.

3 The influence of the *Grand Siècle* is especially evident in de Wailly's taste for a degree of ornamental richness. See *Charles de Wailly, peintre architecte dans l'Europe des Lumières*, exhibition catalogue of the Caisse Nationale des Monumens Historiques et des Sites (Paris, 1979), pp. 40, 41.

4 On the emergence of the idea of nature in the eighteenth century, read J. Ehrard, *L'idée de nature en France dans la première moitié du XVIIIe siècle* (1963; Geneva, 1981).

5 In his 'Discourse on the nature of the principles of civil architecture', which constituted the second part of his work, Leroy distinguished three successive states in Greek Doric: a first state, corresponding to a column height less than six times the module, as at the Temple of Corinth, a second state, where this height was roughly equal to six times the module, as at the Temple of Theseus, and a final state, which was close to modern conceptions and which was illustrated by the Temple of Augustus at Athens. Finally, on the question of proportion: 'the best one can do in this area [of architecture] is to look at all the fragments of ancient monuments from Greece; all those which one can find in Asia Minor or in Syria, together with those which still remain in Rome; the precepts of Vitruvius on the proportions of the orders; and finally the views of the most famous architects regarding these proportions, as so many elements which may serve to compose the best possible orders'. As can be seen, this position is close to that held by Patte, *Les ruines des plus beaux monuments de la Grèce* (Paris, 1758), part 2, p. vi. On the influence of travel in Greece at the end of the eighteenth century, see *Paris–Rome–Athènes. Le voyage en Grèce des architectes français aux XIXe et XXe siècle*, exhibition catalogue of the Ecole Nationale Supérieure des Beaux-Arts (Paris, 1982); D. Wiebenson, *Sources of Greek revival architecture.*

6 C.-F. Volney, *Les ruines ou méditations sur les révolutions des empires* (Paris, 1791). Written after his *Voyage en Syrie et en Egypte* (1787), the *Ruines*, which are, to borrow Jean Gaulmier's phrase, a kind of 'rationalist apocalypse', enjoyed remarkable success at the beginning of the Revolution. The Middle East, which had formerly been rich and densely populated, and which was now a desert, served as a framework for a meditation upon human history. See J. Gaulmier, *Un grand témoin de la Révolution*, pp. 112ff. Cassas, who was both an architect and an artist, had met Volney during his journey of 1782–5. His *Voyage pittoresque de la Syrie, de la Phénicie, de la Palestine et de la Basse-Egypte* appeared from 1799 on, with engravings whose aesthetic values often seemed close to those of revolutionary architecture. On Cassas, whose collection of maquettes, later incorporated into the collections of the Musée des Beaux-Arts made him one of the pioneers of the

'musée d'architecture', read W. Szambien, *Le musée d'architecture (1776–1836). Un projet inachevé* (SRA research report, Paris, 1984), pp. 19, 20, 65–9

7 Jean-Louis Viel de Saint-Maux, the author of the *Lettres sur l'architecture des anciens et des modernes*, was brother to Charles-François Viel. On him, read J.-R. Mantion, 'La solution symbolique. Les lettres sur l'architecture de Viel de Saint-Maux', in *VRBI* (Brussels), 9, 1984, xlvi–lviii; J.-M. Pérouse de Montclos, 'Charles-François Viel, architecte de l'Hôpital général et Jean-Louis Viel de Saint-Maux, architecte, peintre et avocat au Parlement de Paris', in *Bulletin de la Société de l'Histoire de l'Art Français* (Paris, 1966), 257–69.

8 W. Oechslin, 'L'intérêt archéologique et l'expérience architecturale avant et après Piranèse', *Piranèse et les français*, proceedings of a conference held at the Villa Medici, on 12–14 May 1976 (Rome, 1978), pp. 395–410. See also M. Tafuri, *Projet et utopie.*

9 The first important discoveries at Pompeii were in fact made in the 1760s and 1770s. See P. Pinon, 'Pompéï, une archéologie du néo-classicisme', in *Archives d'Architecture Moderne* (Brussels), 22, 1982, 53–68.

10 See M. Gallet, 'Jean-Charles Delafosse, architecte', in *Gazette des Beaux-Arts* (Paris), 61, 1963, 157–64.

11 See P. Duboy, *Jean-Jacques Lequeu. Une énigme* (1986; Paris, 1987), and J. Guillerme, 'Lequeu, entre l'irrégulier et l'éclectique', in *Dix-Huitième Siècle* (Paris), 6, 1974, 167–80.

12 G. Teyssot, Introduction to E. Kaufmann, *Trois architectes révolutionnaires*, p. 27 in particular.

13 C.-N. Ledoux, *L'architecture*, p. 19.

14 Quoted by G. Gusdorf, *Dieu, la nature, l'homme au siècle des Lumières*, p. 309.

15 It was the materialists, such as d'Holbach, who worked the hardest to oust what was, to their contemporaries, so monstrous a disproportion, in opposition to the deists, who compared nature to a book in which the favourable intentions of the creator could be read directly. D'Holbach wrote as follows, in a style which seems curiously to anticipate Ledoux's effusions: 'Suns are extinguished and are encrusted; planets perish and are scattered across the plains of the air; new suns are illuminated; new planets are formed in order to carry out their own revolutions, or in order to describe new paths; and so man, who is but an infinitely small portion of the globe, who is himself but an imperceptible point in the immensity, believes that it was for him that the universe was made, imagines that he should place his trust in nature, flatters himself that he is eternal, and calls himself the king of the universe' (P.-H.-D. d'Holbach, *Système de la nature, ou des lois du monde physique et du monde moral* (1770; Paris, 1821), vol. 1, p. 105. This perception of a nature in movement, whose rhythms crush man, was further exacerbated at the end of the eighteenth century. The only possible response was therefore to discover whatever laws were common to 'the physical world and the

moral world' and to conform strictly to them. The reader may have recognised in this alliance between the physical and the moral one of the themes dear to the *idéologues*, such as Volney and, above all, Cabanis, a former protégé of Helvétius and of d'Holbach, whose *Rapports du physique et du moral de l'homme* were published in 1802.

16 J. Starobinski, *1789. Les emblèmes de la raison.*

17 Quoted by J. Starobinski, ibid., pp. 164, 165.

18 See, for example, A. Soboul, *La révolution française* (Paris, 1984), pp. 57–116.

19 See A. Vidler, 'The architecture of the lodges: ritual form and associational life in the late Enlightenment', in *Oppositions* (New York), 5, 1976, 75–7.

20 R. Darnton, *Mesmerism and the end of the Enlightenment in France.* Consult also C. Gillispie, *Science and polity in France*, pp. 261ff.

21 E.-L. Boullée, *Architecture. Essai sur l'art*, pp. 78, 79.

22 The letter in which Beckford related this episode is reproduced in M. Gallet, *Ledoux et Paris*, pp. 146–8.

23 Ledoux, *L'architecture*, pp. 195, 196.

24 The saltworks had originally been conceived on its own. We know that Ledoux would later present the ideal town of Chaux as his initial project, which was mutilated through the incomprehension of the bureaux de la Ferme and of the royal administration. See M. Gallet, *Ledoux*, pp. 222ff.

25 Ledoux often referred to the alternation of day and night, to the succession of the seasons and to the rhythms of nature, at once changing and eternal. Although he referred directly to Fontenelle's *Entretiens sur la pluralité des mondes*, the architect would hardly seem to have retained anything other from the philosopher's lesson in physics and relativism but the image of these Cartesian vortices which populated the void with their vertiginous figures and which were organised in terms of a kind of universal meteorology. The astronomical metaphor then recalled yet again the ambiguity which was still attached to the term 'revolution', which was an upheaval, a breach and a founding act but also the beginning of a cycle. In which sense, indeed, did late eighteenth-century architecture consider itself to be 'revolutionary'?

26 Some passages in Ledoux bear comparison to the lyricism of Volney in *Les ruines ou méditation sur les révolutions des empires*. The 'Elevation of Chaux cemetery', which represents the earth as emerging from the clouds, is in fact reminiscent of plate 2 of *Les ruines*, in which the planet features in the form of a huge whirling billiard ball.

27 Boullée, *Architecture. Essai sur l'art*, p. 34.

28 I. Kant, *Critique of the faculty of judgement* (1790; translated by J. C. Meredith, Oxford, 1952), p. 168.

29 See M. Mosser and D. Rabreau, 'Nature et architecture parlante: Soufflot, De

Wailly et Ledoux touchés par les Lumières', in *Cahiers de la recherche architecturale*, proceedings of the conference 'Soufflot et l'architecture des Lumières', pp. 222–39.

30 See H. Damisch, 'Ledoux avec Kant', Preface to E. Kaufmann, *De Ledoux à Le Corbusier*, pp. 11-21.

31 Boullée, *Architecture. Essai sur l'art*, p. 49.

32 Le Camus de Mézières went on to give an analysis of the various rooms in a dwelling even more detailed than that of Blondel.

33 Condillac, *Traité des sensations* (1754; Paris, 1984), p. 265.

34 Boullée again rehearsed the dispute between Perrault and Blondel, so as to show how inane it had been. 'You can only be wholly certain that you have identified the principles upon which an art is founded when there is nothing whatsoever that would allow you to distance yourself from these same principles'. Experience, however, had shown that the proportions had always been the object of contradictory interpretations. The founding principles of architecture would therefore have to be sought elsewhere. (*Architecture*, pp. 60, 61, 65–9.)

35 Ibid., p. 62.

36 Ibid., p. 64.

37 'We call beautiful those objects which are most analogous to our own sensations'; 'we reject those which, deprived of such an analogy, do not suit our way of being'. A kind of analogy, a law of congruence, therefore corresponded to each regular figure (Boullée, *Architecture*, p. 61).

38 Ibid., p. 35.

39 On the notion of character in Boullée, see W. Szambien, *Reflet de la connaissance*, pp. 96–129.

40 Boullée, *Architecture*, p. 73.

41 N. Le Camus de Mézières, *Le génie de l'architecture*, p. 3.

42 Ibid., pp. 6, 7. On the relationship between scenography and architectural debate, read R. Rabreau, 'Des scènes figurées à la mise en scène du monument urbain. Notes sur le dessin "théatral" et la création architecturale en France après 1750', *Piranèse et les français*, pp. 443–66.

43 Boullée, *Architecture*, p. 59.

44 J.-V. Viel de Saint-Maux, *Lettres sur l'architecture des anciens, et celle des modernes* (Paris, 1787), p. ix. On Viel de Saint-Maux's doctrine, one should read J.-R. Mantion, 'La solution symbolique'.

45 See M. Mosser and D. Rabreau, 'Nature et architecture parlante'.

46 Ledoux, *L'architecture*, p. 194.

47 Boullée, *Architecture*, pp. 79–97. However, it is as well, in one's interpretation of revolutionary design, to err on the side of caution. Jean-Marie Pérouse de Montclos has, for example, shown how the cenotaph for Newton, which has been the object of so many commentaries, may perhaps derive from an

archaeological fantasy rather than from any very worked-out philosophical speculations. J.-M. Pérouse de Montclos, 'De nova stelli anni 1784', in *Revue de l'Art* (Paris), 58/59, 1982–3, 75–84. In the case of Boullée and Ledoux, symbolism, mythology and archaeology almost invariably coexisted. Even if one ignores the evocation of the three kingdoms of nature, Boullée's cathedral also resembles certain cross-sections of diagrams of the temple of Baalbek. J.-M. Pérouse de Montclos, *Etienne-Louis Boullée*, p. 158.

48 We do not know precisely when the *Essai* was composed, but it was first published in 1781, in a volume of Rousseau's writings on music, by Du Peyrou of Geneva.

49 A. Court de Gebelin, *Le monde primitif analysé et comparé avec le monde moderne* (Paris, 1773–81). In this huge work, which was regarded at the time as a monument of erudition, the author was especially concerned to rediscover the original language, which he saw as being a naïve and faithful depiction of nature, resembling imitative harmony and music.

50 J. Derrida, *Of grammatology*.

51 A. Rossi, *Introduction à Boullée* (Nantes, 1982), p. 2.

52 Ledoux, *L'architecture*, p. 135.

53 According to the descrption given in the *Prospectus* of 1802, these four volumes were to have been devoted to the town of Chaux, then to the houses of the town, the services, the monuments and châteaux, the country houses, the palaces and, finally, to the propylaea of Paris. C.-N. Ledoux, *L'Architecture considérée sous le rapport de l'art, des mœurs et de la législation, par C.-N. Ledoux, Prospectus* (1802; Arc-et-Senans, 1985).

54 D. Ramée, advance notice for the *Architecture de C.-N. Ledoux* (1847; Princeton, 1983). I differentiate between these two editions below, referring to *L'architecture* (1804) and to *Architecture* (1847).

55 See B. Didier, 'Ledoux écrivain', in *Cahiers de la recherche architecturale*, proceedings of the conference 'Soufflot et l'architecture des Lumières', pp. 253–9.

56 Emil Kaufmann, as is well known, saw in Ledoux 'a quite personal breach with tradition', so that in more than one respect his work anticipated the modern movement of the 1920s, hence his famous title *Von Ledoux bis Le Corbusier.*

57 In both cases, the text represented a kind of rewriting of the work that had been built, designed therefore to emphasise its coherence and its prophetic scope.

58 J.-F. Blondel, *Cours d'architecture*, vol. 4, p. iii.

59 A. Vidler, 'The theatre of production, Claude-Nicolas Ledoux and the architecture of social reform', p. 56.

60 Ledoux, *L'architecture*, p. 135.

61 Ibid., p. 113.

62 Ledoux often takes his reader aside and exhorts him to 'contemplate', to 'see' or to 'look'. The abandonment of the rules of *convenance* can also be seen as a

means of attracting attention, of seducing at any price, and sometimes also of shocking.

63 Ledoux, *L'architecture*, pp. 43ff.

64 Ibid., p. 6.

65 Ibid., pp. 7, 8

66 Ibid., p. 8.

67 'Enlightened art is revived, and the great principles are adapted even to familiar customs, and it is then that its celestial benefits unfold and are generalised' (ibid.,pp. 8, 9).

68 Ibid., plate 6.

69 Ledoux, *Architecture*, plates 288–91. Comparison between the registers of the natural and of human involvement constituted a frequent theme in late eighteenth-century architecture. This theme was generally seen as an association between rock and column, such as features both in Delafosse's *Nouvelle iconologie historique* and in Ledoux's designs. See M. Mosser, 'Le rocher et le colonne, un thème d'iconographie architecturale au XVIIIe siècle', in *Revue de l'art* (Paris), 58/59, 1982–3, 53–74.

70 Ledoux, *L'architecture*, plates 61, 62.

71 Ibid., plate 107.

72 See M. Gallet, *Ledoux et Paris*, pp. 151–86.

73 Ledoux, *L'architecture*, plate 88.

74 Ibid., plate 63.

75 Ledoux, *Architecture*, plate 55.

76 This was true, the reader will recall, of the staircase of the Director's house at the saltworks.

77 See chapter 4.

78 Ledoux, *Architecture*, plate 176.

79 Ibid., plate 270.

80 Ibid., plate 35. The opposition between the grotto and the columns of the peristyle also refers to the transition between the kingdom of nature and the work of man, as M. Mosser emphasises, 'Le rocher et la colonne', p. 58.

81 Boullée, *Architecture*, p. 35.

82 Ibid., p. 91.

83 Ledoux, *L'architecture*, plate 74.

84 Ibid., p. 135.

85 See chapter 4.

86 Ledoux, *Architecture*, plate 159.

87 Ibid., plate 177.

88 See A. Fortier, 'L'invention de la maison. Un nouvel espace de projet', in *Architecture–Mouvement–Continuité* (Paris), 51, 1980, 29–35.

89 Ledoux, *L'architecture*, plates 90, 91.

90 Ibid., plate 76.
91 Ledoux, *Architecture*, plates 45–51.
92 Ledoux, *L'architecture*, plate 33.
93 Ibid., p. 104.
94 Ibid., p. 106.
95 Ibid., p. 119.
96 Ibid., plate 104.
97 Ibid., p. 199. For a study of the social meaning of this design, see A. Vidler, 'Asylums of libertinage, Sade, Fourier, Ledoux', in *Lotus International* (Milan), 44, 1984, 28–38.
98 S. Conrad, 'De l'architecture de Claude-Nicolas Ledoux considérée dans ses rapports avec Piranèse', *Piranèse et les français*, pp. 161–80.
99 Ledoux, *L'architecture*, p. 77.
100 See chapter 7.
101 See Jean Starobinski's observation on the transition towards Romanticism: 'The circle was now no longer the certain form of the world, nor the emblem of the presence of God: it was one of the mental schemas, one of the practical categories through which man left his pinpoint of solitude; it was the first of the interpretational models which enabled man to organise the place in which he dwelled' (Introduction to G. Poulet, *Les métamorphoses du cercle* (1961; Paris, 1979), p. 20.
102 Ledoux, *L'architecture*, plate 113.
103 Boullée was particularly explicit on this point, in *Architecture*, pp. 97–107.
104 See M. Ozouf, *La fête révolutionnaire 1789–1799* (Paris, 1976).
105 Boullée, *Architecture*, p. 119. This design is reproduced in J.-M. Pérouse de Montclos, *Etienne-Louis Boullée*, ills. 119, 120. The theme of the colosseum was also very prominent in reflections on town-planning during the revolutionary period. See M. Mosser and D. Rabreau, 'Circus, amphitheatre, colosseum: revolutionary Paris as a new Rome', in *Lotus International* (Milan), 39, 1983, 108–18.
106 This was the commentary on the Cénobie design, a fable which tended to demonstrate the absurdity of brutal revolutions. 'Sixteen families lived together in the calm of the woods . . . A modern philosopher, an economist appeared: happiness fled, anxiety began, each person was troubled; the interpretation of a new social system preoccupied everyone's mind . . . Tired by a journey as arduous as it had been unsatisfying, our coenobites looked in vain in the forest of Chaux for the places where they had dwelled in their childhood; everything had disappeared, all that was left were immense ruins: it was then they recognised that a bird in the hand is worth two in the bush' (*L'architecture*, pp. 181–3).
107 B. Stoloff, *L'affaire Claude-Nicolas Ledoux: autopsie d'un mythe* (Brussels, 1977).
108 Popular unrest in fact resulted in Ledoux being dismissed in 1787, the campaign

against him being summarised in the famous distych: 'Le mur murant rend Paris murmurant' ['The walling-up wall puts all Paris in a murmur'].

109 Ledoux, *L'architecture*, p. 184 and plate 92.

110 Ibid., plate 43.

111 Ibid., p. 185. On this type of civic monument, which was so characteristic of the revolutionary period, see M.-K. Deming and C. de Vaulchier, 'La loi et ses monuments en 1791', in *Dix-Huitième Siècle* (Paris), 14, 1982, 117–30.

112 B. Fortier, 'Logiques de l'équipement. Notes pour une histoire du projet', p. 83.

113 Ledoux, *L'architecture*, p. 117.

114 For a more detailed account of Ledoux's handling of this problem, see M. Ozouf, *L'école de la France. Essais sur la Révolution, l'utopie et l'enseignement* (Paris, 1984), pp. 286–320.

115 Compare L. Girard, *Les libéraux français 1814–1875* (Paris, 1985). See also M. Regaldo, 'Lumières, élite, democratie: la difficile position des idéologues', in *Dix-Huitième Siècle* (Paris), 6, 1974, 193-207.

116 Ledoux, *L'architecture*, pp. 72ff.

117 Boullée, *Architecture*, p. 35.

118 Ibid., p. 37.

119 Ibid.

120 See chapter 2.

121 On the primitive hut and Laugier's interpretations of it, see J. Rykwert, *Adam's house in Paradise*, pp. 46ff.

122 M.-A. Laugier, *Essai sur l'architecture*, pp. 8ff.

123 See chapter 7.

124 Boullée, *Architecture*, pp. 126ff. This design is reproduced in J.-M. Pérouse de Montclos, *Etienne-Louis Boullée*, ills. 98–102.

125 Boullée, *Architecture*, p. 49.

126 Ledoux, *L'architecture*, p. 15.

127 See J.-M. Pérouse de Montclos, ill. 102.

128 They are featured in numerous engravings in *L'architecture*, especially plate 35.

129 A.-L.-C. Destutt de Tracy, *Elémens d'idéologie. Première partie. Idéologie proprement dite*, pp. 302ff.

130 Boullée and Ledoux's designs did, for example, feature isolated columns supporting an entablature. *L'architecture* and *Architecture, essai sur l'art* likewise presented bridges in much the same way as the engineers were doing. For a more precise discussion of this point, see below.

131 On the rediscovery of Delorme's trusses, and on Legrand and Molinos' design, see M.-K. Deming, *La halle au blé de Paris*, pp. 175ff.

132 J.-M. Pérouse de Montclos, 'Innovation technique et archéologie des techniques dans l'architecture néo-classique', in *Les Cahiers de la Recherche Architecturale* (Marseilles), 18, 1985, 44–9.

133 J.-N.-L. Durand, *Précis des leçons d'architecture données à l'Ecole Polytechnique (1802–1805)* (Paris, 1817–19), vol. 1, pp. 9ff. This work consisted of two volumes, the first of which was republished in 1809, 1813, 1819, 1823, 1840. The second volume was reprinted in 1817, 1823, 1825 and 1840. I shall refer in what follows to the 1819 edition of the first volume and to the 1817 edition of the second. For Durand's criticisms of the Abbé Laugier, see W. Szambien, '"Régulas" architecture: imitation in Durand', in *Lotus International* (Milan), 32, 1981, 102–11.

134 A.-J.-B. Rondelet, *Traité théorique et pratique de l'art de bâtir* (Paris, 1802–17), vol. 4, pp. 1ff, plate xciii.

135 See chapter 7.

136 I use the term in Derrida's sense, in his *Of grammatology*.

137 C.-F. Mandar, *Détails de construction d'une maison donnés pour l'instruction de MM. les élèves de l'Ecole Royale des Ponts et Chaussées* (Paris, 1818). This design also featured in Krafft's *Recueil d'architecture civile, contenant les plans, coupes et élévations des châteaux, maisons de campagne et habitations rurales . . . situés aux environs de Paris et dans les départemens voisins.* Charles-François Mandar (1757–1844) had begun his career as a military engineer, but had joined the corps of the Ponts et Chaussées in 1798 and had turned to architecture in 1804. See chapter 6. Having gone into semi-retirement in 1820, Mandar continued both to work as a marine engineer and to teach. He built some twenty houses in Paris and its environs, sixteen of which were in the street which now bears his name. No general study has been devoted to this man, whose influence upon the engineers of the Ponts et Chaussées was almost equal to that of Durand. In the meantime, one should consult his engineer's file, which is stored in the Archives Nationales under F14 2274/2.

138 There were a number of issues, however, upon which Mandar took his distance from Durand. For instance, he recommended the use of perspective; one has the sense that he was none too convinced by the attempts of the professor of architecture at the Polytechnique to standardise the orders. See below. More generally, Mandar's teaching seemed to lead to a kind of constructivism, somewhat removed from Durand's theoretical preoccupations.

139 C.-F. Mandar, *Etudes d'architecture civile* (Paris, 1826), plates 117, 118, 121. Following a similar set of ideas, Mandar also built a Chinese billiard room for the Inspector-General of ports in the Cochin sea. There is a reproduction of this in J.-C. Krafft, *Recueil d'architecture civile*, notebook 9, plate 54.

140 What could be more telling in this respect than the reasons which led Boullée to abandon the theory of proportions? 'One can only be wholly sure of the principles upon which an art is founded when it becomes altogether out of the question to deviate from these same principles' (*Architecture*, p. 67). Through being imprecise, classical proportions were seen as in effect arbitrary – an arbitrariness which had nothing natural about it – and this meant that it was

condemned as roundly as were the false systems of the physicists prior to the Newtonian revolution. It was a revolution of the same order as that which Boullée was in the last analysis proposing with his theory of simple bodies. Around the same period, Prony only allowed 'Greek proportions', which were the only ones 'subject to strict and constant rules' (*Réflexions*).

The imprecision of the proportions was subsequently to constitute a constant cause of complaint against architecture. Around 1830, Alexandre Corréard's *Journal du Génie Civil* published articles on this topic in virtually every issue, written by engineers, and sometimes even by architects, such as the curious figure Louis Lebrun, the author of a *Cours écrit de l'architecture grecque et romaine*, who wished to restore the true principles 'of form, proportion and stability', which modern architecture, in imitation of the monuments of Antiquity, sought to obey.

141 E. Kaufmann, 'Die Architekturtheorie der französischen Klassik und des Klassizismus'.

142 See W. Szambien, *Le musée d'architecture*, pp. 21–5.

143 See W. Szambien, *Jean-Nicolas-Louis Durand*, ill. 9.

144 See chapter 2.

145 P. Madec, 'Le plan architectural entre classicisme et modernisme', in *Raison–projet–représentation*, pp. 19–210, especially pp. 73–89.

146 As far as Blondel was concerned, sublimity in architecture, a quality appropriate to basilicas and public buildings, could be classically defined, and subject to the rules of *convenance* (*Cours d'architecture*, vol. 1, pp. 377–80). He saw Boullée and Ledoux's work as characterised by 'licentious' qualities, which seemed to opt quite freely for oddity.

147 See J. Guillerme, 'Notes pour l'histoire de la régularité', in *Revue d'Esthétique* (Paris), 3/4, 1970, 383–94. On the birth of mineralogy and crystallography at the end of the eighteenth century, see also D. Longeot, 'La genèse de la cristallographie et de la minéralogie scientifique', in *Dix-Huitième Siècle* (Paris), 3, 1971, 253–64.

148 A.-L.-C. Destutt de Tracy, *Elémens d'idéologie*, Preface, p. xv.

149 See chapter 6.

150 See chapter 7.

151 H. Damisch, 'Ledoux avec Kant', p. 17.

152 J.-J. Rousseau, *The social contract* (1762; translated by M. Cranston, Harmondsworth, 1968), p. 60.

153 A.-L.-C. Destutt de Tracy, *Elémens d'idéologie*, pp. 107ff. Condillac, who considered touch to be at the origin of perception, was partly responsible for this concept. If one should in fact 'think Ledoux with Rousseau, if not with Kant', as Hubert Damisch has argued, one should also think revolutionary architecture with *idéologie*, with which it was exactly contemporary. We have already noted

the possible comparison between the functioning of the town of Chaux, which was both centralised and decentralised, and the alternation in Sieyès' schemes of phases of assembly and division of the social order. Speaking now in more general terms, one can discern some affinity, or at least a shared unease, between those thinkers who had willingly jettisoned metaphysics and who were in search of those principles directing nature, and architects who rejected both the certainties of classicism and the immediate imitation of the productions of nature. In either case, there was a search for principles, and a bid to organise them into a system, these principles and systems only being recoverable through their actual deployment. The *idéologues* and the revolutionary architects argued that this deployment could be made to coincide with that of nature, so that one might accompany it in its endless becoming, rather than try to fix it arbitrarily. This accounts for Ledoux's constant reference to the movements of the planets.

154 P. Foulquié and R. Saint-Jean, *Dictionnaire de la langue philosophique* (Paris, 1969), p. 747.

155 Boullée, *Architecture*, p. 145. This design is reproduced in J.-M. Pérouse de Montclos, *Etienne-Louis Boullée*, ill. 103.

156 Ledoux, *L'architecture*, plates 3, 4.

157 'I have conceived the decoration of this bridge by going back to the first ideas, which consist in establishing, by means of boats, a passage across the water' (Boullée, *Architecture*, p. 146).

158 See chapter 9.

159 See M. Foucault, *The birth of the clinic*, translated by A. M. Sheridan Smith (London, 1973), pp. 94–5.

160 See chapter 7.

161 Ledoux was responding to this exigency in particular, when he was designing the Arc-et-Senans saltworks.

162 See chapter 7.

163 A. Koyré, *Etudes d'histoire de la pensée scientifique*, p. 14.

164 O. Hamelin, *Le système d'Aristote* (1920; Paris, 1976), pp. 353–70.

165 See F. Jacob, *La logique du vivant. Une histoire de l'hérédité* (1970; Paris, 1978).

166 See Koyré's studies of Galileo, Gassendi or Pascal, *Etudes d'histoire de la pensée scientifique*.

167 See, for example, R. Lenoble, *Mersenne ou la naissance du mécanisme* (1943; Paris, 1971).

168 See chapter 2. Towards the end of his life, Newton also speculated on the dimensions of the Temple of Solomon. From Villalpande to Newton, science, esotericism and architecture were still closely bound together in the classical age. See J. Rukwert, *The first moderns. The architects of the eighteenth century* (1980; Cambridge, Mass., 1983), p. 157.

169 C.-A. d'Aviler, *Dictionnaire d'architecture civile et hydraulique*, p. 32.

170 B. Forest de Bélidor, *Architecture hydraulique ou l'art de conduire, d'élever et de ménager les eaux pour les différens besoins de la vie.*

171 B. Forest de Bélidor, *La science des ingénieurs*, book 5.

172 A.-F. Frézier, *Dissertation historique et critique sur les ordres d'architecture* (Paris, 1769).

173 G.-G.-A. de l'Hôpital, *Analyse des infiniments petits, pour l'intelligence des lignes courbes* (Paris, 1696). This was used as a textbook at the Ecole des Ponts et Chaussées up to the Revolution.

174 See, for example, R. Woodhouse, *A history of the calculus of variations in the eighteenth century* (1810; New York, Chelsea, n.d.).

175 S. Ince and H. Rouse, *History of hydraulics*, pp. 89–112.

176 'The reader will find absolutely no illustrations in this work. The methods which I expound here require neither constructions nor geometrical or mechanical reasonings, but solely algebraical operations, subject to a uniform and regular procedure. Those who love analysis will be glad to see mechanics become a new branch of it, and will be grateful to me for having thereby extended its domain' (J.-L. Lagrange, *Mécanique analytique* (1788; Paris, 1965), vol. 1, pp. i, ii). On Lagrange's work, read R. Dugas, *A history of mechanics*, pp. 332ff.

177 *Histoire générale des sciences*, vol. 2, pp. 478–80.

178 S. Ince and H. Rouse, *History of hydraulics*, pp. 113–38.

179 The bases of analysis are rendered particularly obscure by the question of the 'infinitely small', which are endowed with paradoxical qualities but are easily handled. Newton and Leibniz had introduced them in order to make calculations easier, and the marquis de l'Hôpital had systematised their usage. In 1734, however, a devastating criticism of the 'infinitely small', Berkeley's *The analyst*, appeared. Nevertheless, this did not stop Euler from using the concept in his own work. For d'Alembert, on the other hand, the 'infinitely small' had no real existence. Prony, who was aware of the importance of this question, drafted a *Réflexions sur la métaphysique de l'infini*, for the attention of his colleagues at the Ponts et Chaussées, in which he tried to link the 'infinitely small' to the notion of limit used by d'Alembert (G. Riche de Prony, *Réflexions sur la métaphysique de l'infini*, ENPC manuscript, MS 1668). The theoretical bases of analysis would not really be clarified until Cauchy published his *Analyse algébrique* (1821). Compare B. Belhoste, *Cauchy 1789–1857. Un mathématicien légitimiste au XIXe siècle* (Paris, 1985), pp. 107ff.

180 Further information on the Ecoles Centrales may be found in D. Julia, 'La fréquentation des Ecoles Centrales: quelques hypothèses', in *Sciences et Techniques en Perspective* (Nantes), 2, 1983, 141–59; A. Léon, *La Révolution française et l'éducation technique* (Paris, 1968); F. Picavet, *Les idéologues*, pp. 37–66.

181 Appointed assistant professor in applied analysis alongside Lagrange, Prony was to teach at the Ecole Polytechnique to 1814–15. See M. Bradley, 'Prony the bridge-builder', pp. 117–49. On the basis of his teaching, Prony published in the Year

VIII his *Mécanique philosophique ou analyse raisonnée de diverses parties de la science de l'équilibre et du mouvement*, and in 1810 and 1815 the two volumes of his *Leçons de mécanique analytique*.

The most exhaustive study of the founding of the Ecole Polytechnique is still A. Fourcy, *Histoire de l'Ecole Polytechnique* (Paris, 1828). It is also worth looking at the school's *Livre du centenaire*, together with the more recent work by P. Callot, *Histoire de l'Ecole Polytechnique, ses légendes, ses traditions, ses gloires* (Paris, 1958).

182 G. Riche de Prony, *Nouvelle architecture hydraulique* (Paris, 1790–6), vol. 1, pp. vi, vii.

183 See chapter 6.

184 B. Forest de Bélidor, *Architecture hydraulique*, vol. 2, pp. 308ff.

185 See J. Payen, *Capital et machine à vapeur au XVIIIe siècle. Les frères Périer et l'introduction en France de la machine à vapeur de Watt* (Paris, 1969).

186 G. Riche de Prony, *Nouvelle architecture hydraulique*, vol. 2, p. 152.

187 Ibid., p. 153.

188 L. Carnot, *Essai sur les machines en général* (Paris, 1788); J.-N.-P Hachette, *Traité élémentaire des machines* (Paris, 1811); N.-L. Sadi Carnot, *Réflexions sur la puissance motrice du feu et sur les machines propres à développer cette puissance* (Paris, 1824). On Sadi Carnot, the heir to the science of the Enlightenment engineers, see T.-S. Kuhn, 'L'œuvre de Sadi Carnot et la tradition de la science des ingénieurs', in *Culture Technique* (Paris), 12, 1984, and above all. P. Redondi, *L'accueil des idées de Sadi Carnot* (Paris, 1980).

189 'We are assured that in England there are over five thousand of these machines; in France, on the other hand, we have barely two hundred of them' (J.-C. Périer, *Sur les machines à vapeur* (Paris, 1810), p. 3). Since it contained an implicit criticism of the economic policies of the Empire, this essay was seized on the orders of the Emperor. See J. Payen, *Capital et machine à vapeur*, p. 175. In this essay, Périer developed the at the time original idea that the genuine strength of a nation could be measured in terms of the horse power that was available to it, whether this power derived from men, from animals, from the water, the wind or from machines.

190 A figure quoted by J.-P. Rioux, *La révolution industrielle 1780–1880* (Paris, 1971), p. 68. On the vexed question of the discrepancy between France and England, one should consult F. Crouzet, *Britain ascendant: comparative studies in Franco-British economic history* (Cambridge, 1990); D. S. Landes, *The unbound Prometheus: technological change and industrial development in western Europe from 1750 to the present* (Cambridge, 1969).

191 More generally, the French engineers attained a deeper mastery than their English counterparts of questions having to do with the rational exposition of technical knowledge, in connection with the development of physico-mathematical science. Telford and Rennie learned French in order to read treatises which had

not been translated into English. Having been trained on the job, in a business context, the English engineers were on the other hand more empirical, and more alert to questions of profit than were the French, who were to some extent the prisoners of their own theoretical formation. The difference between English and French forms of training at the beginning of the nineteenth century refers to one of the central problems in the history of technology, namely, that of the part played by the organisation of knowledge in innovatory processes, a part that varied in importance depending upon the period. One has to allow that this organisation proved less crucial in England than in France, where the rationalising tradition of the state and of its main agents weighed very heavily.

192 Compare F. Braudel and E. Labrousse, *Histoire économique et sociale de la France* (Paris, 1970–82), vol. III, part 1, pp. 257ff.

193 G. Riche de Prony, 'Notice biographique sur Navier', in *Annales des Ponts et Chaussées* (Paris), First quarter, 1837, 1–2.

194 See note 133. My information on Durand is also derived from the exhaustive study by W. Szambien, *Jean-Nicholas-Louis Durand*.

195 J.-N.-L. Durand, *Précis*, vol. 1, p. 3.

196 The *convenance* to which Durand referred no longer had much in common with Blondel's spirit of *convenance*: 'for a building to have *convenance*, it must be stable, salubrious and convenient'. The use of the term 'convenient' lends *convenance* a functional air here. J.-N.-L. Durand, *Précis*, vol. 1, p. 6.

197 Ibid., p. 26.

198 Ibid., p. 31.

199 Ibid., p. 30.

200 There is a detailed analysis of the proportions presented by Durand in W. Szambien, *Jean-Nicholas-Louis Durand*, pp. 74–80.

201 J.-N.-L. Durand, *Précis*, vol. 1, p. 29.

202 Ibid., pp. 7, 8.

203 Durand was not the first to use a grid layout in order to position the elements of a design. The Italian architect Vittone had resorted to this device in his *Instruzioni Elementari* (1760). Compare A. Pérez-Gomez, *Architecture and the crisis of modern science*, p. 109.

204 G. Monge, *Géométrie descriptive. Leçons données aux écoles normales de l'an 3 de la République* [Year VII] (Paris, 1827), p. 1. On Monge's life and work, read L. de Launay, *Monge, fondateur de l'Ecole Polytechnique* (Paris, n.d.); R. Taton, *L'œuvre scientifique de Gaspard Monge* (Paris, 1951).

205 G. Monge, *Géométrie descriptive*, pp. 1, 2.

206 W. Szambien, *Jean-Nicholas-Louis Durand*, p. 89.

207 See chapter 3.

208 It was undoubtedly this sense of a common purpose which had made it possible for Navier, in the course of republishing Bélidor's *Science des ingénieurs* in 1813, to

append Durand's doctrine in note form to the chapter on the architectural orders, whereas elsewhere he added more and more references to rational mechanics, to the resistance of materials and to the theory of machines. B. Forest de Bélidor, *La science des ingénieurs* (Paris, 1813), 'with notes by M. Navier, engineer with the Ponts et Chaussées', book 5. Navier writes in particular that the

> maxims which Bélidor gives here have been consecrated and developed by M. Durand in some lectures the memory of which will always be precious to the pupils of the Ecole Polytechnique . . . M. Durand was the first to establish the principles of architecture upon a firm foundation. He recognises no other principle than that of *convenance*, that is to say, the establishment of a perfect relationship between the disposition of a building and the use to which it is to be put; in such a way that to plan a building is to resolve a problem which turns upon the conditions of stability, economy and utility to which it is subject.
> (pp. 497, 498)

209 J.-J. Winkelmann, *Histoire de l'art chez les anciens* (1764; Paris, 1790–2), pp. xi, xii.

210 D. Leroy, *Les ruines des plus beaux monuments de la Grèce*, part 2, p. ii.

211 We should bear in mind that Volney was a friend of Cassas, one of the precursors of the Musée d'Architecture. History, as Volney conceived of it, consisted of the moral and physical system of a people, just as the history of art embraced both the meticulous description of works and the analysis of the sensations and ideas which were related to them.

212 See W. Szambin, *Le musée d'architecture*.

213 On the emergent history of art, and upon Quatremère de Quincy's doctrine, see A. Vidler, 'The idea of type: The transformation of the academic ideal, 1750–1830', in *Oppositions* (New York), 8, 1977, 95–115.

214 F. Béguin, *Architectures de la modernité* (SRA research report (Paris, 1982), p. 115).

215 See F. Picavet, *Les idéologues*. Joseph-Michel Dutens, an engineer with the Ponts et Chaussées, even went so far as to draft a report in the Year VIII on public education which seemed to be inspired by Condorcet's views, as expressed in his *Rapport et projet de décret sur l'organisation générale de l'instruction publique* (1792). Like Destutt de Tracy or Prony, he recommended 'the rigorous observation of the descent of ideas' (J.-M. Dutens, *Des moyens de nationaliser l'instruction, de sa doctrine* (Evreux, Year VIII), p. 10).

216 W. Szambien, *Jean-Nicholas-Louis Durand*, pp. 35-55.

217 Compare R. Chafée, 'The teaching of architecture at the Ecole des Beaux-Arts', pp. 61–77.

218 On Charles de Wailly's works on town-planning, see *Charles de Wailly, peintre architecte dans l'Europe des Lumières*, pp. 64ff.

219 See chapter 6.

220 See J. Hubbel Weiss, *The making of technological mind* (Cambridge, Mass., 1982, pp. 31–3).

221 The decree of 7 Fructidor Year XII (25 August 1804), which is reproduced in A. Brunot and R. Coquand, *Le corps des Ponts et Chaussées*, pp. 670–1. The General Council of the Ponts et Chaussées replaced the former Assembly of the Ponts et Chaussées, a more informal gathering of the Ponts et Chaussées engineers present in Paris which Trudaine had established. On the powers enjoyed by the Council, see A. Guillerme, 'La formation des nouveaux édiles: ingénieurs des Ponts et Chaussées et architectes 1804–1815', a paper given to the conference 'Ville et territoire pendant la periode napoléonienne', Rome, Ecole Française de Rome, 3–5 May 1984.

222 On the Conseil des Bâtiments Civils, read G. Teyssot, 'Planning and building in towns: the systems of the Bâtiments Civils in France 1795–1848', in *The Beaux-Arts and nineteenth-century French architecture* (London, 1982), pp. 35–50.

223 The status of engineer and scientist was, however, more and more clearly distinguished. A scientific career was acquiring its own independence, and it was becoming difficult to keep abreast of high level research and applied work. In order to pursue his studies in mathematics, Cauchy had to abandon his career in the Ponts et Chaussées. But the links between science and the world of technology were growing stronger, as the composition of the Institute serves to show. Both engineers and scientists founded the Ecole Polytechnique, and this common mould therefore predisposed them to dialogue. Thus, Arago, Gay-Lussac or Cauchy were interested in technology, even if they did not practise the profession of engineer. These contacts between science and technology also helped to form Prony, Girard or Navier, authors of many built works.

224 J.-M. Dutens, *Mémoires sur les travaux publics de l'Angleterre* (Paris, 1819), pp. xviff. This work belongs to a long sequence of reports written by French engineers on English technology. In the first half of the nineteenth century, alongside study voyages by such figures as Dupin or Navier, pupils from the Ponts et Chaussées were regularly sent on missions on the other side of the Channel, on an itinerary decided upon by the Council of the Ecole des Ponts et Chaussées.

225 Navier went to England in 1821 and 1823 to study suspension bridges. He opted for the new system in his *Rapport et mémoire sur les ponts suspendus* (1823), and the conclusions of this work were immediately accepted by the administration of the Ponts et Chaussées. Two years later, the Séguin brothers built the first French suspension bridge at Tournon, which lies to the south of Lyons. Many similar structures followed. On the introduction of suspension bridges into France, see D. Amouroux and B. Lemoine, 'L'age d'or des ponts suspendus en France 1823–1850', in *Annales des Ponts et Chaussées* (Paris), special issue, 1981, 53–63.

226 See F. Etner, 'Le calcul économique et le corps des Ponts et Chaussées entre les deux Napoléon'.

227 P.-S. Laplace, *Essai philosophique sur les probabilités* (Paris, 1814), p. 2.

228 On the advent of civil engineers, see J. Hubbel Weiss, 'Les changements de

structure dans la profession d'ingénieur en France de 1800 à 1850', in *L'ingénieur dans la société française*, pp. 19–38; and *The making of technological man*; G. Ribeill, 'Des ingénieurs civils enquête d'un titre: le cas de l'Ecole des Ponts et Chaussées (1851–1934)', a paper given to the conference on the title of engineer, CERTES-ENPC, Paris, 1984; 'Profils des ingénieurs civils au XIXe siècle. Le cas des centraux', in *L'ingénieur dans la société française*, pp. 11–125.

INDEX

For EU product safety concerns, contact us at Calle de José Abascal, 56–1°,
28003 Madrid, Spain or eugpsr@cambridge.org.

www.ingramcontent.com/pod-product-compliance
Lightning Source LLC
LaVergne TN
LVHW080007210926
848481LV00011B/2964
* 9 7 8 0 5 2 1 1 2 3 6 9 3 *